LÉVY PROCESSES IN LIE GROUPS

The theory of Lévy processes in Lie groups is not merely an extension of the theory of Lévy processes in Euclidean spaces. Because of the unique structures possessed by noncommutative Lie groups, these processes exhibit certain interesting limiting properties that are not present for their counterparts in Euclidean spaces. These properties reveal a deep connection between the behavior of the stochastic processes and the underlying algebraic and geometric structures of the Lie groups themselves.

The purpose of this work is to provide an introduction to Lévy processes in general Lie groups, the limiting properties of Lévy processes in semi-simple Lie groups of noncompact type, and the dynamical behavior of such processes as stochastic flows on certain homogeneous spaces. The reader is assumed to be familiar with Lie groups and stochastic analysis, but no prior knowledge of semi-simple Lie groups is required.

Ming Liao is a Professor of Mathematics at Auburn University.

162   Lévy Processes in Lie Groups

LÉVY PROCESSES IN LIE GROUPS

MING LIAO

Auburn University

CAMBRIDGE
UNIVERSITY PRESS

PUBLISHED BY THE PRESS SYNDICATE OF THE UNIVERSITY OF CAMBRIDGE
The Pitt Building, Trumpington Street, Cambridge, United Kingdom

CAMBRIDGE UNIVERSITY PRESS
The Edinburgh Building, Cambridge CB2 2RU, UK
40 West 20th Street, New York, NY 10011-4211, USA
477 Williamstown Road, Port Melbourne, VIC 3207, Australia
Ruiz de Alarcón 13, 28014 Madrid, Spain
Dock House, The Waterfront, Cape Town 8001, South Africa

http://www.cambridge.org

First published 2004

Printed in the United States of America

Typeface Times 10.25/13 pt.   *System* LaTeX 2_ε [TB]

A catalog record for this book is available from the British Library.

Library of Congress Cataloging in Publication Data
Liao, Ming.
Lévy processes in Lie groups / Ming Liao.
p. cm – (Cambridge tracts in mathematics ; 162)
Includes bibliographical references and index.
ISBN 0-521-83653-0
1. Lévy processes.   2. Lie groups.   I. Title.   II. Series.

QA274.73.L43 2004
512′.482–dc22

2003066662

ISBN 0 521 83653 0 hardback

Contents

Preface

The present volume provides an introduction to Lévy processes in general Lie groups, and hopefully an accessible account on the limiting and dynamical properties of such processes in semi-simple Lie groups of non-compact type. Lévy processes in Euclidean spaces, including the famous Brownian motion, have always played a central role in probability theory. In recent times, there has been intense research activity in exploring the probabilistic connections of various algebraic and geometric structures, therefore, the study of stochastic processes in Lie groups has become increasingly important. This book is aimed at serving two purposes. First, it may provide a foundation to the theory of Lévy processes in Lie groups, as this is perhaps the first book written on the subject, and second it will present some important results in this area, revealing an interesting connection between probability and Lie groups.

Please note: when referring to a result in a referenced text, the chapter and enunciation numbering system in that text has been followed.

David Applebaum, Olav Kallenberg and Wang Longmin read portions of the manuscript and provided useful comments. Part of the book was written during the author's visit to Nankai University, Tianjin, China in the fall of 2002. I wish to take this opportunity to thank my hosts, Wu Rong and Zhou Xingwei, for their hospitality. It would be hard to imagine this work ever being completed without my wife's support and understanding.

List of Symbols

$\mathfrak{a}$ and A: the maximal abelian subspace and the associated abelian Lie group, 105

$\mathfrak{a}_+$: the Weyl chamber, 107

Ad: adjoint action of a Lie group, 40

ad: $\mathrm{ad}(X)Y = [X, Y]$, 40

$\mathcal{B}(G)$ and $\mathcal{B}(G)_+$: Borel σ-algebra and space of nonnegative Borel functions on G, 8

c_g: conjugation, 9

$C^k, C_c^k, C_b, C_0, C_u$: function spaces, 11

$C_0^{k,l}$ and $C_0^{k,r}$: function spaces, 11

$GL(d, \mathbb{R})$, $\mathfrak{gl}(d, \mathbb{R})$, and $GL(d, \mathbb{R})_+$: the general linear group, its Lie algebra, and its identity component, 27, 119, 246

$GL(n, \mathbb{C})$: the complex general linear group, 81, 247

G_μ: the closed subgroup generated by a Lévy process, 146

$\mathfrak{g}_\alpha$: the root space of the root α, 106

g_t^e: the Lévy process g_t starting at the identity element e of a Lie group G, 7

H^+: the rate vector of a Lévy process, 186, 197

H_ρ: the element of $\mathfrak{a}$ representing the half sum of positive roots, 134

I_d: the $d \times d$ identity matrix, 27

$\mathrm{Irr}(G, \mathbb{C})_+$: the set of equivalent classes of nontrivial irreducible representations, 82

K: fixed point set of Θ (with $\mathfrak{k}$ as Lie algebra), 104

$\mathfrak{k}$: eigenspace of θ corresponding to eigenvalue $+1$ (Lie algebra of K), 104

L_g: left translation, 9

$\mathfrak{m}$ and M: the centralizers of A in $\mathfrak{k}$ and K, denoted by $\mathfrak{u}$ and U in Chapter 8, 106

M': normalizer of A in K, denoted by U' in Chapter 8, 111

$\mathfrak{n}^+$ and $\mathfrak{n}^-$: the nilpotent Lie algebras spanned by positive and negative root spaces, 109

N^+ and N^-: the nilpotent Lie groups generated by $\mathfrak{n}^+$ and $\mathfrak{n}^-$, 109

$O(d)$ and $o(d)$: the orthogonal group and its Lie algebra, 120

$O(1, d)$ and $o(1, d)$: the Lorentz group and its Lie algebra, 124

$\mathfrak{p}$: the eigenspace of θ corresponding to eigenvalue -1, 104

R_g: right translation, 9

$SL(d, \mathbb{R})$ and $\mathfrak{sl}(d, \mathbb{R})$: the special linear group and its Lie algebra, 119

$SO(d)$: the special orthogonal group, 120

$SO(1, d)_+$: the identity component of $O(1, d)$, 125

T_μ: the closed semigroup generated by a Lévy process, 149

U, U', and $\mathfrak{u}$: centralizer, normalizer of A in K, and their common Lie algebra, 205

$U(n)$ and $SU(n)$: unitary group and special unitary group, 82

U^δ: unitary representation of class δ, 82

W: the Weyl group, 112

X^l and X^r: the left and right invariant vector fields on G induced by $X \in \mathfrak{g}$, 11, 244

χ_δ: character of class δ, 83

Π: the Lévy measure, 12

ψ_δ: normalized character of class δ, 83

Θ and θ: the Cartan involutions on G and on $\mathfrak{g}$, 104

θ_t: the time shift, 201

Introduction

Like the simple additive structure on an Euclidean space, the more complicated algebraic structure on a Lie group provides a convenient setting under
which various stochastic processes with interesting properties may be defined and studied. An important class of such processes are Lévy processes
that possess translation invariant distributions. Since a Lie group is in general
noncommutative, there are two different types of Lévy processes, left and
right Lévy processes, defined respectively by the left and right translations.
Because the two are in natural duality, for most purposes, it suffices to study
only one of them and derive the results for the other process by a simple
transformation. However, the two processes play different roles in applications. Note that a Lévy process may also be characterized as a process that
possesses independent and stationary increments.

The theory of Lévy processes in Lie groups is not merely an extension of the
theory of Lévy processes in Euclidean spaces. Because of the unique structures
possessed by the noncommutative Lie groups, these processes exhibit certain
interesting properties that are not present for their counterparts in Euclidean
spaces. These properties reveal a deep connection between the behavior of the
stochastic processes and the underlying algebraic and geometric structures of
the Lie groups.

The study of Lie group–valued processes can be traced to F. Perrin's work
in 1928. Itô's article on Brownian motions in Lie groups was published in
1951. Hunt [32] in 1956 obtained an explicit formula for the generator of
a continuous convolution semigroup of probability measures that provides
a complete characterization, in the sense of distribution, of Lévy processes
in a general Lie group. More recently, Applebaum and Kunita [3] in 1993
proved that a Lévy process is the unique solution of a stochastic integral
equation driven by a Brownian motion and a Poisson random measure. This
corresponds to the Lévy –Itô representation in the Euclidean case and provides
a pathwise characterization of Lévy processes in Lie groups.

The celebrated Lévy–Khintchine formula provides a useful Fourier transform characterization of infinite divisible laws, or distributions of Lévy processes, on Euclidean spaces. For Lévy processes in Lie groups, there is also a natural connection with Fourier analysis. Gangolli [22] in 1964 obtained a type of Lévy–Khintchine formula for spherically symmetric infinite divisible laws on symmetric spaces, which may be regarded as a result on Lévy processes in semi-simple Lie groups. More generally, Fourier methods were applied to study the probability measures on locally compact groups (see, for example, Heyer [28] and Siebert [55]) and to study the distributional convergence of random walks to Haar measures on finite groups and in some special cases Lie groups (see Diaconis [15] and Rosenthal [53]). Very recently, the author [43] studied the Fourier expansions of the distribution densities of Lévy processes in compact Lie groups based on the Peter–Weyl theorem and used it to obtain the exponential convergence of the distributions to Haar measures.

Perhaps one of the deepest discoveries in probability theory in connection with Lie groups is the limiting properties of Brownian motions and random walks in semi-simple Lie groups of noncompact type, both of which are examples of Lévy processes. Dynkin [16] in 1961 studied the Brownian motion x_t in the space X of Hermitian matrices of unit determinant. He found that, as time $t \to \infty$, x_t converges to infinity only along certain directions and at nonrandom exponential rates. The space X may be regarded as the homogeneous space $SL(d, \mathbb{C})/SU(d)$, where $SL(d, \mathbb{C})$ is the group of $d \times d$ complex matrices of determinant 1 and $SU(d)$ is the subgroup of unitary matrices. The Brownian motion x_t in X may be obtained as the natural projection of a continuous Lévy process in $SL(d, \mathbb{C})$. Therefore, the study of x_t in X may be reduced to that of a Lévy process in the Lie group $SL(d, \mathbb{C})$.

The matrix group $SL(d, \mathbb{C})$ belongs to a class of Lie groups called semi-simple Lie groups of noncompact type and the space X is an example of a symmetric space. Such a symmetric space possesses a polar decomposition under which any point can be represented by "radial" and "angular" components (both of which may be multidimensional). Dynkin's result says that the Brownian motion in X has a limiting angular component and its radial component converges to ∞ at nonrandom exponential rates. This result is reminiscent of the fact that a Brownian motion in a Riemannian manifold of pinched negative curvature has a limiting "angle" (see Prat [49]). However, the existence of the angular limit is not implied by this fact because a symmetric space of noncompact type may have sections of zero curvature.

Part of Dynkin's result, the convergence of the angular component, was extended by Orihara [47] in 1970 to Brownian motion in a general symmetric space of noncompact type. Malliavin and Malliavin [44] in 1972 obtained a complete result for the limiting properties of a horizontal diffusion in a general semi-simple Lie group of noncompact type. See also Norris, Rogers, and Williams [46], Taylor [56, 57], Babillot [6], and Liao [38] for some more recent studies of this problem from different perspectives.

In a different direction, Furstenberg and Kesten [21] in 1960 initiated the study of limiting properties of products of iid (independent and identically distributed) matrix or Lie group–valued random variables. Such processes may be regarded as random walks or discrete-time Lévy processes in Lie groups. The study of these processes was continued in Furstenberg [20], Tutubalin [58], Virtser [59], and Raugi [51]. In Guivarc'h and Raugi [24], the limiting properties of random walks on semi-simple Lie groups of noncompact type were established under a very general condition. These methods could be extended to a general Lévy process. This extension was made in Liao [40] and was applied to study the asymptotic stability of Lévy processes in Lie groups viewed as stochastic flows on certain homogeneous spaces.

The discrete-time random walks in Lie groups exhibit the same type of limiting properties as the more general Lévy processes. However, the limiting properties of continuous-time Lévy processes do not follow directly from them. The Lévy processes also include diffusion processes in Lie groups that possess translation invariant distributions. The limiting properties of such processes can be studied in connection with their infinitesimal characterizations, namely, their infinitesimal generators or the vector fields in the stochastic differential equations that drive the processes. For example, the limiting exponential rates mentioned here may be expressed in terms of the generator or the vector fields, which allows explicit evaluation in some special cases.

The limiting properties of Lévy processes may also be studied from a dynamic point of view. If the Lie group G acts on a manifold M, then a right Lévy process g_t in G may be regarded as a random dynamical system, or a stochastic flow, on M. The limiting properties of g_t as a Lévy process imply interesting ergodic and dynamical behaviors of the stochastic flow. For example, the Lyapunov exponents, which are the nonrandom limiting exponential rates at which the tangent vectors on M are contracted by the stochastic flow, and the stable manifolds, which are random submanifolds of M contracted by the stochastic flow at the fixed exponential rates, can all be determined explicitly in terms of the group structure.

The purpose of this work is to provide an introduction to Lévy processes in general Lie groups and, hopefully, an accessible account on the limiting and dynamical properties of Lévy processes in semi-simple Lie groups of noncompact type. The reader is assumed to be familiar with Lie groups and stochastic analysis. The basic definitions and facts in these subject areas that will be needed are reviewed in the two appendices at the end of this work. However, no prior knowledge of semi-simple Lie groups will be assumed.

Because the present work is not intended as a comprehensive treatment of Lévy processes in Lie groups, many interesting and related topics are not mentioned, some of which may be found in the survey article by Applebaum [1]. The distribution theory and limiting theorems of convolution products of probability measures in more general topological groups and semigroups can be found in Heyer [28] and in Högnäs and Mukherjea [29]. The theory of Lévy processes in Euclidean spaces has been well developed and there remains enormous interest in these processes (see Bertoin [9], Sato [54], and the proceedings in which [1] appears).

This work is organized as follows: The first chapter contains the basic definitions and results for Lévy processes in a general Lie group, including the generators of Lévy processes regarded as Markov processes, the Lévy measures that are the counting measures of the jumps of Lévy processes, and the stochastic integral equations satisfied by Lévy processes. In Chapter 2, we study the processes in homogeneous spaces induced by Lévy processes in Lie groups as one-point motions, we will discuss the Markov property of these processes and establish some basic relations among various invariance properties, and we will show that a Markov process in a manifold invariant under the action of a Lie group G is the one-point motion of a Lévy process in G. We will also discuss Riemannian Brownian motions in Lie groups and homogeneous spaces. Chapter 3 contains the proofs of some basic results stated in the previous two chapters, including Hunt's result on the generator [32] and the stochastic integral equation characterization of Lévy processes due to Applebaum and Kunita [3]. In Chapter 4, we study the Fourier expansion of the distribution densities of Lévy processes in compact Lie groups based on the Peter–Weyl theorem and obtain the exponential convergence of the distribution to the normalized Haar measure. The results of this chapter are taken from Liao [43]. In the first four chapters, only a basic knowledge of Lie groups will be required.

In the second half of the book, we will concentrate on limiting and dynamical properties of Lévy processes in semi-simple Lie groups of noncompact type. Chapter 5 provides a self-contained introduction to semi-simple Lie groups of noncompact type as necessary preparation for the next three chapters. The basic limiting properties of Lévy processes in such a Lie group are established in Chapters 6. To establish these results, we follow closely the basic ideas in Guivarc'h and Raugi [24], but much more detail is provided with considerable modifications. We also include in this chapter a simple and elementary proof of the limiting properties for certain continuous Lévy processes. Additional limiting properties under an integrability condition, including the existence of nonrandom exponential rates of the "radial components," are established in Chapter 7. In Chapter 8, the dynamical aspects of the Lévy processes are considered by viewing Lévy processes as stochastic flows on certain compact homogeneous spaces. We will obtain explicit expressions for the Lyapunov exponents, the associated stable manifolds, and a clustering pattern of the stochastic flows in terms of the group structures. The main results of this chapter are taken from Liao [40, 41, 42].

1

Lévy Processes in Lie Groups

This chapter contains an introduction to Lévy processes in a general Lie group. The left and right Lévy processes in a topological group G are defined in Section 1.1. They can be constructed from a convolution semigroup of probability measures on G and are Markov processes with left or right invariant Feller transition semigroups. In the next two sections, we introduce Hunt's theorem for the generator of a Lévy process in a Lie group G and prove some related results for the Lévy measure determined by the jumps of the process. In Section 1.4, the Lévy process is characterized as a solution of a stochastic integral equation driven by a Brownian motion and an independent Poisson random measure whose characteristic measure is the Lévy measure. Some variations and extensions of this stochastic integral equation are discussed. The proofs of the stochastic integral equation characterization, due to Applebaum and Kunita, and of Hunt's theorem, will be given in Chapter 3. For Lévy processes in matrix groups, a more explicit stochastic integral equation, written in matrix form, is obtained in Section 1.5.

1.1. Lévy Processes

The reader is referred to Appendices A and B for the basic definitions and facts on Lie groups, stochastic processes, and stochastic analysis.

We will first consider Lévy processes in a general topological group G. A topological group G is a group and a topological space such that both the product map, $G \times G \ni (g, h) \mapsto gh \in G$, and the inverse map, $G \ni g \mapsto g^{-1} \in G$, are continuous. Starting from the next section, we will exclusively consider Lévy processes in Lie groups unless explicitly stated otherwise. A Lie group G is a group and a manifold such that both the product and the inverse maps are smooth. In this work, a manifold is always assumed to be smooth (i.e., C^∞) with a countable base of open sets.

Let G be a topological group and let g_t be a stochastic process in G. For $s < t$, since $g_t = g_s g_s^{-1} g_t = g_t g_s^{-1} g_s$, we will call $g_s^{-1} g_t$ the right increment and $g_t g_s^{-1}$ the left increment of the process g_t over the time interval $(s, \ t)$. The process g_t is said to have independent right (resp. left) increments if these increments over nonoverlapping intervals are independent, that is, if for any $0 < t_1 < t_2 < \cdots < t_n$,

$$g_0, \ g_0^{-1} g_{t_1}, \ g_{t_1}^{-1} g_{t_2}, \ldots, \ g_{t_{n-1}}^{-1} g_{t_n} \ \left(\text{resp. } g_0, \ g_{t_1} g_0^{-1}, \ g_{t_2} g_{t_1}^{-1}, \ldots, g_{t_n} g_{t_{n-1}}^{-1} \right)$$

are independent. The process is said to have stationary right (resp. left) increments if $g_s^{-1} g_t \overset{d}{=} g_0^{-1} g_{t-s}$ (resp. $g_t g_s^{-1} \overset{d}{=} g_{t-s} g_0^{-1}$) for any $s < t$, where $x \overset{d}{=} y$ means that the two random variables x and y have the same distribution.

A stochastic process x_t in a topological space is called *càdlàg (continu à droite, limites à gauche)* if almost all its paths $t \mapsto g_t$ are right continuous on $\mathbb{R}_+ = [0, \ \infty)$ and have left limits on $(0, \ \infty)$.

A càdlàg process g_t in G is called a left Lévy process if it has independent and stationary right increments. At the moment it may seem more natural to call a left Lévy process a right Lévy process because it is defined using its right increments. However, we call it a left Lévy process because its transition semigroup and generator are invariant under left translations, as will be seen shortly. Similarly, a càdlàg process g_t in G is called a right Lévy process if it has independent and stationary left increments.

Given a filtration $\{\mathcal{F}_t\}$, a left Lévy process g_t in G is called a left Lévy process under $\{\mathcal{F}_t\}$, or a left $\{\mathcal{F}_t\}$-Lévy process, if it is $\{\mathcal{F}_t\}$-adapted and, for any $s < t$, $g_s^{-1} g_t$ is independent of $\mathcal{F}_s$. A right $\{\mathcal{F}_t\}$-Lévy process is defined similarly. Evidently, a left (resp. right) Lévy process is always a left (resp. right) Lévy process under its natural filtration $\{\mathcal{F}_t^0\}$.

If g_t is a left Lévy process, then g_t^{-1} is a right Lévy process, and vice versa. This is a one-to-one correspondence between left and right Lévy processes. There are other ways to establish such a correspondence; for example, if G is a matrix group and g' denotes the matrix transpose of $g \in G$, then $g_t \leftrightarrow g_t'$ gives another one-to-one correspondence between left and right Lévy processes. Because of the duality between the left and right Lévy processes, any result for the left Lévy process has a counterpart for the right Lévy process, and vice versa. We can concentrate only on one of these two processes and derive the results for the other process by a suitable transformation. In the following, we will mainly concentrate on left Lévy processes, except in Chapter 8 and a few other places where it is more natural to work with right Lévy processes.

Let g_t be a left Lévy process in G. Define

$$g_t^e = g_0^{-1} g_t. \tag{1.1}$$

Then g_t^e is a left Lévy process in G starting at the identity element e of G, that is, $g_0^e = e$, and is independent of g_0. Note that, for $t > s$, $(g_s^e)^{-1} g_t^e = g_s^{-1} g_t$.

It is clear that if g_t is a left Lévy process under a filtration $\{\mathcal{F}_t\}$ and if s is a fixed element of $\mathbb{R}_+$, then $g_t' = g_s^{-1} g_{s+t}$ is a left Lévy process identical in distribution to the process g_t^e and independent of $\mathcal{F}_s$. The following proposition says that s may be replaced by a stopping time.

Proposition 1.1. *Let g_t be a left Lévy process under a filtration $\{\mathcal{F}_t\}$. If τ is an $\{\mathcal{F}_t\}$ stopping time with $P(\tau < \infty) > 0$, then under the conditional probability $P(\cdot \mid \tau < \infty)$, the process $g_t' = g_\tau^{-1} g_{\tau+t}$ is a left Lévy process in G that is independent of $\mathcal{F}_\tau$ and has the same distribution as the process g_t^e under P.*

Proof. First assume τ takes only discrete values. Fix $0 < t_1 < t_2 < \cdots < t_k$, $\phi \in C_c(G^k)$ and $\xi \in (\mathcal{F}_\tau)_+$, where $(\mathcal{F}_\tau)_+$ is the set of nonnegative $\mathcal{F}_\tau$-measurable functions. Because $\xi 1_{[\tau=t]} \in (\mathcal{F}_t)_+$, we have

$$E\left[\phi\big(g_\tau^{-1} g_{\tau+t_1}, \ldots, g_\tau^{-1} g_{\tau+t_k}\big) \xi \mid \tau < \infty\right]$$

$$= \sum_{t<\infty} E\left[\phi\big(g_t^{-1} g_{t+t_1}, \ldots, g_t^{-1} g_{t+t_k}\big)\xi; \tau = t\right]/P(\tau < \infty)$$

$$= \sum_{t<\infty} E\left[\phi\big(g_t^{-1} g_{t+t_1}, \ldots, g_t^{-1} g_{t+t_k}\big)\right] E(\xi; \tau = t)/P(\tau < \infty)$$

$$= E\left[\phi\big(g_0^{-1} g_{t_1}, \ldots, g_0^{-1} g_{t_k}\big)\right] E(\xi \mid \tau < \infty). \tag{1.2}$$

Setting $\xi = 1$ yields $E[\phi(g_\tau^{-1} g_{\tau+t_1}, \ldots, g_\tau^{-1} g_{\tau+t_k}) \mid \tau < \infty] = E[\phi(g_0^{-1} g_{t_1}, \ldots, g_0^{-1} g_{t_k})]$. Therefore, for a general $\xi \in (\mathcal{F}_\tau^0)_+$, the expression in (1.2) is equal to

$$E[\phi(g_\tau^{-1} g_{\tau+t_1}, \ldots, g_\tau^{-1} g_{\tau+t_k}) \mid \tau < \infty] \, E(\xi \mid \tau < \infty).$$

This proves the desired result for a discrete stopping time τ.

For a general stopping time τ, let $\tau_n = (k+1)2^{-n}$ on the set $[k \cdot 2^{-n} \leq \tau < (k+1)2^{-n}]$ for $k = 0, 1, 2, \ldots$. Then τ_n are discrete stopping times and $\tau_n \downarrow \tau$ as $n \uparrow \infty$. The result for τ follows from the discrete case and the right continuity of g_t. $\qquad\square$

Let $\mathcal{B}(G)$ be the Borel σ-algebra on G and let $\mathcal{B}(G)_+$ be the space of nonnegative Borel functions on G. For $t \in \mathbb{R}_+, g \in G$, and $f \in \mathcal{B}(G)_+$, define

$$P_t f(g) = E\left[f\big(g g_t^e\big)\right]. \tag{1.3}$$

Because g_t is a left Lévy process, for $t > s$ and $f \in \mathcal{B}(G)_+$,

$$E\big[f(g_t) \mid \mathcal{F}_s^0\big] = E\big[f\big(g_s g_s^{-1} g_t\big) \mid \mathcal{F}_s^0\big] = E\big[f\big(h g_{t-s}^e\big)\big]_{h=g_s} = P_{t-s} f(g_s),$$

almost surely. If $g = g_0$, then taking the expectation of this expression, we obtain $P_t f(g) = P_s P_{t-s} f(g)$. This shows that $\{P_t; t \in \mathbb{R}_+\}$ is a semigroup of probability kernels on G and that g_t is a Markov process with transition semigroup P_t (see Appendix B.1).

For any $g \in G$, let $L_g \colon G \ni g' \mapsto gg' \in G$ and $R_g \colon G \ni g' \mapsto g'g \in G$ be, respectively, the left and right translations on G. Let $c_g \colon G \ni g' \mapsto gg'g^{-1} \in G$ be the conjugation map on G.

Let H be a subgroup of G. A linear operator T with domain $D(T)$, operating in some function space on G, is called left H-invariant if it is invariant under L_h for all $h \in H$, that is, if

$$\forall h \in H \quad \text{and} \quad f \in D(T), \quad f \circ L_h \in D(T) \quad \text{and} \quad T(f \circ L_h) = (Tf) \circ L_h.$$

Similarly, a right H-invariant operator T is defined using R_h instead of L_h. A left (resp. right) G-invariant operator will simply be called left (resp. right) invariant. A semigroup of probability kernels $\{P_t\}$ will be called left (resp. right) (resp. H-) invariant if each P_t is such an operator on G with domain $\mathcal{B}_b(G)$, the space of all the bounded Borel functions on G. A Markov process with such a transition semigroup will be called left (resp. right) (resp. H-) invariant.

Now assume the topological group G is locally compact and has a countable base of open sets. Then by (1.3) and the right continuity of the process g_t, it can be shown that P_t is a Feller semigroup and is left invariant on G. Therefore, g_t is a Feller process and is left invariant on G. See Appendix B.1 for the definition of Feller processes.

For any measure μ and measurable function f on a measurable space, the integral $\int f \, d\mu$ may be written as $\mu(f)$. In the following, the measurability consideration on a topological space will always refer to the Borel σ-algebra of the space unless explicitly stated otherwise. The convolution of two measures μ and ν on G is a measure $\mu * \nu$ on G defined by

$$\mu * \nu(f) = \int f(gh)\mu(dg)\nu(dh) \tag{1.4}$$

for $f \in \mathcal{B}(G)_+$. A convolution semigroup of probability measures on G is a family $\{\mu_t; t \in \mathbb{R}_+\}$ of probability measures on G such that $\mu_0 = \delta_e$ (the unit point mass at e) and $\mu_t * \mu_s = \mu_{t+s}$ for $s, t \in \mathbb{R}_+$. It will be called continuous if $\mu_t \to \delta_e$ weakly as $t \to 0$. Then $\mu_t \to \mu_s$ weakly as $t \downarrow s$ for any $s \in \mathbb{R}_+$.

Let g_t be a left Lévy process in G and let $\{\mu_t; t \in \mathbb{R}_+\}$ be the family of the marginal distribution of the process g_t^e; that is, μ_t is the distribution of g_t^e for each $t \in \mathbb{R}_+$. Note that $\mu_t = P_t(e, \cdot)$. Then $\{\mu_t; t \in \mathbb{R}_+\}$ is a continuous convolution semigroup of probability measures on G and

$$P_t f(g) = \int f(gh)\mu_t(dh). \tag{1.5}$$

Conversely, let $\{\mu_t; t \in \mathbb{R}_+\}$ be a continuous convolution semigroup of probability measures on G. Then P_t defined by (1.5) is a left invariant Feller semigroup. By the discussion in Appendix B.1, there is a càdlàg Markov process g_t in G with transition semigroup P_t and an arbitrary initial distribution. By the Markov property of the process g_t, for $s < t$,

$$E\left[f\left(g_s^{-1}g_t\right)|\mathcal{F}_s^0\right] = P_{t-s}(f \circ L_g)(g_s)\,|_{g=g_s^{-1}} = \mu_{t-s}(f),$$

almost surely, where $\{\mathcal{F}_t^0\}$ is the natural filtration of the process g_t. This shows that the process g_t has independent and stationary right increments; therefore, it is a left Lévy process in G. Note that if g_t is a càdlàg Markov process with a left invariant transition semigroup P_t, then $\mu_t = P_t(e, \cdot)$ is a continuous convolution semigroup of probability measures on G satisfying (1.5), and hence g_t is a left Lévy process.

To summarize, we record the following result:

Proposition 1.2. *Let G be a locally compact topological group with a countable base of open sets.*

>   *(a) A left Lévy process g_t in G is a Markov process with a left invariant Feller transition semigroup P_t given by (1.3). Moreover, the marginal distributions μ_t of the process $g_t^e = g_0^{-1}g_t$ form a continuous convolution semigroup of probability measures on G satisfying (1.5).*
>   *(b) If $\{\mu_t; t \in \mathbb{R}_+\}$ is a continuous convolution semigroup of probability measures on G and ν is a probability measure on G, then there is a left Lévy process g_t in G with initial distribution ν such that μ_t is the distributions of g_t^e for each $t \in \mathbb{R}_+$.*
>   *(c) A left invariant càdlàg Markov process g_t in G is a left Lévy process in G.*

1.2. Generators of Lévy Processes

Let M be a manifold. For any integer $k \geq 0$, let $C^k(M)$ be the space of the real- or complex-valued functions on M that have continuous derivatives up to order k with $C(M) = C^0(M)$ being the space of continuous functions on

M and $C^\infty(M) = \bigcap_{k>0} C^k(M)$. Let $C^k_c(M)$ denote the space of the functions in $C^k(M)$ with compact supports. Let $C_b(M)$, $C_0(M)$, and $C_u(M)$ be, respectively, the space of bounded continuous functions, the space of continuous functions convergent to 0 at ∞, and the space of uniformly continuous functions on M with respect to the topology of the one-point compactification of M. The spaces $C(M)$, $C_c(M)$, $C_b(M)$, $C_0(M)$, and $C_u(M)$ are also defined for a locally compact Hausdorff space M with a countable base of open sets.

From now on, we will assume that G is a Lie group of dimension d with Lie algebra $\mathfrak{g}$ unless explicitly stated otherwise.

Hunt [32] found a complete characterization of left invariant Feller semigroups of probability kernels on G, or equivalently, left Lévy processes in G, by their generators. To state this result, we will fix a basis $\{X_1, X_2, \ldots, X_d\}$ of $\mathfrak{g}$. There are functions $x_1, x_2, \ldots, x_d \in C^\infty_c(G)$ such that $x_i(e) = 0$ and $X_j x_k = \delta_{jk}$. These functions may be used as local coordinates in a neighborhood of the identity element e of G with $X_i = (\partial/\partial x_i)$ at e, and, hence, will be called the coordinate functions associated to the basis $\{X_1, \ldots, X_d\}$. In a neighborhood U of e, which is the diffeomorphic image of some open subset of $\mathfrak{g}$ under the exponential map exp of the Lie group G, x_i may be defined to satisfy $g = \exp[\sum_i x_i(g)X_i]$ for $g \in U$. Note that the coordinate functions are not uniquely determined by the basis, but if $x'_1, \ldots, x'_d$ form another set of coordinate functions associated to the same basis, then

$$x'_i = x_i + O(|x|^2) \tag{1.6}$$

on some neighborhood of e, where $|x|^2 = \sum_{i=1}^n x_i^2$.

Any $X \in \mathfrak{g}$ induces a left invariant vector field X^l on G defined by $X^l(g) = DL_g(X)$, where DL_g is the differential map of L_g. It also induces a right invariant vector field X^r on G defined by $X^r(g) = DR_g(X)$. For any integer $k \geq 0$, let $C^{k,l}_0(G)$ be the space of $f \in C^k(G) \cap C_0(G)$ such that

$$Y_1^l f \in C_0(G), \quad Y_1^l Y_2^l f \in C_0(G), \quad \ldots, \quad Y_1^l Y_2^l \cdots Y_k^l f \in C_0(G)$$

for any $Y_1, Y_2, \ldots, Y_k \in \mathfrak{g}$. We may call $C^{k,l}_0(G)$ the space of functions on G with continuous derivatives vanishing at ∞ taken with respect to left invariant vector fields up to order k. The space $C^{k,r}_0(G)$ of functions on G with continuous derivatives vanishing at ∞ taken with respect to right invariant vector fields up to order k is defined similarly with Y_i^l replaced by Y_i^r.

Theorem 1.1. *Let L be the generator of a left invariant Feller semigroup of probability kernels on a Lie group G. Then its domain $D(L)$ contains $C^{2,l}_0(G)$,* *and $\forall f \in C_0^{2,l}(G)$ and $g \in G$,*

$$Lf(g) = \frac{1}{2} \sum_{j,k=1}^d a_{jk} X_j^l X_k^l f(g) + \sum_{i=1}^d c_i X_i^l f(g)$$

$$+ \int_G \left[f(gh) - f(g) - \sum_{i=1}^d x_i(h) X_i^l f(g) \right] \Pi(dh), \quad (1.7)$$

where a_{jk}, c_i are constants with $\{a_{jk}\}$ being a nonnegative definite symmetric matrix, and Π is a measure on G satisfying

$$\Pi(\{e\}) = 0, \qquad \Pi\left(\sum_{i=1}^d x_i^2 \right) < \infty, \qquad \text{and} \quad \Pi(U^c) < \infty \quad (1.8)$$

for any neighborhood U of G with U^c being the complement of U in G.

Conversely, if the matrix $\{a_{jk}\}$ and the measure Π satisfy the conditions here and c_i are arbitrary constants, then there exists a unique left invariant Feller semigroup P_t of probability kernels on G whose generator L restricted to $C_0^{2,l}(G)$ is given by (1.7).

Note that the condition (1.8) on Π is independent of the choice of the basis $\{X_1, \ldots, X_d\}$ of $\mathfrak{g}$ and the associated coordinate functions $x_1, \ldots, x_d \in C_c^\infty(G)$. A measure Π on G satisfying this condition will be called a Lévy measure on G. It will be called the Lévy measure of a Feller semigroup P_t if it is associated to P_t as in Theorem 1.1, and it will be called the Lévy measure of a Lévy process g_t if P_t is the transition semigroup of g_t. Clearly, any finite measure on G that does not charge e is a Lévy measure.

In some of the literature, the functions $x_1, \ldots, x_d$ are not assumed to be compactly supported; then one should replace $\Pi(\sum_{i=1}^d x_i^2) < \infty$ in (1.8) by $\int_U \sum_{i=1}^d x_i^2 d\Pi < \infty$ and require U to be relatively compact. However, in the present work, we will always assume that the coordinate functions $x_1, \ldots, x_d$ are contained in $C_c^\infty(G)$.

The proof of Theorem 1.1 will be given in Chapter 3. The reader is referred to theorem 5.1 in [32] for Hunt's original proof. Some obscure points in Hunt's paper was clarified by Ramaswami [50]. A complete proof of Hunt's result is also given in Heyer [28, chapter IV]. Note that Hunt's original result was stated in a slightly different form. He regarded the generator L as an operator on $C_u(G)$, the space of uniformly continuous functions on G under a metric compatible with the one-point compactification topology on G, and he proved that its domain contains the space of functions $f \in C_u(G) \cap C^2(G)$ such that $X^l f$, $X^l Y^l f \in C_u(G)$ for any $X, Y \in \mathfrak{g}$ and established (1.7) for such functions. Because such a function f can be written as $f = f_0 + c$

for some $f_0 \in C_0^{2,l}(G)$ and $c \in \mathbb{R}$ (the set of real numbers), it follows that Theorem 1.1 is equivalent to Hunt's original result.

Let U be a neighborhood of e that is a diffeomorphic image of the exponential map. Assume the coordinate functions x_i satisfy $g = \exp[\sum_{i=1}^d x_i(g)X_i]$ for $g \in U$. Let $f \in C^2(G)$. Applying Taylor's expansion to $\phi(t) = f(g \exp(t \sum_{i=1}^d x_i X_i))$, we obtain

$$f(gh) - f(g) - \sum_{i=1}^d x_i(h)X_i^l f(g) = \frac{1}{2} \sum_{j,k=1}^d x_j(h)x_k(h)X_j^l X_k^l f(gh') \quad (1.9)$$

for $g \in G$ and $h \in U$, where $h' = \exp[t \sum_{i=1}^d x_i(h)X_i]$ for some $t \in [0, 1]$. In particular, the left-hand side of (1.9) is $O(|x|^2)$, and it is $O(|x|^2)$ even when the x_is are arbitrary coordinate functions associated to the same basis due to (1.6). Therefore, by the condition (1.8) imposed on Π, the integral in (1.7) is well defined and $Lf \in C_0(G)$ for $f \in C_0^{2,l}(G)$.

Proposition 1.3. *The differential operator* $D = (1/2)\sum_{j,k=1}^d a_{jk}X_j^l X_k^l$ *on* $C^2(G)$ *and the Lévy measure* Π *given in Theorem 1.1 are completely determined by the generator* L *and are independent of the basis* $\{X_1, \dots, X_d\}$ *of* $\mathfrak{g}$ *and the associated coordinate functions* x_i *and coefficients* a_{ij}.

Proof. By (1.7), for any $f \in C_c^\infty(G)$ that vanishes in a neighborhood of e, $Lf(e) = \Pi(f)$. Since $\Pi(\{e\}) = 0$, this proves that Π is determined by L.

Since $x_i(e) = 0$ and $X_j^l x_k(e) = \delta_{jk}$, $L(x_j x_k)(e) = a_{jk} + \Pi(x_j x_k)$. Let $x_1', \dots, x_d'$ be another set of coordinate functions associated to the same basis $\{X_1, \dots, X_d\}$ of $\mathfrak{g}$ and let $D' = (1/2)\sum_{j,k=1}^d a_{jk}' X_j^l X_k^l$ be the corresponding differential operator. Then

$$a_{jk} + \Pi(x_j' x_k') = L(x_j' x_k')(e) = a_{jk}' + \Pi(x_j' x_k').$$

This implies $a_{jk} = a_{jk}'$; hence, D is independent of the choice of the coordinate functions if the basis of $\mathfrak{g}$ is fixed.

Now let $\{Y_1, \dots, Y_d\}$ be another basis of $\mathfrak{g}$ with $Y_j = \sum_{p=1}^d b_{jp}X_p$. Let $\{c_{jk}\}$ be the inverse matrix of $\{b_{jk}\}$. Then $y_k = \sum_{q=1}^d c_{qk}x_q$ for $1 \le k \le d$ can be used as coordinate functions associated to the new basis $\{Y_1, \dots, Y_d\}$. Let $D' = (1/2)\sum_{j,k=1}^d a_{jk}' Y_j^l Y_k^l$ be the operator corresponding to the new basis. Then

$$a_{jk}' = L(y_j y_k)(e) - \Pi(y_j y_k)$$

$$= \sum_{p,q=1}^d c_{pj}c_{qk}[L(x_p x_q)(e) - \Pi(x_p x_q)] = \sum_{p,q=1}^d a_{pq}c_{pj}c_{qk}$$

and

$$D' = \frac{1}{2} \sum_{j,k=1}^{d} a'_{jk} Y^l_j Y^l_k = \frac{1}{2} \sum_{j,k,p,q,u,v=1}^{d} a_{pq} c_{pj} c_{qk} b_{ju} b_{kv} X_u X_v$$

$$= \frac{1}{2} \sum_{p,q=1}^{d} a_{pq} X_p X_q = D.$$

This proves that D is independent of the basis $\{X_1, \ldots, X_d\}$. $\square$

The second-order differential operator $D = (1/2) \sum_{i,j=1}^{d} a_{ij} X^l_i X^l_j$ will be called the diffusion part of the generator L. Note that the coefficients c_i in (1.7) in general depend on the choice of the basis of $\mathfrak{g}$ and the associated coordinate functions.

We note that the uniqueness stated in the second half of Theorem 1.1 may be slightly strengthened as follows: There is at most one left invariant Feller semigroup P_t of probability kernels on G whose generator L restricted to $C_c^\infty(G)$ is given by (1.7). To prove this, note that, by the proof of Proposition 1.3, Π and a_{ij} are determined by this restriction, and it is easy to show the same for c_i. This will be formally stated in Theorem 3.1.

If the Lévy measure Π satisfies the following finite first moment condition:

$$\int \sum_{i=1}^{d} |x_i(g)| \, \Pi(dg) < \infty, \tag{1.10}$$

then the integral $\int_G [f(gh) - f(g)] \, \Pi(dh)$ exists and the formula (1.7) simplifies to:

$$Lf(g) = \frac{1}{2} \sum_{j,k=1}^{d} a_{jk} X^l_j X^l_k f(g) + \sum_{i=1}^{d} b_i X^l_i f(g) + \int_G [f(gh) - f(g)] \Pi(dh) \tag{1.11}$$

for $f \in C_0^{2,l}(G)$, where $b_i = c_i - \int_G x_i(h) \Pi(dh)$. In this case, there is no need to introduce the coordinate functions $x_1, \ldots, x_d$. Note that the condition (1.10) is independent of the choice of the basis and the associated coordinate functions and is satisfied if Π is finite.

1.3. Lévy Measure

Let g_t be a càdlàg process taking values in a Lie group G. Since $g_t = g_{t-} g_{t-}^{-1} g_t = g_t g_{t-}^{-1} g_{t-}$, we will call $g_{t-}^{-1} g_t$ a right jump and $g_t g_{t-}^{-1}$ a left jump of the process at time t.

Let g_t be a left Lévy process in a Lie group G with Lévy measure Π and let N be the counting measure of the right jumps; that is, N is the random measure on $\mathbb{R}_+ \times G$ defined by

$$N([0, t] \times B) = \#\{s \in (0, t]; \quad g_{s-}^{-1}g_s \neq e \text{ and } g_{s-}^{-1}g_s \in B\} \qquad (1.12)$$

for $t \in \mathbb{R}_+$ and $B \in \mathcal{B}(G)$. Since $g_{s-}^{-1}g_s = (g_{s-}^e)^{-1}g_s^e$, $N([0, t] \times B)$ is $\mathcal{F}_t^e$-measurable, where $\mathcal{F}_t^e$ is the natural filtration of the process g_t^e.

Proposition 1.4. *Let g_t be a left Lévy process in G. Then the random measure N defined by (1.12) is a Poisson random measure on $\mathbb{R}_+ \times G$ associated to the filtration $\{\mathcal{F}_t^e\}$ and its characteristic measure is the Lévy measure Π of g_t.*

See Appendix B.3 for the definition of Poisson random measures.

Proof. For $B \in \mathcal{B}(G)$ lying outside a neighborhood of e, let $N_t^B = N([0, t] \times B)$. This is a right continuous increasing process that increases only by jumps almost surely equal to 1. Because of the independent and stationary right increments of g_t, N_t^B also has independent and stationary increments. It is well known (see [52, chapter XII, proposition 1.4]) that such a process is a Poisson process of rate $c_B = E(N_1^B)$. For disjoint sets $B_1, B_2, \ldots, B_k$ in $\mathcal{B}(G)$ lying outside a neighborhood of e, any two of the processes $N_t^{B_1}, N_t^{B_2}, \ldots, B_t^{B_k}$ cannot jump at the same time; it follows from [52, XII, proposition 1.7] that these are independent Poisson processes. This proves that N is a Poisson random measure on $\mathbb{R}_+ \times G$, which is easily seen to be associated to the filtration $\{\mathcal{F}_t^e\}$. It remains to prove $c_B = \Pi(B)$. For any $\phi \in \mathcal{B}(G)_+$, let $F_\phi = \int_0^1 \int_G \phi(g)N(dt\, dg)$ and let $c_\phi = E(F_\phi)$. It suffices to show that $c_\phi = \Pi(\phi)$ for any $\phi \in C_c^\infty(G)$ such that $0 \le \phi \le 1$ and $\phi = 0$ in a neighborhood of e. Let $0 = t_0 < t_1 < t_2 < \cdots < t_n = 1$ with $t_{i+1} - t_i = \delta = 1/n$ for all $0 \le i \le n-1$ and let

$$f_\delta = \sum_{i=0}^{n-1} \phi(g_{t_i}^{-1}g_{t_{i+1}}).$$

Then $f_\delta \to F_\phi$ almost surely as $\delta \to 0$. However,

$$E(f_\delta^2) = \sum_{i=0}^{n-1}\sum_{j=0}^{n-1} E\left[\phi(g_{t_i}^{-1}g_{t_{i+1}})\phi(g_{t_j}^{-1}g_{t_{j+1}})\right]$$

$$= \sum_{i \neq j} E\left[\phi(g_\delta^e)\right] E\left[\phi(g_\delta^e)\right] + \sum_{i=0}^{n-1} E\left[\phi(g_\delta^e)^2\right]$$

$$= \frac{1-\delta}{\delta^2}[P_\delta\phi(e)]^2 + \frac{1}{\delta}P_\delta(\phi^2)(e) \to [L\phi(e)]^2 + L(\phi^2)(e) \quad \text{as } \delta \to 0$$

$$= \Pi(\phi)^2 + \Pi(\phi^2) < \infty,$$

where the first half of Theorem 1.1 is used to justify the convergence and the equality. This proves that the family $\{f_\delta; \delta > 0\}$ is $L^2(P)$-bounded; hence, it is uniformly integrable. It follows that

$$c_\phi = E(F_\phi) = \lim_{\delta \to 0} E(f_\delta) = \lim_{\delta \to 0} \frac{1}{\delta} P_\delta \phi(e) = L\phi(e) = \Pi(\phi). \qquad \Box$$

In some literature, the Lévy measure Π is defined to be a measure on $G - \{e\}$ and N a Poisson random measure on $\mathbb{R}_+ \times (G - \{e\})$. However, we find it more convenient to regard Π as a measure on G not charging $\{e\}$ and N as a Poisson random measure on $\mathbb{R}_+ \times G$ not charging $\mathbb{R}_+ \times \{e\}$.

By the discussion in Appendix B.3, a nontrivial Poisson random measure N on $\mathbb{R}_+ \times G$ with a finite characteristic measure is determined by a sequence of random times $T_n \uparrow \infty$, with iid differences $T_n - T_{n-1}$ of an exponential distribution (setting $T_0 = 0$) and an independent sequence of iid G-valued random variables σ_n. More precisely,

$$N([0, t] \times B) = \#\{n;\ n > 0,\ T_n \leq t,\ \text{and}\ \sigma_n \in B\}$$

for $t \in R_+$ and $B \in \mathcal{B}(G)$.

Jumps, determined by an independent Poisson random measure N' of finite characteristic measure, may be added to g_t to form a new Lévy process y_t, as described in the following proposition. The process y_t will be called the Lévy process obtained by interlacing g_t with jumps determined by N'.

Proposition 1.5. *Let g_t be a left Lévy process in G with generator L restricted to $C_0^{2,l}(G)$ given by (1.7) and Lévy measure Π, and let N' be an independent, nontrivial Poisson random measure on $\mathbb{R}_+ \times G$ with a finite characteristic measure Π' satisfying $\Pi'(\{e\}) = 0$. Suppose N' is determined by the random times $T_n \uparrow \infty$ and G-valued random variables σ_n. Define $y_t = g_t$ for $0 \leq t < T_1$ and inductively $y_t = y(T_n-)\sigma_n g(T_n)^{-1} g_t$ for $T_n \leq t < T_{n+1}$, that is,*

$$y_t = g(T_1-)\sigma_1 g(T_1)^{-1} g(T_2-)\sigma_2 g(T_2)^{-1} g(T_3-) \cdots g(T_n-)\sigma_n g(T_n)^{-1} g_t \tag{1.13}$$

for $T_n \leq t < T_{n+1}$ with $T_0 = 0$. Then y_t is a left Lévy process in G whose generator restricted to $C_0^{2,l}(G)$ is given by (1.7) with Π replaced by $\Pi + \Pi'$ and c_i replaced by $c_i + \int x_i d\Pi'$.

Proof. It is easy to see that the natural filtration of y_t is given by

$$\mathcal{F}_t^y = \sigma\{g_s \text{ and } N'([0, s] \times B);\ 0 \leq s \leq t \text{ and } B \in \mathcal{B}(G)\}.$$

For $t > s$, $y_s^{-1} y_t$ can be constructed from $\{g_s^{-1} g_u$ and $N'([s, u] \times \cdot);$ $u \in [s,\ t]\}$ in the same way as $y_0^{-1} y_{t-s}$ from $\{g_u^e$ and $N'([0, u) \times \cdot);$

$u \in [0, \ t - s]\}$; it follows that $y_s^{-1} y_t$ is independent of $\mathcal{F}_s^y$ and its distribution depends only on $t - s$. This proves that y_t is a left Lévy process in G. Let L^y be the generator of y_t. For $f \in C_0^{2,l}(G)$ and $g \in G$,

$$
\begin{aligned}
L^y f(g) &= \lim_{t \to 0} \frac{1}{t} \{ E[f(g y_t^e)] - f(g) \} \\
&= \lim_{t \to 0} \frac{1}{t} \{ E[f(g g_t^e); \, t < T_1] \\
&\quad + E\left[f(g g^e(T_1-) \sigma_1 g^e(T_1)^{-1} g_t^e; \, T_1 \le t < T_2 \right] \\
&\quad + E\left[f(g y_t^e); \, t > T_2 \right] - f(g) \}.
\end{aligned}
$$

Since T_1 is an exponential random variable of rate $c = \Pi'(G)$ independent of the process g_t^e and $P(t \ge T_2) = O(t^2)$, we have

$$
\begin{aligned}
L^y f(g) &= \lim_{t \to 0} \frac{1}{t} \{ E\left[f(g g_t^e) \right] e^{-ct} \\
&\quad + E\left[f(g g^e(T_1-) \sigma_1 g^e(T_1)^{-1} g_t^e; T_1 \le t \right] - f(g) \} \\
&= \lim_{t \to 0} \frac{1}{t} \{ E\left[f(g g_t^e) \right] - f(g) \} e^{-ct} \\
&\quad + \lim_{t \to 0} \frac{1}{t} \int_0^t \left\{ E\left[f\left(g g_{s-}^e \sigma_1 (g_s^e)^{-1} g_t^e \right) \right] - f(g) \right\} c e^{-cs} ds \\
&= L f(g) + c E[f(g \sigma_1) - f(g)] \\
&= L f(g) + \int_G [f(g \sigma) - f(g)] \Pi'(d\sigma).
\end{aligned}
$$

Note that only the first half of Theorem 1.1 is used in this proof. $\qquad\square$

Note that, in Proposition 1.5, the process g_t does not jump at time $t = T_n$ almost surely because it is independent of N'.

Let A be a Borel subset of G with a finite $\Pi(A)$. The next proposition shows that the jumps contained in A may be removed from a Lévy process g_t to obtain a new Lévy process x_t whose Lévy measure is supported by A^c. Thus the original process g_t may be regarded as x_t, which has only jumps contained in A^c, interlaced with jumps contained in A. For any measure ν on a measurable space $(S, \mathcal{S})$ and $A \in \mathcal{S}$, the restriction of ν to A is the measure $\nu|_A$ defined by $\nu|_A(B) = \nu(A \cap B)$ for $B \in \mathcal{S}$.

Proposition 1.6. *Suppose g_t is a left Lévy process in G with generator L restricted to $C_0^{2,l}(G)$ given by (1.7) and Lévy measure Π, and suppose A is a Borel subset of G such that $0 < \Pi(A) < \infty$. Let $T_1 = \inf\{t > 0; \, g_{t-}^{-1} g_t \in A\}$ and inductively let $T_{n+1} = \inf\{t > T_n; \, g_{t-}^{-1} g_t \in A\}$ (set $\inf \emptyset = \infty$). Define $x_t = g_t$ for $t < T_1$ and inductively $x_t = x(T_n-) g(T_n)^{-1} g_t$ for $T_n \le t < T_{n+1}$;*

that is,

$$x_t = g(T_1-)g(T_1)^{-1}g(T_2-)g(T_2)^{-1}g(T_3-)\cdots g(T_n-)g(T_n)^{-1}g_t \quad (1.14)$$

for $T_n \leq t < T_{n+1}$ with $T_0 = 0$. Then $\infty > T_n \uparrow \infty$ almost surely and x_t is a left Lévy process whose generator restricted to $C_0^{2,l}(G)$ is given by (1.7) with Π replaced by $\Pi|_{A^c}$ and c_i replaced by $c_i - \int x_i d(\Pi|_A)$. Moreover, $\{T_1, T_2 - T_1, T_3 - T_2, \ldots\}$ is a sequence of independent exponential random variables of rate $\Pi(A)$ and

$$\{g(T_1-)^{-1}g(T_1), \quad g(T_2-)^{-1}g(T_2), \quad g(T_3-)^{-1}g(T_3), \ldots\}$$

is a sequence of independent G-valued random variables with a common distribution given by $(\Pi|_A)/\Pi(A)$. Furthermore, these two sequences and the process x_t are independent.

Note that $N_A = N|_{\mathbb{R}_+ \times A}$ is a Poisson random measure with a finite characteristic measure $\Pi|_A$, and it is determined by the two sequences $\{T_n\}$ and $\{g(T_n-)^{-1}g(T_n)\}$ given here. By the present proposition, the process x_t and N_A are independent.

Proof. By Theorem 1.1, there is a left Lévy process x_t' in G whose generator is given by (1.7) with Π replaced by $\Pi|_{A^c}$ and c_i replaced by $c_i - \int x_i d(\Pi|_A)$. By Proposition 1.5, jumps may be added to x_t' to obtain a left Lévy process g_t' whose generator restricted to $C_0^{2,l}(G)$ is given by (1.7) with the original Π and c_i. By the uniqueness stated in Theorem 1.1, the two processes g_t and g_t' must have the same distribution. The claims of the present proposition follow from a comparison between these two processes. $\qquad\square$

It is clear that a left Lévy process g_t is continuous if and only if $\Pi = 0$. In this case, it is called a left invariant diffusion process in G.

All the discussion thus far, including Theorem 1.1 and all the propositions, applies also to a right Lévy process g_t and a right invariant Feller semigroup P_t with appropriate changes: For example, the words "left" and "right" should be switched, and in (1.1), (1.3) and (1.5), (1.7), (1.9), and (1.11), one should replace $g_t^e = g_0^{-1}g_t$, gg_t^e, gh, X_i^l, and $C_0^{2,l}(G)$ by $g_t^e = g_t g_0^{-1}$, $g_t^e g$, hg, X_i^r, and $C_0^{2,r}(G)$, respectively. In Proposition 1.4, N should be regarded as the counting measure of the left jumps of g_t, defined by

$$N([0, t] \times B) = \#\{s \in (0, t]; \; g_s g_{s-}^{-1} \neq e \text{ and } g_s g_{s-}^{-1} \in B\} \quad (1.15)$$

for $t \in \mathbb{R}_+$ and $B \in \mathcal{B}(G)$. In Propositions 1.5 and 1.6, one should replace the right jump $g_{t-}^{-1}g_t$ by the left jump $g_t g_{t-}^{-1}$ and redefine y_t and x_t as

$y_t = g_t g(T_n)^{-1} \sigma_n y(T_n-)$ and $x_t = g_t g(T_n)^{-1} x(T_n-)$ for $T_n \le t < T_{n+1}$; hence, the order of the products in (1.13) and (1.14) should be reversed. A continuous right Lévy process is called a right invariant diffusion process in G. In the following such changes will not always be mentioned explicitly.

1.4. Stochastic Integral Equations

Let $\{X_1, \ldots, X_d\}$ be a basis of $\mathfrak{g}$ and let $x_1, \ldots, x_d$ be the associated coordinate functions introduced earlier. The reader is referred to Appendix B for the definitions of a d-dimensional Brownian motion with covariance matrix $\{a_{ij}\}$, the compensated random measure $\tilde{N}$ of a Poisson random measure N on $\mathbb{R}_+ \times G$, and the independence of $\{B_t\}$ and N under a filtration.

The basic results in stochastic analysis often require that the underlying probability space is equipped with a filtration that is right continuous. In the following, two types of filtrations will be mentioned: the natural filtration of a Lévy process and that generated by a Brownian motion and an independent Poisson random measure (possibly including an independent random variable as well) as defined in Section B.1. It is not hard to show that both filtrations are right continuous after completion. In the rest of this work, these filtrations will automatically be assumed to be completed, and all the filtrations will be assumed to be right continuous and complete.

The following result, due to Applebaum and Kunita [3], characterizes a Lévy process in G by a stochastic integral equation involving stochastic integrals with respect to a Brownian motion and a Poisson random measure.

Theorem 1.2. *Let g_t be a left Lévy process in G. Assume its generator L restricted to $C_0^{2,l}(G)$ is given by (1.7) with coefficients a_{jk}, c_i and the Lévy measure Π. Let N be the counting measure of the right jumps of g_t given by (1.12), and let $\{\mathcal{F}_t^e\}$ be the natural filtration of the process $g_t^e = g_0^{-1} g_t$. Then there exists a d-dimensional $\{\mathcal{F}_t^e\}$-Brownian motion $B_t = (B_t^1, \ldots, B_t^d)$ with covariance matrix a_{ij} such that it is independent of N under $\{\mathcal{F}_t^e\}$ and, $\forall f \in C_0^{2,l}(G)$,*

$$
\begin{aligned}
f(g_t) = f(g_0) &+ \sum_{i=1}^{d} \int_0^t X_i^l f(g_{s-}) \circ dB_s^i + \sum_{i=1}^{d} c_i \int_0^t X_i^l f(g_{s-}) \, ds \\
&+ \int_0^t \int_G [f(g_{s-}h) - f(g_{s-})] \tilde{N}(ds\,dh) \\
&+ \int_0^t \int_G \left[f(g_{s-}h) - f(g_{s-}) - \sum_{i=1}^{d} x_i(h) X_i^l f(g_{s-}) \right] ds\,\Pi(dh).
\end{aligned}
$$

$$(1.16)$$

Conversely, given a G-valued random variable u, a d-dimensional Brownian motion B_t with covariance matrix $\{a_{ij}\}$, some constants c_i, and a Poisson random measure N on $\mathbb{R}_+ \times G$ whose characteristic measure Π is a Lévy measure, such that u, $\{B_t\}$, and N are independent, then there is a unique càdlàg process g_t in G with $g_0 = u$, adapted to the filtration $\{\mathcal{F}_t\}$ generated by u, $\{B_t\}$, and N, such that (1.16) is satisfied for any $f \in C_0^{2,l}(G)$. Moreover, g_t is a left Lévy process in G whose generator restricted to $C_0^{2,l}(G)$ is given by (1.7).

For an integral taken over an interval on the real line with respect to any measure, the convention $\int_s^t = \int_{(s,\,t]}$ is used here and throughout the rest of this work.

The proof of Theorem 1.2 will be given in Chapter 3. In fact, a more complete result as stated in [3] will be proved. For example, it will also be shown that the pair of $\{B_t\}$ and N is uniquely determined by the Lévy process g_t and that the filtration generated by $\{B_t\}$ and N is equal to $\{\mathcal{F}_t^e\}$.

Note that g_{s-} may be replaced by g_s for the two integrals taken with respect to the Lebesgue measure ds on $\mathbb{R}_+$ in (1.16). See Appendix B.2 for the existence of the Stratonovich stochastic integral $\int_0^t X_i^l f(g_{s-}) \circ dB_s^i$. The existence of the stochastic integral taken with respect to $\tilde{N}(ds\,dh)$ is guaranteed by the discussion in Appendix B.3 and the condition (1.8). This condition also ensures that the last integral in (1.16) exists and is finite. Therefore, all the integrals in (1.16) exist and are finite.

We note that if g_t is a left Lévy process that satisfies (1.16) for any $f \in C_c^\infty(G)$, then for any $f \in C^1(G)$,

$$\int_0^t f(g_{s-}) \circ dB_s^j = \int_0^t f(g_{s-})dB_s^j + \frac{1}{2}\sum_{k=1}^d \int_0^t X_k^l f(g_s)a_{jk}\,ds, \quad (1.17)$$

where the first integral on the right-hand side is an Itô stochastic integral. To prove this, let $y = (y^1, \ldots, y^d)$ be a set of local coordinates on a relatively compact open subset U of G such that each y^i is extended to be a function in $C_c^\infty(G)$. Then $y_t = y(g_t)$ is a d-tuple of semi-martingale. Suppose $f \in C^1(G)$ is supported by U. Then $\exists \tilde{f} \in C_c^1(R^d)$ such that $f(g) = \tilde{f}(y(g))$ for $g \in G$. By (1.16) applied to y^i, (B.3), and (B.11), we have

$$\int_0^t f(g_{s-}) \circ dB_s^j = \int_0^t \tilde{f}(y_{s-}) \circ dB_s^j$$

$$= \int_0^t \tilde{f}(y_{s-})dB_s^j + \frac{1}{2}[\tilde{f}(y.), B^j]_t^c$$

$$= \int_0^t \tilde{f}(y_{s-})dB_s^j + \frac{1}{2}[f(g.), B^j]_t^c$$

$$= \int_0^t f(g_{s-})dB_s^j + \frac{1}{2}\sum_{k=1}^d \int_0^t X_k^l f(g_s)d[B^j, B^k]_s^c.$$

This proves (1.17) because $[B^j, B^k]_t = a_{jk}t$. For a general $f \in C^1(G), (1.17)$ can be proved by a standard argument using a partition of unity on G as in Appendix B.2.

Replacing f by $X_j^l f$ in (1.17), we see that the stochastic integral equation (1.16) can also be written as

$$f(g_t) = f(g_0) + M_t^f + \int_0^t Lf(g_s)\,ds, \tag{1.18}$$

where

$$M_t^f = \sum_{j=1}^d \int_0^t X_j^l f(g_{s-})\,dB_s^j + \int_0^t \int_G [f(g_{s-}h) - f(g_{s-})]\tilde{N}(ds\,dh)$$

is an L^2-martingale.

The two processes B_t and N in Theorem 1.2 will be called, respectively, the driving Brownian motion and the driving Poisson random measure of the left Lévy process g_t. It can be shown from the explicit constructions in Propositions 1.5 and 1.6 that the three left Lévy processes g_t, x_t, and y_t in these two propositions have the same driving Brownian motion $\{B_t\}$. Indeed, the same $\{B_t\}$ for g_t and y_t is a consequence of Lemma 3.10 in Chapter 3, and from which it also follows that x_t has the same $\{B_t\}$ as g_t.

In fact, the stochastic integral equation (1.16) holds for f contained in a larger function space. Let $f \in C_b(G) \cap C^2(G)$, the space of bounded functions on G possessing continuous second-order derivatives. Then the first integral in (1.16), a Stratonovich stochastic integral, exists and is finite by the discussion in Appendix B.2. The same is true for the second integral because it is a pathwise Lebesgue integral. For the last two integrals, let U be a relatively compact open neighborhood of e and let

$$\tau_U = \inf\{t > 0; \ g_t \in U^c\},$$

the first exit time from U. Then $UU = \{gh; g, h \in U\}$ is relatively compact and $\tau_U \uparrow \infty$ as $U \uparrow G$. Each of the last two integrals in (1.16), taken with respect to $\tilde{N}(ds\,dh)$ and $ds\,\Pi(dh)$, can be written as a sum $\int_0^t \int_U + \int_0^t \int_{U^c}$. The first term of this sum is finite when t is replaced by $t \wedge \tau_U = \min(t, \tau_U)$ because of the Taylor expansion (1.9) and also because $g_{s-}h \in UU$, and so is the second term because f is bounded and $\Pi(U^c) < \infty$. Therefore, all the integrals in (1.16) exist and are finite for $f \in C_b(G) \cap C^2(G)$.

Proposition 1.7. *The stochastic integral equation (1.16) for a left Lévy process g_t holds for any $f \in C_b(G) \cap C^2(G)$.*

Proof. Let $R_t(f)$ denote the right-hand side of (1.16). Fix $f \in C_b(G) \cap C^2(G)$. Let U be a relatively compact open neighborhood of e and let τ_U be the first exit time from U such that the x_i vanish on U^c. Choose $\phi \in C_c^\infty(G)$ such that $0 \le \phi \le 1$ and $\phi = 1$ on UU. Since $f\phi \in C_c^2(G) \subset C_0^{2,l}(G)$, (1.16) holds when f is replaced by $f\phi$. If $t < \tau_U$, $Y \in \mathfrak{g}$, and $h \in U$, then $(f\phi)(g_t) = f(g_t)$, $Y^l(f\phi)(g_{t-}) = Y^l f(g_{t-})$, and $(f\phi)(g_{t-}h) = f(g_{t-}h)$. It follows that, for $t < \tau_U$,

$$f(g_t) = f(g_0) + \sum_{i=1}^d \int_0^t X_i^l f(g_{s-}) \circ dB_s^i + \sum_{i=1}^d c_i \int_0^t X_i^l f(g_s)\, ds$$

$$+ \int_0^t \int_G [f(g_{s-}h) - f(g_{s-})]\tilde{N}(ds\, dh)$$

$$+ \int_0^t \int_{U^c} [(f\phi)(g_{s-}h) - f(g_{s-}h)]\tilde{N}(ds\, dh)$$

$$+ \int_0^t \int_G \left[f(g_s h) - f(g_s) - \sum_{i=1}^d x_i(h) X_i^l f(g_s) \right] ds\, \Pi(dh)$$

$$+ \int_0^t \int_{U^c} [(f\phi)(g_s h) - f(g_s h)]\, ds\, \Pi(dh)$$

$$= R_t(f) + \int_0^t \int_{U^c} [(f\phi)(g_{s-}h) - f(g_{s-}h)]N(ds\, dh)$$

$$= R_t(f) + \sum_{T_n \le t} [(f\phi)(g_{T_n-}\sigma_n) - (f\phi)(g_{T_n-})],$$

where the random times $T_n \uparrow \infty$ and G-valued random variables σ_n are determined by the Poisson random measure $N|_{\mathbb{R}_+ \times U^c}$. Since $\tau_U \uparrow \infty$ and $T_1 \uparrow \infty$ as $U \uparrow G$, it follows that $f(g_t) = R_t(f)$ almost surely for any $t \in \mathbb{R}_+$. □

In Theorem 1.2, if the Lévy measure Π satisfies the finite first moment condition (1.10), then the integral

$$\int_0^t \int_G [f(g_{s-}h) - f(g_{s-})]N(ds\, dh)$$

exists and the stochastic integral equation (1.16) can be simplified as follows:

$$f(g_t) = f(g_0) + \sum_{i=1}^{d} \int_0^t X_i^l f(g_{s-}) \circ dB_s^i + \sum_{i=1}^{d} b_i \int_0^t X_i^l f(g_s) \, ds$$

$$+ \int_0^t \int_G [f(g_{s-}h) - f(g_{s-})] N(ds \, dh), \qquad (1.19)$$

where $b_i = c_i - \int_G x_i(h) \Pi(dh)$. The generator L of g_t, restricted to $C_0^{2,l}(G)$, is now given by (1.11).

The stochastic integral equation (1.19) in fact holds for any $f \in C^2(G)$. To see this, note that the first two integrals in (1.19) clearly exist and are finite for $f \in C^2(G)$. The third integral can be written as a sum $\int_0^t \int_U + \int_0^t \int_{U^c}$, where U is a relatively compact open neighborhood of e. The first term of this sum is finite when t is replaced by $t \wedge \tau_U$ because of (1.10) and also because $g_{s-}h \in UU$ for $h \in U$. The second term is

$$\int_0^t \int_{U^c} [f(g_{s-}h) - f(g_{s-})] N(ds \, dh) = \sum_{T_n \leq t} [f(g_{T_n-}\sigma_n) - f(g_{T_n-})], \quad (1.20)$$

where T_n and σ_n are given in the proof of Proposition 1.7. The sum on the right-hand side of (1.20) contains only finitely many terms. Therefore, all the integrals in (1.19) exist and are finite. As in the proof of Proposition 1.7, it can be shown that (1.19) holds for any $f \in C^2(G)$.

Remark. Let $C_b^{2,l}(G)$ be the space of functions $f \in C^2(G)$ such that $X^l f \in C_b(G)$ and $X^l Y^l f \in C_b(G)$ for any $X, Y \in \mathfrak{g}$. The function space $C_b^{2,r}(G)$ is defined similarly with X^l and Y^l replaced by X^r and Y^r. We note that, for $f \in C_b^{2,l}(G)$, the expression on the right-hand side of (1.7) is bounded. Therefore, Lf may be defined for $f \in C_b^{2,l}(G)$ as this expression or as the simpler expression on the right-hand side of (1.11) when the Lévy measure Π has a finite first moment. Note that, for $f \in C_b^{2,l}(G)$, M_t^f in (1.18) is still an L^2-martingale. Taking the expectation on (1.18) yields $E[f(g_t)] = E[f(g_0)] + E[\int_0^t Lf(g_s) \, ds]$. Taking the derivative of $P_t f = E[f(g_t)]$ at $t = 0$, we obtain

$$\forall f \in C_b^{2,l}(G) \quad \text{and} \quad g \in G, \qquad \frac{d}{dt} P_t f(g) \mid_{t=0} = Lf(g). \qquad (1.21)$$

A similar conclusion holds for a right Lévy process g_t in G and $f \in C_b^{2,r}(G)$.

Sometimes it is convenient to work with stochastic integral equations driven by a standard Brownian motion. Let $a = \{a_{jk}\}$ be the covariance matrix of the Brownian motion $B_t = (B_t^1, \ldots, B_t^d)$ in (1.16). Suppose $\sigma = \{\sigma_{ij}\}$ is an $m \times d$ matrix such that $a = \sigma'\sigma$, where σ' is the matrix transpose of σ. Note that such a matrix σ always exists for some integer $m > 0$, in particular, for $m = d$. Then there is an m-dimensional standard Brownian motion $W_t = (W_t^1, \ldots, W_t^m)$ such that $B_t = W_t\sigma$. When a is invertible and $m = d$, one may let $W_t = B_t\sigma^{-1}$, but in general one may have to extend the original probability space Ω (see chapter II, theorem 7.1' on p. 90 in [33]; not theorem 7.1 on p. 84). More precisely, the standard Brownian motion W_t may have to be defined on an extension of the original probability space $(\Omega, \mathcal{F}, P)$, that is, on $(\Omega \times \Omega', \mathcal{F} \times \mathcal{F}', P \times P')$, where $(\Omega', \mathcal{F}', P')$ is some probability space. Then a process x_t on the original probability space may be naturally regarded as a process on the extended probability space by setting $x_t(\omega, \omega') = x_t(\omega)$ for $(\omega, \omega') \in \Omega \times \Omega'$. Let

$$Y_j = \sum_{i=1}^{d} \sigma_{ji} X_i \quad \text{for } 1 \leq j \leq m \quad \text{and} \quad Z = \sum_{i=1}^{d} c_i X_i. \tag{1.22}$$

Then the stochastic integral equation (1.16) becomes

$$f(g_t) = f(g_0) + \sum_{j=1}^{m} \int_0^t Y_j^l f(g_{s-}) \circ dW_s^j + \int_0^t Z^l f(g_s) \, ds$$

$$+ \int_0^t \int_G [f(g_{s-}h) - f(g_{s-})] \tilde{N}(ds\,dh)$$

$$+ \int_0^t \int_G \left[f(g_{s-}h) - f(g_{s-}) - \sum_{i=1}^{d} x_i(h) X_i^l f(g_{s-}) \right] ds\,\Pi(dh) \tag{1.23}$$

for $f \in C_b(G) \cap C^2(G)$. Moreover, the generator L of g_t restricted to $C_0^{2,l}(G)$, given by (1.7), becomes

$$Lf(g) = \frac{1}{2} \sum_{j=1}^{m} Y_j^l Y_j^l f(g) + Z^l f(g)$$

$$+ \int_G \left[f(gh) - f(g) - \sum_{i=1}^{d} x_i(h) X_i^l f(g) \right] \Pi(dh). \tag{1.24}$$

However, Equation (1.23) can be converted to (1.16) using (1.22) and $B_t = W_t\sigma$. As a direct consequence of Theorem 1.2, we obtain the following result.

Corollary 1.1. *Let g_t be a left Lévy process in G with Lévy measure Π, let N be its counting measure of right jumps given by (1.12), and let $\{\mathcal{F}_t^e\}$ be the natural filtration of the process $g_t^e = g_0^{-1}g_t$. Then there are $Y_1, \ldots, Y_m, Z \in \mathfrak{g}$*

and an m-dimensional standard Brownian motion $W_t = (W_t^1, \ldots, W_t^m)$, *adapted to* $\{\mathcal{F}_t^e\}$ *and independent of* N *under* $\{\mathcal{F}_t^e\}$, *possibly defined on an extension of the original probability space, such that (1.23) holds for any* $f \in C_b(G) \cap C^2(G)$.

Conversely, given $Y_1, \ldots, Y_m, Z \in \mathfrak{g}$, *a G-valued random variable* u, *an m-dimensional standard Brownian motion* $\{W_t\}$, *and a Poisson random measure* N *on* $\mathbb{R}_+ \times G$ *with characteristic measure being a Lévy measure, such that* u, $\{W_t\}$, *and* N *are independent, there exists a unique càdlàg process* g_t *in G with* $g_0 = u$, *adapted to the filtration generated by* u, $\{B_t\}$, *and* N, *such that (1.23) holds for any* $f \in C_0^{2,l}(G)$. *Moreover,* g_t *is a left Lévy process in G whose generator L restricted to* $C_0^{2,l}(G)$ *is given by (1.24).*

Note that the integer m in the first half of this corollary can always be taken to be $d = \dim(G)$ and, in this case, if $\{a_{ij}\}$ is nondegenerate, then no extension of probability space is required.

If Π satisfies the finite first moment condition (1.10), then (1.23) simplifies to

$$f(g_t) = f(g_0) + \sum_{i=1}^{m} \int_0^t Y_i^l f(g_{s-}) \circ dW_s^i + \int_0^t Y_0^l f(g_s)\, ds$$

$$+ \int_0^t \int_G [f(g_{s-}h) - f(g_{s-})] N(ds\, dh) \qquad (1.25)$$

for $f \in C^2(G)$, where $Y_0 = Z - \sum_{i=1}^{d} [\int_G x_i(h)\Pi(dh)] X_i$. In this case, the generator L restricted to $C_0^{2,l}(G)$ has the following simpler form. For $f \in C_0^{2,l}(G)$,

$$Lf(g) = \frac{1}{2} \sum_{i=1}^{m} Y_i^l Y_i^l f(g) + Y_0^l f(g) + \int_G [f(gh) - f(g)]\Pi(dh). \quad (1.26)$$

If we assume the stronger condition that the Lévy measure Π is finite, then N is determined by a sequence of random times $T_n \uparrow \infty$ with iid differences $T_n - T_{n-1}$ of an exponential distribution and an independent sequence of iid G-valued random variables σ_n. In this case, the last term in (1.25) may be written as $\sum_{T_n \leq t} [f(g_{T_n-}\sigma_n) - f(g_{T_n-})]$. Therefore, the stochastic integral equation (1.25) is equivalent to the stochastic differential equation

$$dg_t = \sum_{i=1}^{m} Y_i^l(g_t) \circ dW_t^i + Y_0^l(g_t)\, dt \qquad (1.27)$$

on G together with the jump conditions $g_t = g_{t-}\sigma_n$ at $t = T_n$ for $n = 1, 2, \ldots$. More precisely, the following result (whose simple proof is omitted) holds.

Proposition 1.8. *Let g_t be a left Lévy process in G with a finite Lévy measure Π, let $u = g_0$, and let random times $T_n \uparrow \infty$ and G-valued random variables σ_n be the two sequences determined by N, the counting measure of right jumps of g_t defined by (1.12). Then the process g_t may be obtained by successively solving the stochastic differential equation (1.27) on the random intervals $[T_n, T_{n+1})$ with initial conditions $g(T_n) = g(T_n-)\sigma_n$ for $n = 0, 1, 2, \ldots,$ where $T_0 = 0$, $g_{0-} = e$, and $\sigma_0 = u$.*

By this proposition, a left Lévy process g_t with a finite Lévy measure may be regarded as a left invariant diffusion process, determined by (1.27), interlaced with iid random jumps at exponentially distributed random time intervals. It can be shown (see Applebaum [2]) that a general left Lévy process can be obtained as a limit of such processes.

If $Y_0 = Y_1 = \cdots = Y_m = 0$ and Π is finite, then the Lévy process g_t is a discrete process consisting of a sequence of iid jumps at exponentially spaced random time intervals. More precisely, $g_t = g_0$ for $0 \leq t < T_1$, and

$$g_t = g_0\sigma_1\sigma_2\cdots\sigma_i \quad \text{for } T_i \leq t < T_{i+1} \quad \text{and} \quad i \geq 1, \tag{1.28}$$

where the two sequences T_n and σ_n are determined by the driving Poisson random measure N of g_t. Therefore, the random walks in G, which are defined as the products of iid G-valued random variables, can be regarded as discrete-time Lévy processes.

To end this section, we present a form of stochastic integral equation that holds for a general left Lévy process g_t and any $f \in C^2(G)$. Let U be a relatively compact open neighborhood of e. The stochastic integral equation (1.23) may be written as

$$f(g_t) = f(g_0) + \sum_{j=1}^{m} \int_0^t Y_j^l f(g_{s-}) \circ dW_s^j + \int_0^t Z_0^l f(g_s)\, ds$$

$$+ \int_0^t \int_U [f(g_{s-}h) - f(g_{s-})] \tilde{N}(ds\, dh)$$

$$+ \int_0^t \int_{U^c} [f(g_{s-}h) - f(g_{s-})] N(ds\, dh)$$

$$+ \int_0^t \int_U \left[f(g_{s-}h) - f(g_{s-}) - \sum_{i=1}^{d} x_i(h) X_i^l f(g_{s-}) \right] ds\, \Pi(dh),$$
$$\tag{1.29}$$

where $Z_0 = Z - \sum_{i=1}^{d} \int_{U^c} x_i(h)\Pi(dh) X_i$. By (1.20), it is easy to show that (1.29) holds for any $f \in C^2(G)$. This is essentially equation (3.6) in Applebaum and Kunita [3].

1.5. Lévy Processes in $GL(d, \mathbb{R})$

In this section, let G be the general linear group $GL(d, \mathbb{R})$, the group of the $d \times d$ real invertible matrices. This is a d^2-dimensional Lie group with Lie algebra $\mathfrak{g}$ being the space $\mathfrak{gl}(d, \mathbb{R})$ of all the $d \times d$ real matrices and the Lie bracket given by $[X, Y] = XY - YX$. See Appendix A.1 for more details.

We may identify $\mathfrak{g} = \mathfrak{gl}(d, \mathbb{R})$ with the Euclidean space $\mathbb{R}^{d^2}$ and $G = GL(d, \mathbb{R})$ with a dense open subset of $\mathbb{R}^{d^2}$. For any $X = \{X_{ij}\} \in \mathbb{R}^{d^2}$, its Euclidean norm

$$|X| = \left(\sum_{i,j} X_{ij}^2 \right)^{1/2} = [\text{Trace}(XX')]^{1/2}$$

satisfies $|XY| \leq |X| |Y|$ for any $X, Y \in \mathbb{R}^{d^2}$, where X' is the transpose of X and XY is the matrix product.

Let E_{ij} be the matrix that has 1 at place (i, j) and 0 elsewhere. Then the family $\{E_{ij}; i, j = 1, 2, \ldots, d\}$ is a basis of $\mathfrak{g}$. Let $\{x_{ij}; i, j = 1, 2, \ldots, d\}$ be a set of associated coordinate functions and let $x = \{x_{ij}\}$. It is easy to show that one may take $x(g) = g - I_d$ for g contained in a neighborhood of $e = I_d$ (the $d \times d$ identity matrix). In general, by (1.6), the coordinate functions satisfy $x(g) = g - I_d + O(|g - I_d|^2)$ for g contained in a neighborhood of e.

For $g \in G = GL(d, \mathbb{R})$, the tangent space $T_g G$ can be identified with $\mathbb{R}^{d^2}$; therefore, any element X of $T_g G$ can be represented by a $d \times d$ real matrix $\{X_{ij}\}$ in the sense that

$$\forall f \in C(G), \qquad Xf = \sum_{i,j=1}^{d} X_{ij} \frac{\partial}{\partial g_{ij}} f(g),$$

where g_{ij}, for $i, j = 1, 2, \ldots, d$, are the standard coordinates on $\mathbb{R}^{d^2}$. It can be shown that for $g, h \in GL(d, \mathbb{R})$ and $X \in T_e G = \mathfrak{gl}(d, \mathbb{R})$, $DL_g \circ DR_h(X)$ is represented by the matrix product gXh, where X is identified with its matrix representation $\{X_{ij}\}$. Therefore, we may write gXh for $DL_g \circ DR_h(X)$. Thus, $X^l(g) = gX$ and $X^r(g) = Xg$. This shorthand notation may even be used for a general Lie group G.

Let g_t be a left Lévy process in $G = GL(d, \mathbb{R})$. Then it satisfies the stochastic integral equation (1.23) for any $f \in C_b(G) \cap C^2(G)$. Let f be the matrix-valued function on G defined by $f(g) = g_{ij}$ for $g \in G$. Although f

is not contained in $C_b(G) \cap C^2(G)$, at least formally, (1.23) leads to the following stochastic integral equation in matrix form:

$$g_t = g_0 + \sum_{i=1}^{m} \int_0^t g_{s-} Y_i \circ dW_s^i + \int_0^t g_s Z \, ds + \int_0^t \int_G g_{s-}(h - I_d) \tilde{N}(ds \, dh)$$

$$+ \int_0^t \int_G g_s [h - I_d - x(h)] \, ds \, \Pi(dh). \tag{1.30}$$

For any process y_t taking values in a Euclidean space, let

$$y_t^* = \sup_{0 \le s \le t} |y_s|.$$

Theorem 1.3. *Let g_t be a left Lévy process in $G = GL(d, \mathbb{R})$ with Lévy measure Π and let N be the counting measure of right jumps of g_t as defined by (1.12). Assume*

$$E[|g_0|^2] < \infty \quad \text{and} \quad \int_G |h - I_d|^2 \Pi(dh) < \infty. \tag{1.31}$$

Then, for any $t > 0$,

$$E\left[(g_t^*)^2\right] < \infty.$$

Moreover, there are $Y_1, \ldots, Y_m, Z \in \mathfrak{g} = \mathfrak{gl}(d, \mathbb{R})$ and an m-dimensional standard Brownian motion $W_t = (W_t^1, \ldots, W_t^m)$, adapted to the filtration $\{\mathcal{F}_t^e\}$ generated by the process $g_t^e = g_0^{-1} g_t$ and independent of N under $\{\mathcal{F}_t^e\}$, possibly defined on an extension of the original probability space, such that (1.30) holds.

Conversely, given $Y_1, \ldots, Y_m, Z \in \mathfrak{g}$, a G-valued random variable g_0, an m-dimensional standard Brownian motion $\{W_t\}$, and a Poisson random measure N on $\mathbb{R}_+ \times G$ with characteristic measure Π being a Lévy measure such that (1.31) is satisfied, there is a unique càdlàg process g_t in G with g_0 as given, adapted to the filtration generated by g_0, $\{W_t\}$, and N, such that (1.30) holds. Moreover, g_t is a left Lévy process in G with Lévy measure Π.

Note that (1.8) implies $\int_U |h - I_d|^2 \Pi(dh) < \infty$ for any compact subset U of G; therefore, the integrability condition for Π in Theorem 1.3 is automatically satisfied if Π is supported by a compact subset of G. Because $[h - I_d - x(h)] = O(|h - I_d|^2)$ for h contained in a neighborhood of I_d, the finiteness of $E[(g_t^*)^2]$ implies the existence of the last integral in (1.30); hence, all the integrals in (1.30); exist and are finite. If the coordinate functions are chosen so that $x(g) = g - I_d$ for g contained in a neighborhood of I_d, then the integrand of the last integral in (1.30) vanishes for h contained in a neighborhood of I_d. We also note that, as in Corollary 1.1, the integer m in the first half of Theorem 1.3 may always be taken to be $\dim(G) = d^2$.

Proof. Let g_t be a left Lévy process in G. Since g_0 and the process $g_t^e = g_0^{-1} g_t$ are independent, for simplicity, we may assume $g_0 = I_d$. Fix two positive integers $n < m$. Let f be a G-valued function with components contained in $C_c^2(G)$ such that $f(g) = g$ for any $g \in G$ with $|g| \leq m$ and $|f(g) - f(h)| \leq C|g - h|$ for any $g, h \in G$ and some $C > 0$, let $\tau = \inf\{t > 0; |g_t| > n\}$, and let

$$U = \{h \in G;\ |gh| \leq m \text{ for any } g \in G \text{ with } |g| \leq n\}.$$

We may assume that the x_{ij} are supported by U. Let $t \wedge \tau = \min(t, \tau)$. By (1.23),

$$
f(g_{t\wedge\tau}) = I_d + \sum_{i=1}^{m} \int_0^{t\wedge\tau} g_{s-} Y_i \circ dW_s^i + \int_0^{t\wedge\tau} g_s Z\, ds
$$
$$
+ \int_0^{t\wedge\tau} \int_U g_{s-}(h - I_d)\tilde{N}(ds\, dh)
$$
$$
+ \int_0^{t\wedge\tau} \int_{U^c} [f(g_{s-}h) - f(g_{s-})]\tilde{N}(ds\, dh)
$$
$$
+ \int_0^{t\wedge\tau} \int_U g_s[h - I_d - x(h)]\, ds\, \Pi(dh)
$$
$$
+ \int_0^{t\wedge\tau} \int_{U^c} [f(g_s h) - f(g_s)]\, ds\, \Pi(dh).
$$

Since $|f(gh) - f(g)| \leq C|gh - g| \leq C|g|\,|h - I_d|$,

$$
E\left\{ \left| \int_0^{t\wedge\tau} \int_{U^c} [f(g_{s-}h) - f(g_{s-})]\tilde{N}(ds\, dh) \right|^2 \right\}
$$
$$
= E \int_0^{t\wedge\tau} \int_{U^c} |f(g_{s-}h) - f(g_{s-})|^2\, ds\, \Pi(dh)
$$
$$
\leq n^2 C^2 t \int_{U^c} |h - I_d|^2 \Pi(dh) \to 0
$$

as $m \uparrow \infty$ because $U \uparrow G$. Similarly,

$$
E\left\{ \left| \int_0^{t\wedge\tau} \int_{U^c} [f(g_s h) - f(g_s)]\, ds\, \Pi(dh) \right| \right\} \leq nCt \int_{U^c} |h - I_d| \Pi(dh) \to 0
$$

as $m \uparrow \infty$. Since $f(g) \to g$ as $m \uparrow \infty$ for any $g \in G$, we have, for any $t \in \mathbb{R}_+$,

$$g_{t \wedge \tau} = I_d + J_t + K_t + L_t,$$

where

$$J_t = \sum_{i=1}^{m} \int_0^{t \wedge \tau} g_{s-} Y_i \, dW_s^i,$$

$$K_t = \int_0^{t \wedge \tau} \int_G g_{s-}(h - I_d) \tilde{N}(ds \, dh),$$

and

$$L_t = \int_0^{t \wedge \tau} \left\{ \frac{1}{2} \sum_{i=1}^{m} g_s Y_i Y_i + g_s Z + \int_G g_s[h - I_d - x(h)] \Pi(dh) \right\} ds.$$

Applying Doob's norm inequality (B.1) to the R^{d^2}-valued martingale J_t and using the basic property of stochastic integrals, we obtain

$$E\left[(J_t^*)^2\right] \le 4E\left[|J_t|^2\right] \le C_1 E\left[\sum_{i=1}^{m} \int_0^{t \wedge \tau} |g_s Y_i|^2 \, ds\right] \le C_2 E\left[\int_0^{t \wedge \tau} (g_s^*)^2 ds\right]$$

for positive constants C_1 and C_2. Similarly, because $\int_G |h - I_d|^2 \Pi(dh)$ is finite,

$$E\left[(K_t^*)^2\right] \le 4E\left[\int_0^{t \wedge \tau} \int_G |g_s(h - I_d)|^2 \Pi(dh) \, ds\right] \le C_3 E\left[\int_0^{t \wedge \tau} (g_s^*)^2 ds\right]$$

for some constant $C_3 > 0$. Using the Schwartz inequality and the fact that

$$\int_G |h - I_d - x(h)| \Pi(dh) < \infty,$$

we obtain

$$E\left[(L_t^*)^2\right] \le C_4 t E\left[\int_0^{t \wedge \tau} (g_s^*)^2 ds\right]$$

for some constant $C_4 > 0$. It follows that

$$E\left[(g_{t \wedge \tau}^*)^2\right] \le B + C_t E\left[\int_0^{t \wedge \tau} (g_s^*)^2 ds\right] \le B + C_t t n^2$$

for some positive constants B and C_t, with the latter depending on t. Hence, $E[(g_{t \wedge \tau}^*)^2]$ is finite. Moreover,

$$E\left[(g_{t \wedge \tau}^*)^2\right] \le B + C_t \int_0^t E\left[(g_{s \wedge \tau}^*)^2\right] ds.$$

The Gronwall inequality (see, for example, [34, lemma 18.4], or simply apply the inequality displayed here repeatedly to itself) yields $E[(g_{t\wedge\tau}^*)^2] \le Be^{C_t t}$. Since $\tau \uparrow \infty$ as $n \uparrow \infty$, this proves that $E[(g_t^*)^2] < \infty$ and (1.30) holds for any $t \in \mathbb{R}_+$.

If we repeat the preceding argument with $g_0 = 0$ (the origin in R^{d^2}), then $B = 0$. This proves the uniqueness of the process g_t satisfying (1.30). The existence of such a process follows from Corollary 1.1, but it can also be established directly using the usual successive approximation method for proving the existence of the strong solution of a stochastic differential equation (see, for example, [33, sec. 3 in ch. 4]). In view of Corollary 1.1, the theorem is proved.   □

Note that if the Lévy measure Π in Theorem 1.3 also satisfies $\int_G |h - I_d| \Pi(dh) < \infty$, then the stochastic integral equation (1.30) becomes

$$g_t = g_0 + \sum_{i=1}^m \int_0^t g_{s-} Y_i \circ dW_s^i + \int_0^t g_s Y_0 \, ds + \int_0^t \int_G g_{s-}(h - I_d) N(ds \, dh),$$
(1.32)

where $Y_0 = Z - \int_G x(h) \Pi(dh)$.

By (1.29), we see that a general left Lévy process g_t in $G = GL(d, \mathbb{R})$ satisfies the following stochastic integral equation obtained by Holevo [30]:

$$g_t = g_0 + \sum_{i=1}^m \int_0^t g_{s-} Y_i \circ dW_s^i + \int_0^t g_s Z_0 \, ds + \int_0^t \int_U g_{s-}(h - I_d) \tilde{N}(ds \, dh)$$

$$+ \int_0^t \int_{U^c} g_{s-}[h - I_d] N(ds \, dh)$$

$$+ \int_0^t \int_U g_s[h - I_d - x(h)] \, ds \, \Pi(dh),$$
(1.33)

where $Z_0 = Z - \int_{U^c} x(h) \Pi(dh)$.

2

Induced Processes

In this chapter, we consider the processes in a homogeneous space of a Lie group G induced by Lévy processes in G. In Section 2.1, these processes are introduced as one-point motions of Lévy processes in Lie groups. We derive the stochastic integral equations satisfied by these processes and discuss their Markov property. In Section 2.2, we consider the Markov processes in a homogeneous space of G that are invariant under the action of G. We study the relations among various invariance properties, we present Hunt's result on the generators of G-invariant processes, and we show that these processes are one-point motions of left Lévy processes in G that are also invariant under the right action of the isotropy subgroup. The last section of this chapter contains a discussion of Riemannian Brownian motions in Lie groups and homogeneous spaces.

2.1. One-Point Motions

Let G be a Lie group that acts on a manifold M on the left and let g_t be a process in G. For any $x \in M$, the process $x_t = g_t x$ will be called the one-point motion of g_t in M starting from x. In general, the one-point motion of a Markov process in G is not a Markov process in M. However, if g_t is a right Lévy process in G, then its one-point motions are Markov processes in M with a common transition semigroup P_t^M given by

$$P_t^M f(x) = E\left[f\left(g_t^e x\right)\right] \qquad \text{for } f \in C_0(M). \tag{2.1}$$

This can be proved as follows. Let $\mathcal{F}_t$ be the natural filtration of the process g_t^e. For $s < t$ and $f \in \mathcal{B}(M)_+$,

$$E[f(x_t) \mid \mathcal{F}_s] = E[f(g_t x) \mid \mathcal{F}_s] = E\left[f\left(g_t g_s^{-1} g_s x\right) \mid \mathcal{F}_s\right]$$
$$= E\left[f\left(g_t g_s^{-1} x_s\right) \mid \mathcal{F}_s\right] = E\left[f\left(g_{t-s}^e z\right)\right] \mid_{z=x_s}.$$

From (2.1), it is easy to see that P_t^M is a Feller semigroup. By the duality between the left and right Lévy processes, and that between the left and right

32

actions, we see that this conclusion holds for the one-point motion of a left Lévy process in a Lie group that acts on M on the right. To summarize, we record the following simple fact:

Proposition 2.1. *Let G be a Lie group that acts on a manifold M on the left (resp. on the right) and let g_t be a right (resp. left) Lévy process in G. Then $\forall x \in M$, $g_t x$ (resp. $x g_t$) is a Markov process in M with a Feller transition semigroup $P_t^M f(x) = E[f(g_t^e x)]$ (resp. $P_t^M f(x) = E[f(x g_t^e)]$) for $x \in M$ and $f \in C_0(M)$.*

In the rest of this work, an action of a Lie group on a manifold will always mean a left action unless stated otherwise.

Let g_t be a right Lévy process in G starting at e. Then it is the solution of a stochastic integral equation of the following form, which is just the version of (1.23) for a right Lévy process: For $f \in C_b(G) \cap C^2(G)$,

$$
\begin{aligned}
f(g_t) = f(e) + \sum_{j=1}^{m} \int_0^t Y_j^r f(g_{s-}) \circ dW_s^j + \int_0^t Z^r f(g_s)\, ds \\
+ \int_0^t \int_G [f(hg_{s-}) - f(g_{s-})] \tilde{N}(ds\, dh) \\
+ \int_0^t \int_G \left[f(hg_s) - f(g_s) - \sum_{i=1}^{d} x_i(h) X_i^r f(g_s) \right] ds\, \Pi(dh), \quad (2.2)
\end{aligned}
$$

where Y_j, Z, X_i, x_i, and W_t have the same meanings as in (1.23), N is the counting measure of the left jumps, and Π is the Lévy measure of g_t.

Any $X \in \mathfrak{g}$ induces a vector field X^* on M given by

$$
X^* f(x) = \frac{d}{dt} f(e^{tX} x) \mid_{t=0} \quad (2.3)
$$

for any $f \in C^1(M)$ and $x \in M$. Let $\pi_x : G \to M$ be the map $\pi_x(g) = gx$. If $f \in C_c^2(M)$, then $f \circ \pi_x \in C_b(G) \cap C^2(G)$ with $X^r(f \circ \pi_x) = (X^* f) \circ \pi_x$ for $X \in \mathfrak{g}$. From (2.2), we obtain the following stochastic integral equation for the one-point motion $x_t = g_t x$ of g_t in M: For $f \in C_c^2(M)$,

$$
\begin{aligned}
f(x_t) = f(x) + \sum_{j=1}^{m} \int_0^t Y_j^* f(x_{s-}) \circ dW_s^j + \int_0^t Z^* f(x_s)\, ds \\
+ \int_0^t \int_G [f(hx_{s-}) - f(x_{s-})] \tilde{N}(ds\, dh) \\
+ \int_0^t \int_G \left[f(hx_s) - f(x_s) - \sum_{i=1}^{d} x_i(h) X_i^* f(x_s) \right] ds\, \Pi(dh) \quad (2.4)
\end{aligned}
$$

$$= f(x) + \sum_{i=1}^{m} \int_0^t Y_i^* f(x_{s-}) \, dW_s^i$$

$$+ \int_0^t \int_G [f(hx_{s-}) - f(x_{s-})] \tilde{N}(ds \, dh) + \int_0^t L^M f(x_s) \, ds, \quad (2.5)$$

where L^M is the generator of the Feller process x_t whose restriction to $C_c^2(M)$ is given by

$$L^M f(x) = \frac{1}{2} \sum_{j=1}^{m} Y_j^* Y_j^* f(x) + Z^* f(x)$$

$$+ \int_G \left[f(hx) - f(x) - \sum_{i=1}^{d} x_i(h) X_i^* f(x) \right] \Pi(dh). \quad (2.6)$$

This last statement may be easily verified by differentiating $E[f(x_t)]$ using (2.5) at $t = 0$.

If Π satisfies the finite first moment condition (1.10), then (2.4) and (2.6) simplify as follows. For $f \in C_c^2(M)$,

$$f(x_t) = f(x) + \sum_{j=1}^{m} \int_0^t Y_j^* f(x_{s-}) \circ dW_s^j + \int_0^t Y_0^* f(x_s) \, ds$$

$$+ \int_0^t \int_G [f(hx_{s-}) - f(x_{s-})] N(ds \, dh) \quad (2.7)$$

and

$$L^M f(x) = \frac{1}{2} \sum_{j=1}^{m} Y_j^* Y_j^* f(x) + Y_0^* f(x) + \int_G [f(hx) - f(x)] \Pi(dh), \quad (2.8)$$

where $Y_0 = Z - \sum_{i=1}^{d} [\int_G x_i(h) \Pi(dh)] X_i$.

We note that because $f \circ \pi_x \in C_b(G) \cap C^2(G)$ for $f \in C_b(M) \cap C^2(M)$ and $f \circ \pi_x \in C^2(G)$ for $f \in C^2(G)$, (2.4) in fact holds for $f \in C_b(M) \cap C^2(M)$ and (2.7) holds for $f \in C^2(M)$.

By Proposition 1.8, suitably adjusted for right Lévy processes, if Π is finite, then Equation (2.7) is equivalent to the stochastic differential equation

$$dx_t = \sum_{j=1}^{m} Y_j^*(x_t) \circ dW_t^j + Y_0^*(x_t) \, dt \quad (2.9)$$

on M driven only by the standard Brownian motion W_t, together with the jump conditions $x_t = \sigma_n x_{t-}$ at $t = T_n$ for $n = 1, 2, \ldots$, where the random times $T_n \uparrow \infty$ and the G-valued random variables σ_n are determined by N.

We have seen that the one-point motion of a right Lévy process in G with an arbitrary starting point in M is a Markov process. If we only require that the one-point motion with a fixed starting point be a Markov process, the right invariance of g_t may be weakened.

Now let g_t be a Markov process in G starting at e with transition semigroup P_t and assume that G acts transitively on M. Fix a point $p \in M$ and let H be the isotropy subgroup of G at p; that is, $H = \{g \in G; \ gp = p\}$. If g_t is right H-invariant as defined in Section 1.1, then it is easy to show that $P_t(f \circ \pi_p)(g)$ depends only on $\pi_p(g)$ for any $f \in \mathcal{B}(M)_+$; therefore, $Q_t f$ given by $Q_t f(x) = P_t(f \circ \pi_p)(g)$, where $g \in G$ is chosen to satisfy $x = gp$, is well defined and

$$(Q_t f) \circ \pi_p = P_t(f \circ \pi_p). \tag{2.10}$$

From this, it is easy to show that $\{Q_t; \ t \in \mathbb{R}_+\}$ is a semigroup of probability kernels on M.

Let x_t be a Markov process in a manifold M with a transition semigroup Q_t and let G be a Lie group acting on M. The process x_t or the semigroup Q_t will be called G-invariant if $Q_t(f \circ g) = (Q_t f) \circ g$ for any $f \in \mathcal{B}(M)_+$ and $g \in G$, where g is regarded as the map $M \ni x \mapsto gx \in M$. In this case, it is easy to show that gx_t is a Markov process in M with semigroup Q_t for any $g \in G$.

Proposition 2.2. *Let G be a Lie group that acts on a manifold M transitively, let H be the isotropy subgroup of G at a fixed point $p \in M$, and let g_t be a Markov process in G starting at e with transition semigroup P_t. If g_t is right H-invariant, then $x_t = g_t p$ is a Markov process in M with transition semigroup Q_t given by (2.10). Moreover, if g_t is also left invariant, then x_t is G-invariant. Furthermore, if P_t is a left invariant Feller semigroup, then Q_t is a Feller semigroup provided either M or H is compact.*

Proof. Let $\{\mathcal{F}_t\}$ be the natural filtration of g_t. Then, for $f \in \mathcal{B}(M)_+$ and $s < t$,

$$E[f(x_t) \mid \mathcal{F}_s] = E[f(g_t p) \mid \mathcal{F}_s] = P_{t-s}(f \circ \pi_p)(g_s)$$
$$= (Q_{t-s} f) \circ \pi_p(g_s) = Q_{t-s} f(x_s).$$

This proves that $x_t = g_t p$ is a Markov process in M with transition semigroup Q_t. In (2.10), replacing f by $f \circ h$ with $h \in G$, we obtain

$$Q_t(f \circ h)(x) = P_t(f \circ h \circ \pi_p)(g) = P_t(f \circ \pi_p \circ L_h)(g).$$

It follows that Q_t is G-invariant if P_t is left invariant.

Now assume that P_t is a left invariant Feller semigroup. Then $P_t f(g) = E[f(gg_t)]$ for $f \in \mathcal{B}(G)_+$ and $g \in G$. Let $f \in C_b(M)$ and $x_n \to x$ on M. Then $\exists g_n, g \in G$ such that $x_n = g_n p$, $x = gp$, and $g_n \to g$. It follows that $Q_t f(x_n) = P_t(f \circ \pi_p)(g_n) \to P_t(f \circ \pi_p)(g) = Q_t f(x)$; hence, $Q_t f \in C_b(M)$ for $f \in C_b(M)$.

Let $f \in C_0(M)$. We want to show $Q_t f \in C_0(M)$. If M is compact, then $C_0(M) = C_b(M)$ and this is proved in the last paragraph. Assume H is compact. Let $x_n \to \infty$ on M. We will show that $Q_t f(x_n) \to 0$ as $n \to \infty$ for any $t \in \mathbb{R}_+$. There exist $g_n \in G$ such that $x_n = g_n p$. For any $\varepsilon > 0$, there is a compact subset F of G such that $P(g_t \in F^c) < \varepsilon$. Since $Q_t f(x_n) = E[f(g_n g_t p)]$, we need only to show that $g_n g p \to \infty$ on M uniformly for $g \in F$. If not, $\exists g_n' \in F$ such that, by taking a subsequence if necessary, $g_n g_n' p$ is contained in a compact subset of $M = G/H$. By the compactness of H, this implies that g_n must be contained in a compact subset of G, and then x_n is contained in a compact subset of M, which is impossible.

It remains to prove that $\sup_{x \in M} |Q_t f(x) - f(x)| \to 0$ as $t \to 0$. It is well known (see, for example, [34, theorem 17.6]) that, if $Q_t[C_0(M)] \subset C_0(M)$, then this property is equivalent to the apparently weaker property: $\forall f \in C_0(M)$ and $\forall x \in M$, $Q_t f(x) \to f(x)$ as $t \to 0$. This follows from $Q_t f(x) - f(x) = E[f(gg_t p) - f(gp)]$ for $g \in G$ with $gp = x$ and the right continuity of g_t at $t = 0$. $\square$

2.2. Invariant Markov Processes in Homogeneous Spaces

If $F: S \to T$ is a measurable map between two measurable spaces S and T, and if μ is a measure on S, then $F\mu = \mu \circ F^{-1}$ will denote the measure on T defined by $F\mu(B) = \mu(F^{-1}(B))$ for any measurable subset B of T, or equivalently, by $F\mu(f) = \mu(f \circ F)$ for any measurable function f on T whenever the integral exists. In the case when $S = T$, the measure μ will be called invariant under the map F or F-invariant if $F\mu = \mu$.

Let G be a Lie group and H be a subgroup. A measure μ on G will be called left (resp. right) H-invariant if $L_h \mu = \mu$ (resp. $R_h \mu = \mu$) for any $h \in H$. It will be called bi H-invariant if it is both left and right H-invariant. When $H = G$, it will simply be called left (resp. right, resp. bi) invariant.

It is well known (see [27, I.1]) that there is a unique left invariant probability measure on any compact Lie group, called the normalized Haar measure. This measure is in fact bi-invariant.

Let K be a compact subgroup of G. A measure ν on the homogeneous space $M = G/K$ is called H-invariant if $h\nu = \nu$ for any $h \in H$, where h is regarded as the map $h: (G/K) \ni gK \mapsto hgK \in (G/K)$ via the natural action of G on G/K.

Let $\pi: G \to M = G/K$ be the natural projection. It is easy to see that if μ is a left K-invariant measure on G, then $\nu = \pi\mu$ is a K-invariant measure on $M = G/K$. A map $S: M = G/K \to G$ satisfying $\pi \circ S = \mathrm{id}_M$ is called a section on M. Although a smooth section may not exist globally on M, there is always a Borel measurable section.

Proposition 2.3. *If ν is a K-invariant (probability) measure on $M = G/K$, then the measure μ on G, defined by*

$$\forall f \in \mathcal{B}(G)_+, \qquad \mu(f) = \int_M \int_K f(S(x)k)\,\rho_K(dk)\,\nu(dx), \qquad (2.11)$$

is the unique bi K-invariant (probability) measure on G with $\pi\mu = \nu$, where ρ_K is the normalized Haar measure on K and S is a measurable section on M.

Proof. For $g \in G$, $S \circ \pi(g) = gk$ for some $k \in K$. If μ is a bi K-invariant measure on G with $\pi\mu = \nu$, then, using the left invariance of ρ_K on K, for $f \in \mathcal{B}(G)_+$,

$$\mu(f) = \int_K \mu(f \circ R_k)\,\rho_K(dk) = \int_K \mu(f \circ R_k \circ S \circ \pi)\,\rho_K(dk)$$

$$= \int_K \nu(f \circ R_k \circ S)\,\rho_K(dk).$$

This shows that μ satisfies (2.11). Conversely, using the bi K-invariance of ρ_K and the K-invariance of ν, it can be shown that μ given by (2.11) is bi K-invariant with $\pi\mu = \nu$. It is clear that, if ν is a probability measure, then so is μ. $\qquad\square$

For any two K-invariant measures μ and ν on $M = G/K$, let

$$\mu * \nu(f) = \int\int f(S(x)y)\,\mu(dx)\,\nu(dy), \qquad (2.12)$$

for $f \in \mathcal{B}(G)_+$, where S is a measurable section on G/K as in (2.11). It is easy to see that $\mu * \nu(f)$ does not depend on the choice of S and is a K-invariant measure on M, called the convolution of μ and ν. It is clear that $\mu * \nu$ is a probability measure if so are μ and ν. A family $\{\nu_t;\ t \in \mathbb{R}_+\}$ of K-invariant probability measures on M will be called a K-invariant convolution semigroup (of probability measures) on M if $\nu_{t+s} = \nu_t * \nu_s$ for $s, t \in \mathbb{R}_+$ and $\nu_0 = \delta_o$, where o is the point eK in $M = G/K$. It will be called continuous if $\nu_t \to \delta_o$ weakly as $t \to 0$.

A family $\{\mu_t;\ t \in \mathbb{R}_+\}$ of probability measures on G will be called a generalized convolution semigroup (of probability measures) on G if

$\mu_{t+s} = \mu_t * \nu_s$ for $s, t \in \mathbb{R}_+$. It will be called continuous if $\mu_t \to \mu_0$ weakly as $t \to 0$. Note that a convolution semigroup $\{\mu_t;\ t \in \mathbb{R}_+\}$ defined in Section 1.1 is a generalized convolution semigroup with $\mu_0 = \delta_e$.

Recall that a Markov process x_t in M with transition semigroup Q_t is called G-invariant if Q_t is G-invariant; that is, $Q_t(f \circ g) = (Q_t f) \circ g$ for $f \in \mathcal{B}(G)_+$ and $g \in G$. The following proposition provides some basic relations among several different invariance properties.

Proposition 2.4. *Let G be a Lie group and let K be a compact subgroup.*

> (a) *If Q_t is a G-invariant (Feller) semigroup of probability kernels on $M = G/K$, then $\nu_t = Q_t(o, \cdot)$ is a K-invariant (continuous) convolution semigroup on M, where $o = eK$.*
> (b) *If ν_t is a K-invariant (continuous) convolution semigroup on M, then Q_t, defined by*
> $$\forall x \in M \text{ and } f \in \mathcal{B}(M)_+, \quad Q_t f(x) = \int f(S(x)y)\nu_t(dy), \quad (2.13)$$
> *where S is a measurable section on G/K, is the unique G-invariant (Feller) semigroup of probability kernels on M satisfying $\nu_t = Q_t(o, \cdot)$.*
> (c) *The map $\mu_t \mapsto \nu_t = \pi\mu_t$ provides a one-to-one correspondence between the set of the generalized (continuous) convolution semigroups μ_t on G, with bi K-invariant μ_t and $\mu_0 = \rho_K$, and the set of the K-invariant (continuous) convolution semigroups ν_t on M.*

Proof. For $f \in \mathcal{B}(M)_+$, writing $S_x = S(x)$, we have $Q_{s+t} f(o) = \int Q_s(o, dx)Q_t(x, f) = \int Q_s(o, dx)Q_t(o, f \circ S_x) = \nu_s * \nu_t(f)$ with $\nu_t = Q_t(o, \cdot)$. From this, it is easy to show that ν_t is a K-invariant convolution semigroup, and it is continuous if Q_t is Feller. This proves (a).

To prove (b), note that $Q_{t+s} f(x) = \int f(S(x)y)\nu_{s+t}(dy) = \iint f(S(x)S(y)z)\nu_t(dy)\nu_s(dz)$ and $S(x)S(y) = S(S(x)y)k'$ for some $k' \in K$. By the K-invariance of ν_s,

$$Q_{s+t} f(x) = \int\int f(S(S(x)y)z)\nu_t(dy)\nu_s(dz)$$

$$= \int Q_s f(S(x)y)\nu_t(dy) = Q_t Q_s f(x).$$

Moreover, for $g \in G$, $S(gx) = gS(x)k'$ for some $k' \in K$ and $Q_t f(gx) = \int f(S(gx)y)\nu_t(dy) = \int f(gS(x)k'y)\nu_t(dy) = Q_t(f \circ g)(x)$. Therefore, Q_t is a G-invariant semigroup on M. By (2.13), $Q_t(o, \cdot) = \nu_t$. This determines Q_t by the G-invariance. It remains to prove that Q_t is Feller if ν_t is continuous.

Although it may not be possible to choose a globally continuous section $S: M \to G$, it is always possible to choose a continuous or even smooth section locally; therefore, for any $x_0 \in M$, $S(x) = S_1(x)k_x$ for x contained in a neighborhood V of x_0, where S_1 is a section continuous in V and $k_x \in K$. It now is easy to see that, for $f \in C_0(M)$, $Q_t f$ defined by (2.13) is continuous on M, by the K-invariance of v_t, and $Q_t f(x) \to 0$ as $x \to \infty$ under the one-point compactification topology on M. The weak convergence $v_t \to \delta_o$ implies $Q_t f \to f$ as $t \to 0$. Therefore, Q_t is a Feller semigroup. Thus (b) is proved.

For (c), let v_t be a K-invariant convolution semigroup on M and let μ_t be defined by (2.11). Then μ_t is the unique bi K-invariant probability measure on G with $\pi \mu_t = v_t$ and $\mu_0 = \rho_K$. For $f \in \mathcal{B}(G)_+$,

$$
\begin{aligned}
\mu_{s+t}(f) &= \int \int f(S(x)k)\rho_K(dk)v_{s+t}(dx) \\
&= \int \int \int f(S(S(y)z)k)\rho_K(dk)v_t(dy)v_s(dz) \\
&= \int \int \int \int f(S(S(y)hz)k)\rho_K(dh)\rho_K(dk)v_t(dy)v_s(dz) \\
&= \int \int \int f(S(gz)k)\rho_K(dk)v_s(dz)\mu_t(dg) \\
&= \int \int \int f(gS(z)k'k)\rho_K(dk)v_s(dz)\mu_t(dg) \\
&= \int \int f(gh)\mu_t(dg)\mu_s(dh) = \mu_t * \mu_s(f).
\end{aligned}
$$

This shows that μ_t is a generalized convolution semigroup on G. Conversely, if μ_t is a bi K-invariant generalized convolution semigroup on G with $\mu_0 = \rho_K$ and if $v_t = \pi \mu_t$, then v_t is K-invariant and, for $f \in \mathcal{B}(M)_+$,

$$
\begin{aligned}
v_{t+s}(f) &= \mu_{t+s}(f \circ \pi) = \int \int f \circ \pi(gh)\mu_t(dg)\mu_s(dh) \\
&= \int \int f(gy)\mu_t(dg)v_s(dy) = \int \int f(S(\pi(g))k'y)\mu_t(dg)v_s(dy) \\
&= \int \int f(S(x)y)v_t(dx)v_s(dy) = v_t * v_s(f).
\end{aligned}
$$

This proves that v_t is a K-invariant convolution semigroup on M. It is clear that μ_t is continuous if and only if $v_t = \pi \mu_t$ is also. $\qquad \square$

Let $\mathfrak{g}$ be the Lie algebra of G. Recall that, for $g \in G$, c_g is the conjugation map on G defined by $c_g(h) = ghg^{-1}$ for $h \in G$. The differential of c_g at e

will be denoted by $\text{Ad}(g)$,

$$\mathfrak{g} \ni X \mapsto \text{Ad}(g)X = Dc_g(X) = DL_g \circ DR_{g^{-1}}(X) \in \mathfrak{g}. \qquad (2.14)$$

The map

$$G \times \mathfrak{g} \ni (g, X) \mapsto \text{Ad}(g)X \in \mathfrak{g}$$

is an action of G on $\mathfrak{g}$, called the adjoint action of the Lie group G on its Lie algebra $\mathfrak{g}$, and is denoted by $\text{Ad}(G)$. For a subgroup H of G, depending on the context, $\text{Ad}(H)$ will denote either the adjoint action of H on its Lie algebra or the action $\text{Ad}(G)$ restricted to H given by $H \times \mathfrak{g} \ni (h, X) \mapsto \text{Ad}(h)X \in \mathfrak{g}$.

We note that

$$\forall X \in \mathfrak{g}, \qquad \text{Ad}(e^X) = e^{\text{ad}(X)}, \qquad (2.15)$$

where $\text{ad}(X): \mathfrak{g} \to \mathfrak{g}$ is the linear map given by $\text{ad}(X)Y = [X, Y]$ (Lie bracket) and $e^{\text{ad}(X)} = \sum_{n=0}^{\infty}(1/n!)\text{ad}(X)^n$ is the exponential of $\text{ad}(X)$. In particular, $(d/dt)\text{Ad}(e^{tX})Y \mid_{t=0} = [X, Y]$.

Because K is compact, there is an $\text{Ad}(K)$-invariant inner product on $\mathfrak{g}$, that is, an inner product $\langle \cdot, \cdot \rangle$ satisfying $\langle \text{Ad}(k)X, \text{Ad}(k)Y \rangle = \langle X, Y \rangle$ for $X, Y \in \mathfrak{g}$ and $k \in K$. We will fix such an inner product. Let $\mathfrak{k}$ be the Lie algebra of K and let $\mathfrak{p}$ be the orthogonal complement of $\mathfrak{k}$ in $\mathfrak{g}$. It is easy to see that $\mathfrak{p}$ is $\text{Ad}(K)$-invariant; that is, $\text{Ad}(k)\mathfrak{p} \subset \mathfrak{p}$ for $k \in K$.

Let $\{X_1, \ldots, X_d\}$ be an orthonormal basis of $\mathfrak{g}$ such that $X_1, \ldots, X_n$ form a basis of $\mathfrak{p}$ and $X_{n+1}, \ldots, X_d$ form a basis of $\mathfrak{k}$. Consider the map

$$\phi: \quad \mathbb{R}^n \ni y = (y_1, \ldots, y_n) \mapsto \pi(e^{\sum_{i=1}^{n} y_i X_i}) \in M.$$

Restricted to a sufficiently small neighborhood V of 0 in $\mathbb{R}^n$, ϕ is a diffeomorphism and hence $y_1, \ldots, y_n$ may be used as local coordinates on $\phi(V)$, a neighborhood of o in M. For $x = \phi(y)$ in $\phi(V)$, we may write $(\partial/\partial y_i)f(x)$ for $(\partial/\partial y_i)f \circ \phi(y)$ for $f \in C^1(M)$ and $1 \le i \le n$. In this way, $\partial/\partial y_i$ may be regarded as a vector field on $\phi(V)$. For $k \in K$ and $x \in \phi(V), kx = k\phi(y(x)) = \pi\{\exp[\sum_{i=1}^{n} y_i(x) \text{Ad}(k)X_i]\}$; it follows that, for $x \in \phi(V)$,

$$\sum_{i=1}^{n} y_i(x) \text{Ad}(k)X_i = \sum_{i=1}^{n} y_i(kx) X_i. \qquad (2.16)$$

The functions y_i may be extended so that $y_i \in C_c^{\infty}(M)$ and (2.16) holds for any $x \in M$ and $k \in K$, by replacing y_i by ψy_i for a K-invariant $\psi \in C_c^{\infty}(M)$ with $\psi = 1$ in a neighborhood of o. Thus extended, the y_i will be called the canonical coordinate functions on M (associated to the basis $\{X_1, \ldots, X_n\}$ of $\mathfrak{p}$).

By the discussion in Sections 1.2 and 1.3, the generator T of a left invariant diffusion process g_t in G takes the form

$$\forall f \in C_c^\infty(G), \qquad Tf = \frac{1}{2} \sum_{i,j=1}^{d} a_{ij} X_i^l X_j^l f + \sum_{i=1}^{d} c_i X_i^l f, \qquad (2.17)$$

where a_{ij} and c_i are some constants with a_{ij} forming a nonnegative definite symmetric matrix. The operator T is left invariant, that is, $T(f \circ L_g) = (Tf) \circ L_g$ for $f \in C_c^\infty(G)$ and $g \in G$, and will be called a left invariant diffusion generator on G. If T is also right K-invariant, that is, if $T(f \circ R_k) = (Tf) \circ R_k$ for $f \in C_c^\infty(G)$ and $k \in K$, then it induces a differential operator $\tilde{T}$ on M, given by $\tilde{T} f(x) = T(f \circ \pi)(g)$ for $f \in C_c^\infty(M)$ and $x \in M$, where $g \in G$ is chosen to satisfy $x = \pi(g)$. The operator $\tilde{T}$ is well defined by the right K-invariance of T and it satisfies

$$\forall f \in C_c^\infty(M), \qquad (\tilde{T} f) \circ \pi = T(f \circ \pi). \qquad (2.18)$$

The left invariance of T implies that $\tilde{T}$ is G-invariant; that is, $\tilde{T}(f \circ g) = (\tilde{T} f) \circ g$ for $f \in C_c^\infty(M)$ and $g \in G$.

The G-invariance of $\tilde{T}$ implies that $\tilde{T}$ is completely determined by $\tilde{T} f(o)$ for $f \in C_c^\infty(M)$. Note that $X_j^l(f \circ \pi)(e) = 0$ for $j > n$, and

$$X_i^l X_j^l(f \circ \pi)(e) = \frac{\partial^2}{\partial s \partial t} f(e^{sX_i} e^{tX_j} o) \mid_{(s,t)=(0,0)},$$

which vanishes for $j > n$. Since $X_j^l X_i^l = X_i^l X_j^l + [X_j, X_i]^l$, if $j > n$, then $X_j^l X_i^l(f \circ \pi)(e) = [X_j, X_i]^l(f \circ \pi)(e)$, which is a linear combination of $X_i^l(f \circ \pi)(e)$ for $i = 1, 2, \ldots, n$. It follows from (2.17) that

$$\forall f \in C_c^\infty(M), \quad \tilde{T} f(o) = \frac{1}{2} \sum_{i,j=1}^{n} a_{ij} X_i^l X_j^l(f \circ \pi)(e) + \sum_{i=1}^{n} c_i' X_i^l(f \circ \pi)(e),$$

$$(2.19)$$

where a_{ij} and c_i' are some constants with a_{ij} forming an $n \times n$ nonnegative definite symmetric matrix. In fact, the a_{ij} are the same as in (2.17). Because both the inner product $\langle \cdot, \cdot \rangle$ and the space $\mathfrak{p}$ are $\mathrm{Ad}(K)$-invariant, for any $k \in K$, there is an $n \times n$ orthogonal matrix $\{b_{ij}(k)\}$ such that $\mathrm{Ad}(k)X_j = \sum_{i=1}^{n} b_{ij}(k)X_i$ for $j = 1, 2, \ldots, n$. Since $X_i^l(f \circ k \circ \pi)(e) = [\mathrm{Ad}(k)X_i]^l(f \circ \pi)(e)$, the coefficients a_{ij} and c'_i in (2.19) satisfy

$$\forall k \in K, \qquad a_{ij} = \sum_{p,q=1}^{n} a_{pq} b_{ip}(k) b_{jq}(k) \qquad \text{and} \qquad c'_i = \sum_{p=1}^{n} c'_p b_{ip}(k)$$

(2.20)

for $i, j = 1, 2, \ldots, n$. In fact, this is also a sufficient condition for (2.19) to be the expression at o of a G-invariant differential operator $\tilde{T}$ on M. It is easy to see from (2.19) and (2.20) that

$$\forall f \in C_c^{\infty}(G), \qquad Tf = \frac{1}{2} \sum_{i,j=1}^{n} a_{ij} X_i^l X_j^l f + \sum_{i=1}^{n} c'_i X_i^l f \qquad (2.21)$$

is the restriction to $C_c^{\infty}(G)$ of a left invariant and right K-invariant diffusion generator T on G satisfying (2.18).

A G-invariant differential operator $\tilde{T}$ having the expression (2.19) at o with coefficients satisfying (2.20) will be called a G-invariant diffusion generator on M because it is the generator, restricted to $C_c^{\infty}(M)$, of a G-invariant diffusion process in M. In fact, $x_t = \pi(g_t)$ is such a process, where g_t is a left invariant diffusion process in G whose generator restricted to $C_c^{\infty}(G)$ is given by (2.21).

The following result is an extension of Hunt's result Theorem 1.1 to a G-invariant semigroup Q_t of probability kernels on $M = G/K$, obtained also by Hunt [32]. The theorem here is a slightly different version of Hunt's original result, but it is easier to state and prove.

Theorem 2.1. *Let G be a Lie group and let K be a compact subgroup of G. Fix an $\mathrm{Ad}(K)$-invariant inner product on the Lie algebra $\mathfrak{g}$ of G, and choose an orthonormal basis $\{X_1, \ldots, X_d\}$ of $\mathfrak{g}$ and the associated canonical coordinate functions $y_1, \ldots, y_n$ on M as was done previously. If Q_t is a G-invariant Feller semigroup of probability kernels on $M = G/K$ with generator $\tilde{L}$, then the domain $D(\tilde{L})$ of $\tilde{L}$ contains $C_c^{\infty}(M)$ and, for any $f \in C_c^{\infty}(M)$,*

$$\tilde{L}f(o) = \tilde{T}f(o) + \int_M \left[f(x) - f(o) - \sum_{i=1}^{n} y_i(x) \frac{\partial}{\partial y_i} f(o) \right] \tilde{\Pi}(dx), \quad (2.22)$$

where o is the point eK in G/K, $\tilde{T}$ is a G-invariant diffusion generator on M, and $\tilde{\Pi}$ is a K-invariant measure on M satisfying

$$\tilde{\Pi}(\{o\}) = 0, \qquad \tilde{\Pi}(|y|^2) < \infty, \qquad \text{and} \qquad \tilde{\Pi}(U^c) < \infty \qquad (2.23)$$

for any neighborhood U of o, where $|y|^2 = \sum_{i=1}^{n} y_i^2$.

Conversely, given $\tilde{T}$ and $\tilde{\Pi}$ as in the previous paragraph, there is a unique G-invariant Feller semigroup Q_t of probability kernels on M such that its generator $\tilde{L}$ at point o, restricted to $C_c^\infty(M)$, is given by (2.22).

The proof of Theorem 2.1 will be given in Chapter 3. Note that, by using the Taylor expansion of f at o and the condition $\tilde{\Pi}(|y|^2) < \infty$, it is easy to see that the integral in (2.22) exists and is finite.

The following result provides a basic relation between G-invariant Feller processes in M and right K-invariant left Lévy processes in G.

Theorem 2.2. *Let G be a Lie group and let K be a compact subgroup. If g_t is a right K-invariant left Lévy process in G with $g_0 = e$, then its one-point motion from $o = eK$ in $M = G/K$ is a G-invariant Feller process in M. Conversely, if x_t is a G-invariant Feller process in M with $x_0 = o$, then there is a right K-invariant left Lévy process g_t in G with $g_0 = e$ such that its one-point motion in M from o is identical to the process x_t in distribution.*

Equivalently this theorem can also be stated in terms of Feller semigroups as follows: A left invariant Feller semigroup P_t (of probability kernels) on G that is also right K-invariant determines a G-invariant Feller semigroup Q_t on M by

$$\forall f \in \mathcal{B}(M)_+, \qquad (Q_t f) \circ \pi = P_t(f \circ \pi). \qquad (2.24)$$

Conversely, given a G-invariant Feller semigroup Q_t on M, there is a left invariant Feller semigroup P_t on G satisfying (2.24). Note that, for a given Q_t, such P_t may not be unique.

Proof. The first part of the theorem is a direct consequence of Proposition 2.2.

To prove the second part, let x_i be coordinate functions associated to the basis $\{X_1, \ldots, X_d\}$ of $\mathfrak{g}$. We may assume the x_i satisfy $g = \exp[\sum_{i=1}^n x_i(g)X_i] \exp[\sum_{i=n+1}^d x_i(g)X_i]$ for g contained in a neighborhood of e. Then $x_i = y_i \circ \pi$ for $1 \le i \le n$, and by (2.16),

$$\forall k \in K, \qquad \sum_{i=1}^n x_i \, \text{Ad}(k)X_i = \sum_{i=1}^n x_i \circ c_k \, X_i, \qquad (2.25)$$

in a neighborhood of e. We may assume $x_i = y_i \circ \pi$ for $1 \le i \le n$ and Equation (2.25) hold globally on G by replacing x_i by ψx_i for a bi K-invariant $\psi \in C_c^\infty(G)$ with $\psi = 1$ in a neighborhood of e.

Suppose the generator $\tilde{L}$ of x_t, restricted to $C_c^\infty(M)$, is given by (2.22). Let T be the left invariant and right K-invariant diffusion generator on G satisfying (2.18). Let $S: M \to G$ be a measurable section such that $S(x) = \exp[\sum_{i=1}^n y_i(x)X_i]$ for x contained in a K-invariant neighborhood V of o. Define a measure Π on G by

$$\forall f \in B(G)_+, \qquad \Pi(f) = \int\int f(kS(x)k^{-1})\rho_K(dk)\tilde{\Pi}(dx). \qquad (2.26)$$

Then Π is K-conjugate invariant (i.e., $c_k\Pi = \Pi$ for $k \in K$), and because of the K-invariance of $\tilde{\Pi}$, $\pi\Pi = \tilde{\Pi}$. By (2.16), for $k \in K$ and $x \in V$, $kS(x)k^{-1} = \exp[\sum_{i=1}^n y_i(kx)X_i]$. It follows that $x_i(kS(x)k^{-1}) = 0$ for $i > n$. We may assume the coordinate functions x_i are supported by $\pi^{-1}(V)$. Then $\Pi(|x_i|) = 0$ for $i > n$. Since $\Pi(\sum_{i=1}^n x_i^2) = \tilde{\Pi}(|y|^2) < \infty$, we see that Π is a Lévy measure on G.

It now follows that, for $f \in C_c^\infty(G)$,

$$Lf(g) = Tf(g) + \int_G \left[f(gh) - f(g) - \sum_{i=1}^n x_i(h)X_i^l f(g) \right] \Pi(dh) \qquad (2.27)$$

is the restriction to $C_c^\infty(G)$ of the generator L of a left invariant Feller semigroup P_t on G. Such a P_t is unique by the discussion in Section 1.2. Let g_t be a left Lévy process in G with $g_0 = e$ and transition semigroup P_t. By the left and right K-invariance of T, the K-conjugate invariance of Π, and (2.25), it is easy to show that L is K-conjugate invariant; that is, $L(f \circ c_k) = (Lf) \circ c_k$ for any $k \in K$. It follows that the generator of the left Lévy process $g_t' = kg_tk^{-1} = c_k(g_t)$ has the same restriction to $C_c^\infty(G)$ as that of g_t; therefore, kg_tk^{-1} has the same transition semigroup as g_t. This implies that g_t is right K-invariant. By Proposition 2.2, its one-point motion from o, $x_t' = g_t o$, is a G-invariant Feller process in M. Let Q_t' and L' be respectively the transition semigroup and the generator of x_t'. Then, for $f \in C_c^\infty(M)$,

$$L'f(o) = \frac{d}{dt}Q_t'f(o)\,|_{t=0} = \frac{d}{dt}P_t(f \circ \pi)(e)\,|_{t=0} = L(f \circ \pi)(e)$$

$$= T(f \circ \pi)(e) + \int_G \left[(f \circ \pi)(h) - f(o) \right.$$

$$\left. - \sum_{i=1}^n x_i(h)X_i^l(f \circ \pi)(e) \right] \Pi(dh).$$

Since $T(f \circ \pi) = (\tilde{T}f) \circ \pi$, $\tilde{\Pi} = \pi\Pi$, $X_i^l(f \circ \pi)(e) = (\partial/\partial y_i)f(o)$, and $x_i = y_i \circ \pi$, this expression is equal to that in (2.22). The uniqueness in Theorem 2.1 implies that x_t' has the same distribution as x_t. $\qquad\square$

Remark. Let x_t and g_t be as in Theorem 2.2. If f is a function on M such that $f \circ \pi \in C_0^{2,l}(G)$, then $(1/t)(Q_t f - f) \circ \pi = (1/t)[P_t(f \circ \pi) - (f \circ \pi)] \to L(f \circ \pi)$ as $t \to 0$ in $C_0(G)$. It follows that the domain $D(\tilde{L})$ of the generator $\tilde{L}$ of x_t in Theorem 2.1 contains all the functions f on M with $f \circ \pi \in C_0^{2,l}(G)$ and (2.22) holds for such f.

2.3. Riemannian Brownian Motions

Let M be a d-dimensional Riemannian manifold equipped with the Riemannian metric $\{\langle \cdot, \cdot \rangle_x; x \in M\}$, where $\langle \cdot, \cdot \rangle_x$ is an inner product on the tangent space $T_x M$ at $x \in M$, and let $\text{Exp}_x: T_x M \to M$ be the Riemannian exponential map at x, defined by $\text{Exp}_x(v) = \gamma(1)$ for $v \in T_x M$, where $\gamma: [0, 1] \to M$ is the geodesic with $\gamma(0) = x$ and $\gamma'(0) = v$. The Laplace–Beltrami operator Δ on M can be defined by

$$\Delta f(x) = \sum_{i=1}^{d} \frac{d^2}{dt^2} f(\text{Exp}_x(tY_i)) \mid_{t=0} \tag{2.28}$$

for $f \in C^2(M)$ and $x \in M$, where $\{Y_1, \ldots, Y_d\}$ is a complete set of orthonormal tangent vectors at x. This definition is independent of the choice of Y_is. It is well known that, in local coordinates $x_1, \ldots, x_d$, writing ∂_i for $\partial/\partial x_i$, we get

$$\Delta f(x) = \sum_{j,k=1}^{d} g^{jk}(x) \partial_j \partial_k f(x) - \sum_{i,j,k=1}^{d} g^{jk}(x) \Gamma_{jk}^i(x) \partial_i f(x), \tag{2.29}$$

where $\{g^{jk}\} = \{g_{jk}\}^{-1}$, $g_{jk}(x) = \langle \partial_j, \partial_k \rangle_x$, and $\Gamma_{jk}^i = (1/2) \sum_{p=1}^{d} g^{ip} \times (\partial_j g_{pk} + \partial_k g_{pj} - \partial_p g_{jk})$. The matrix $\{g_{ij}\}$ is called the Riemannian metric tensor under the local coordinates.

A diffusion process x_t in M is called a Riemannian Brownian motion if its generator, when restricted to $C_c^{\infty}(G)$, is equal to $(1/2)\Delta$. When $M = \mathbb{R}^d$ is equipped with the standard Euclidean metric, the Laplace–Beltrami operator is just the usual Laplace operator and a Riemannian Brownian motion is a standard d-dimensional Brownian motion.

From its definition, it is easy to see that the Laplace–Beltrami operator with domain $C^2(M)$ or $C_c^{\infty}(M)$ is invariant under any isometry ϕ on M. It follows that if x_t is a Riemannian Brownian motion in M, then so is $\phi(x_t)$ for any isometry ϕ on M.

The Riemannian Brownian motion has the following scaling property: If $c > 0$ is a constant, Δ is the Laplace–Beltrami operator, and x_t is a Riemannian Brownian motion with respect to the metric $\{\langle \cdot, \cdot \rangle_x; x \in M\}$, then $(1/c)\Delta$ is the Laplace–Beltrami operator and $x_{t/c}$ is a Riemannian Brownian motion with respect to the metric $\{c\langle\cdot,\cdot\rangle_x; x \in M\}$.

There is considerable interest in studying Riemannian Brownian motions in connection with the geometric structures of the manifolds (see, for example, Elworthy [17, 18], Ikeda and Watanabe [33], and Hsu [31]). In this section, we consider several cases when Riemannian Brownian motions appear in Lie groups and the associated homogeneous spaces.

Let G be a Lie group and let $\langle\cdot,\cdot\rangle$ be an inner product on the Lie algebra $\mathfrak{g}$ of G. The latter induces a left invariant Riemannian metric $\{\langle\cdot,\cdot\rangle_g; g \in G\}$ on G under which the left translations L_g for $g \in G$ are isometries. Note that a Riemannian Brownian motion g_t in G with respect to this metric is a left invariant diffusion process, that is, a continuous left Lévy process, in G. We will express the Laplace–Beltrami operator Δ in terms of left invariant vector fields on G, and from this expression we will obtain a stochastic differential equation satisfied by g_t.

Let $\{X_1, \ldots, X_d\}$ be an orthonormal basis of $\mathfrak{g}$ with respect to $\langle\cdot,\cdot\rangle$ and let

$$[X_j, X_k] = \sum_{i=1}^{d} C_{jk}^i X_i. \tag{2.30}$$

The numbers C_{jk}^i are called the structure coefficients of G under the basis $\{X_1, \ldots, X_d\}$. Since $[X_j, X_k] = -[X_k, X_j]$, $C_{jk}^i = -C_{kj}^i$.

Lemma 2.1. *Let $x_1, \ldots, x_d$ be the coordinate functions associated to the basis $\{X_1, \ldots, X_d\}$ such that $\mathfrak{g} = \exp[\sum_{i=1}^{d} x_i(g)X_i]$ for $\mathfrak{g}$ close to e. Then for g contained in a neighborhood U of e,*

$$X_k^l(g) = \frac{\partial}{\partial x_k} + \frac{1}{2}\sum_{i,j=1}^{d} x_j(g)C_{jk}^i \frac{\partial}{\partial x_i} + \sum_{p,q=1}^{d} b_{pq}^i(g)\frac{\partial}{\partial x_i},$$

where $b_{pq}^i \in C^\infty(U)$ and $b_{pq}^i = O(|x|^2)$.

Proof. For $X, Y \in \mathfrak{g}$, let Y_X be the tangent vector to the curve $t \mapsto X + tY$ in $\mathfrak{g}$ at $t = 0$. We will denote $D\exp(Y_X)$ by $D\exp_X(Y)$. By Helgason [26, Chapter II, theorem 1.7],

$$D\exp_X(Y) = DL_{\exp(X)}\left\{\left[\frac{\mathrm{id}_\mathfrak{g} - e^{-\mathrm{ad}(X)}}{\mathrm{ad}(X)}\right]Y\right\}, \tag{2.31}$$

where $\mathrm{id}_\mathfrak{g}$ denotes the identity map on $\mathfrak{g}$ and

$$\frac{\mathrm{id}_\mathfrak{g} - e^{-\mathrm{ad}(X)}}{\mathrm{ad}(X)} = \sum_{r=0}^{\infty} \frac{(-1)^r}{(r+1)!}\mathrm{ad}(X)^r = \mathrm{id}_\mathfrak{g} - \frac{1}{2}\mathrm{ad}(X) + \sum_{r=2}^{\infty} \frac{(-1)^r}{(r+1)!}\mathrm{ad}(X)^r$$

is a linear endomorphism on $\mathfrak{g}$, which is invertible when X is sufficiently close to 0. It is easy to show that

$$\left[\frac{\mathrm{id}_\mathfrak{g} - e^{-\mathrm{ad}(X)}}{\mathrm{ad}(X)}\right]^{-1} = \mathrm{id}_\mathfrak{g} + \frac{1}{2}\mathrm{ad}(X) + \sum_{r=2}^{\infty} c_r \mathrm{ad}(X)^r,$$

where the last series converges absolutely in the operator norm for X sufficiently close to 0. Recall that $g = \exp[\sum_{i=1}^{d} x_i(g)X_i]$ for g contained in a sufficiently small neighborhood U of e. For $X = \sum_{j=1}^{n} x_j X_j$, let

$$Z = \left[\frac{\mathrm{id}_\mathfrak{g} - e^{-\mathrm{ad}(X)}}{\mathrm{ad}(X)}\right]^{-1} X_k = X_k + \frac{1}{2}[X, X_k] + O(\|X\|^2)$$

$$= X_k + \frac{1}{2}\sum_{i,j=1}^{d} x_j C_{jk}^i X_i + O(|x|^2).$$

By (2.31), $DL_{\exp(X)}(X_k) = D\exp_X(Z)$. It follows that, for $f \in C^\infty(U)$ and $g = e^X \in U$,

$$X_k^l f(g) = \frac{d}{dt} f(e^X e^{tX_k})\mid_{t=0} = \frac{d}{dt} f(e^{X+tZ})\mid_{t=0}.$$

This proves the claim because

$$X + tZ = \sum_{i=1}^{d}\left[x_i + t\left(\delta_{ik} + \frac{1}{2}\sum_{j=1}^{d} x_j C_{jk}^i + O(|x|^2)\right)\right]X_i. \qquad \square$$

From Lemma 2.1, $\partial_i = X_i^l - (1/2)\sum_{j,k} x_j C_{ji}^k X_k^l + O(|x|^2)$. The Riemannian metric tensor of the left invariant metric on G induced by the inner product $\langle\cdot,\cdot\rangle$ on $\mathfrak{g}$ can be written as

$$g_{ij}(x) = \langle\partial_i, \partial_j\rangle_x = \delta_{ij} - \frac{1}{2}\sum_p x_p C_{pj}^i - \frac{1}{2}\sum_p x_p C_{pi}^j + O(|x|^2)$$

and $\partial_k g_{ij}(e) = -(1/2)C_{kj}^i - (1/2)C_{ki}^j$. Since $g_{ij}(e) = g^{ij}(e) = \delta_{ij}$,

$$\Gamma_{jk}^i(e) = \frac{1}{2}[\partial_j g_{ik}(e) + \partial_k g_{ij}(e) - \partial_i g_{jk}(e)]$$

$$= \frac{1}{4}\left(- C_{jk}^i - C_{ji}^k - C_{ki}^j - C_{kj}^i + C_{ij}^k + C_{ik}^j\right) = \frac{1}{2}\left(C_{ij}^k + C_{ik}^j\right).$$

Because $C_{ii}^k = 0$,

$$X_i^l X_i^l = \left[\partial_i + \frac{1}{2}\sum_j x_j C_{ji}^k \partial_k + O(|x|^2)\right]\left[\partial_i + \frac{1}{2}\sum_j x_j C_{ji}^k \partial_k + O(|x|^2)\right]$$

$$= \partial_i \partial_i + O(|x|).$$

It follows from (2.29) and the previous computation that, for $f \in C^2(G)$,

$$\Delta f(g) = \sum_{i=1}^{d} X_i^l X_i^l f(g) - X_0^l f(g) \qquad \text{with} \qquad X_0 = \sum_{i,j=1}^{d} C_{ij}^j X_i \quad (2.32)$$

holds for $g = e$. Since both sides of (2.32) are left invariant differentiable operators on G, therefore, it holds for all $g \in G$.

We note that X_0 in (2.32) is independent of the choice of the orthonormal basis $\{X_1, \ldots, X_d\}$ of $\mathfrak{g}$ because the operator $\sum_{i=1}^{d} X_i^l X_i^l$ is independent of this basis.

By the discussion in Appendix B.2, the solution g_t of the stochastic differential equation

$$dg_t = \sum_{i=1}^{d} X_i^l(g_t) \circ dW_t^i - \frac{1}{2} X_0^l(g_t) \, dt, \qquad (2.33)$$

where $W_t = (W_t^1, \ldots, W_t^d)$ is a d-dimensional standard Brownian motion, is a diffusion process in G with generator $(1/2)\Delta$; hence, it is a Riemannian Brownian motion in G. To summarize, we have the following result:

Proposition 2.5. *The Laplace–Beltrami operator on G with respect to the left invariant metric induced by an inner product $\langle \cdot, \cdot \rangle$ on $\mathfrak{g}$ is given by (2.32), where $\{X_1, \ldots, X_d\}$ is an orthonormal basis of $\mathfrak{g}$ and X_0 defined there is independent of the choice of this basis. Consequently, if g_t is a solution of the stochastic differential equation (2.33), then it is a Riemannian Brownian motion in G with respect to this metric.*

By a similar computation, it can be shown that if in (2.32) and (2.33), X_i^l is replaced by X_i^r and $-X_0^l$ by $+x_0^r$ then Δ becomes the Laplace-Beltrami operator on G and the process g_t becomes a Riemannian Brownian motion in G with respect to the right invariant metric induced by $\langle \cdot, \cdot \rangle$.

If G is compact, then there is an $\text{Ad}(G)$-invariant inner product $\langle \cdot, \cdot \rangle$ on $\mathfrak{g}$. Then the left invariant metric on G induced by $\langle \cdot, \cdot \rangle$ will also be right invariant; hence, it will be called a bi-invariant metric on G. In this case, it can be shown that $X_0 = 0$ in (2.32) and (2.33). In fact, a more general conclusion holds as in the following proposition. For two subsets $\mathfrak{u}$ and $\mathfrak{v}$ of $\mathfrak{g}$, $[\mathfrak{u}, \mathfrak{v}]$ will denote the space spanned by $[X, Y]$ for $X \in \mathfrak{u}$ and $Y \in \mathfrak{v}$.

Proposition 2.6. *Let G be a Lie group with Lie algebra $\mathfrak{g}$ and let $\langle \cdot, \cdot \rangle$ be an inner product on $\mathfrak{g}$ that is $\text{Ad}(H)$-invariant for some Lie subgroup H of G. If $[\mathfrak{p}, \mathfrak{p}] \subset \mathfrak{h}$, where $\mathfrak{p}$ is the orthogonal complement of the Lie algebra $\mathfrak{h}$ of H in*

$\mathfrak{g}$, *then* $X_0 = 0$ *in (2.32) and (2.33). In particular, if* $\langle \cdot, \cdot \rangle$ *is* Ad(G)-*invariant, then* $X_0 = 0$.

Proof. We note that $\mathfrak{p}$ is Ad(H)-invariant. Since $(d/dt)\mathrm{Ad}(e^{tX})Y\mid_{t=0} = [X, Y]$, $[\mathfrak{h}, \mathfrak{p}] \subset \mathfrak{p}$. Without loss of generality, we may assume that the orthonormal basis $\{X_1, \ldots, X_d\}$ of $\mathfrak{g}$ is chosen such that $X_1, \ldots, X_p$ form a basis of $\mathfrak{p}$ and $X_{p+1}, \ldots, X_d$ form a basis of $\mathfrak{h}$. Since $[\mathfrak{h}, \mathfrak{p}] \subset \mathfrak{p}$ and $[\mathfrak{p}, \mathfrak{p}] \subset \mathfrak{h}$, it is easy to show that $C_{ij}^j = 0$ for $1 \le i \le p$ and any j. Because, for $h \in H$, Ad(h): $\mathfrak{p} \to \mathfrak{p}$ and Ad(h): $\mathfrak{h} \to \mathfrak{h}$ are orthogonal with respect to the restrictions of $\langle \cdot, \cdot \rangle$ to $\mathfrak{p}$ and to $\mathfrak{h}$, it follows that, for $X \in \mathfrak{h}$, the matrix representations of ad(X): $\mathfrak{p} \to \mathfrak{p}$ under the basis $\{X_1, \ldots, X_p\}$ and ad(X): $\mathfrak{h} \to \mathfrak{h}$ under the basis $\{X_{p+1}, \ldots, X_d\}$ are skew-symmetric; hence, $C_{ij}^j = 0$ for $p + 1 \le i \le d$ and any j. This proves that $X_0 = 0$ in (2.32) and (2.33). If the inner product $\langle \cdot, \cdot \rangle$ is Ad(G)-invariant, we may take $H = G$ and then the condition for $X_0 = 0$ is satisfied. □

We will now assume that H is a closed subgroup of G with Lie algebra $\mathfrak{h}$, $\langle \cdot, \cdot \rangle$ is an Ad(H)-invariant inner product on $\mathfrak{g}$, and $\mathfrak{p}$ is the orthogonal complement of $\mathfrak{h}$ in $\mathfrak{g}$. The Ad(H)-invariance of the inner product implies that the space $\mathfrak{p}$ is Ad(H)-invariant. Consider the following stochastic differential equation on G:

$$dg_t = \sum_{i=1}^{p} X_i^l(g_t) \circ dW_t^i + c \sum_{i=p+1}^{d} X_i^l(g_t) \circ dW_t^i, \qquad (2.34)$$

where c is an arbitrary constant and $\{X_1, \ldots, X_d\}$ is an orthonormal basis of $\mathfrak{g}$ with respect to $\langle \cdot, \cdot \rangle$ such that $\{X_1, \ldots, X_p\}$ is a basis of $\mathfrak{p}$ and $\{X_{p+1}, \ldots, X_d\}$ is a basis of $\mathfrak{h}$. Note that, when $c = 1$, the distribution of the process g_t determined by (2.34) is independent of the choice of any orthonormal basis of $\mathfrak{g}$ because a suitable orthogonal transformation maps this basis into another that can be divided into a basis of $\mathfrak{p}$ and a basis of $\mathfrak{h}$ as required here and such a transformation will also transform the standard Brownian motion W_t into another standard Brownian motion. In this case, (2.34) becomes (2.33) with $X_0 = 0$.

We will consider the one-point motions of the process g_t in the homogeneous space G/H. Let $\pi \colon G \to G/H$ be the natural projection. The restriction of $\langle \cdot, \cdot \rangle$ to $\mathfrak{p}$ is an Ad(H)-invariant inner product on $\mathfrak{p}$ that induces a G-invariant Riemannian metric $\{\langle \cdot, \cdot \rangle_x; x \in G/H\}$ on G/H such that $\langle D\pi(X), D\pi(Y) \rangle_{\pi(e)} = \langle X, Y \rangle$ for any $X, Y \in \mathfrak{p}$. By Kobayashi and Nomizu [35, chapter X], if the following condition is satisfied:

$$\langle X, [Z, Y]_{\mathfrak{p}} \rangle + \langle [Z, X]_{\mathfrak{p}}, Y \rangle = 0 \quad \text{for } X, Y, Z \in \mathfrak{p}, \qquad (2.35)$$

where $X_{\mathfrak{p}}$ denotes the $\mathfrak{p}$-component of $X \in \mathfrak{g}$ under the direct sum decomposition $\mathfrak{g} = \mathfrak{h} \oplus \mathfrak{p}$, then, for any $X \in \mathfrak{p}$, $t \mapsto e^{tX}H$ is a geodesic in G/H.

Proposition 2.7. *Let g_t be a left invariant diffusion process in G satisfying (2.34). Then g_t is right H-invariant. Moreover, if $g_0 = e$ and if the restriction of $\langle \cdot, \cdot \rangle$ to $\mathfrak{p}$ satisfies the condition (2.35), then $x_t = g_t o$ is a Riemannian Brownian motion in G/H with respect to the G-invariant metric on G/H defined previously, where o is the point eH in G/H.*

Proof. For any $h \in H$, $\mathrm{Ad}(h)$ maps any orthonormal basis in $\mathfrak{p}$ (resp. in $\mathfrak{h}$) into another orthonormal basis in $\mathfrak{p}$ (resp. in $\mathfrak{h}$), it follows that g_t is right H-invariant; hence, by Proposition 2.2, $x_t = g_t o$ is a Markov process in G/H with a Feller transition semigroup Q_t defined by $Q_t f(x) = P_t(f \circ \pi)(g)$ for $f \in C_0(G/H)$ and $x = go$, where P_t is the transition semigroup of g_t. Let $L_{G/H}$ and L_G be respectively the generators of x_t and g_t. For $X \in \mathfrak{g}$, let $\tilde{X} = D\pi(X) \in T_o(G/H)$. Then $\{\tilde{X}_1, \ldots, \tilde{X}_p\}$ is an orthonormal basis of $T_o(G/H)$ and for any $Y \in \mathfrak{p}$ and $g \in G$, $\mathrm{Exp}_{go}[Dg(\tilde{Y})] = \pi(ge^Y) = ge^Y o$. We have, for $f \in C^\infty(G/H)$ and $x = go \in G/H$,

$$
L_{G/H}f(x) = \frac{d}{dt}Q_t f(x)\,|_{t=0} = \frac{d}{dt}P_t(f \circ \pi)(g)\,|_{t=0} = L_G(f \circ \pi)(g)
$$

$$
= \frac{1}{2}\sum_{i=1}^{p} X_i^l X_i^l (f \circ \pi)(g) + \frac{c^2}{2}\sum_{i=p+1}^{d} X_i^l X_i^l (f \circ \pi)(g)
$$

$$
= \frac{1}{2}\sum_{i=1}^{p} X_i^l X_i^l (f \circ \pi)(g) = \frac{1}{2}\sum_{i=1}^{p} \frac{d^2}{dt^2} f(ge^{tX_i}o)\,|_{t=0}
$$

$$
= \frac{1}{2}\sum_{i=1}^{p} \frac{d^2}{dt^2} f(\mathrm{Exp}_x(tDg(\tilde{X}_i)))\,|_{t=0} = \frac{1}{2}\Delta_{G/H}f(x),
$$

where $\Delta_{G/H}$ is the Laplace-Beltrami operator on G/H.   □

Now suppose G is a compact Lie group that acts on a manifold M transitively. An $\mathrm{Ad}(G)$-invariant inner product $\langle \cdot, \cdot \rangle$ on the Lie algebra $\mathfrak{g}$ of G induces a bi-invariant Riemannian metric on G. Fix $p \in M$ and let H be the isotropy subgroup of G at p with Lie algebra $\mathfrak{h}$. The manifold M may be identified with G/H via the map $G/H \ni gH \mapsto gp \in M$. Because the inner product $\langle \cdot, \cdot \rangle$ is $\mathrm{Ad}(H)$-invariant, it induces a G-invariant Riemannian metric on $M = G/H$, which will be called the G-invariant metric on M induced by the $\mathrm{Ad}(G)$-invariant inner product $\langle \cdot, \cdot \rangle$ on $\mathfrak{g}$. It can be shown that this metric is independent of the point $p \in M$.

The condition (2.35) is satisfied because for $X, Y, Z \in \mathfrak{p}$,

$$\langle X, [Z, Y]_{\mathfrak{p}} \rangle + \langle [Z, X]_{\mathfrak{p}}, Y \rangle = \langle X, [Z, Y] \rangle + \langle [Z, X], Y \rangle$$

$$= \frac{d}{dt} \langle \mathrm{Ad}(e^{tZ})X, \mathrm{Ad}(e^{tZ})Y \rangle \mid_{t=0} = 0$$

by the $\mathrm{Ad}(G)$-invariance of the inner product. It follows from Proposition 2.7 with $c = 1$ in (2.34) that if g_t is a Riemannian Brownian motion in G with $g_0 = e$, then $g_t p$ is a Riemannian Brownian motion in M. Because p can be an arbitrary point in M and g_t is right invariant, $g_t x$ is a Riemannian Brownian motion for any $x \in M$ and even when $g_0 \neq e$. To summarize, we have the following result.

Proposition 2.8. *Let G be a compact Lie group that acts transitively on a manifold M and let $\langle \cdot, \cdot \rangle$ be an $\mathrm{Ad}(G)$-invariant inner product on the Lie algebra $\mathfrak{g}$ of G. If g_t is a Riemannian Brownian motion in G with respect to the bi-invariant metric induced by $\langle \cdot, \cdot \rangle$, then $\forall x \in M$, $x_t = g_t x$ is a Riemannian Brownian motion in M with respect to the G-invariant metric induced by $\langle \cdot, \cdot \rangle$.*

3

Generator and Stochastic Integral Equation of a Lévy Process

This chapter is devoted to the proofs of Theorems 1.1, 1.2, and 2.1. In Section 3.1, we show that the generator L of a left invariant Feller semigroup P_t, restricted to $C_0^{2,l}(G)$, is given by (1.7), which also determines P_t. This is the first half of Theorem 1.1 and the uniqueness stated in the second half. We basically follow the arguments in chapter IV of [28], but some changes are made. In Section 3.2, we prove that the stochastic integral equation (1.16) has a unique solution. This is the second half of Theorem 1.2. As a straightforward consequence, the existence stated in the second half of Theorem 1.1 follows. The first half of Theorem 1.2—that is, a given left Lévy process satisfying the stochastic integral equation (1.16)—is proved in Section 3.3. The main ideas from these two sections are taken from Applebaum and Kunita [3], but more details are provided here. Theorem 2.1 is proved in Section 3.4 mainly by suitably changing the arguments in Section 3.1. The amount of work is reduced by combining the proofs of the three results together.

3.1. The Generator of a Lévy Process

Let $\{\mu_t;\ t \in \mathbb{R}_+\}$ be a continuous convolution semigroup of probability measures on a Lie group G, let $\{P_t;\ t \in \mathbb{R}_+\}$ be the left invariant Feller transition semigroup of probability kernels on G defined by (1.5), that is, $P_t f(g) = \mu_t(f \circ L_g)$ for $f \in \mathcal{B}(G)_+$ and $g \in G$, and let L be its generator. We show in this section that the domain $D(L)$ contains $C_0^{2,l}(G)$ and (1.7) holds for any $f \in C_0^{2,l}(G)$. We also prove that the restriction of L to $C_0^{2,l}(G)$ determines the left invariant Feller semigroup P_t uniquely.

Lemma 3.1. *For any neighborhood V of e,*

$$\sup_{t>0} \frac{1}{t} \mu_t(V^c) < \infty.$$

Proof. Recall that $C_u(G)$ is the space of all the uniformly continuous functions on G under the one-point compactification topology and that any $f \in C_u(G)$

may be expressed as $f = f_0 + c$ for $f_0 \in C_0(G)$ and $c \in \mathbb{R}$. Let $\mathcal{C}$ be the space of $f \in C_u(G)$ such that $(1/t)(P_t f - f)$ converges uniformly on G as $t \to 0$. Since $D(L)$ is dense in $C_0(G)$, it follows that $\mathcal{C}$ is dense in $C_u(G)$. Therefore, for any $f \in C_u(G)$ and $\varepsilon > 0$, there exists $\phi \in \mathcal{C}$ such that $\sup_{g \in G} |f(g) - \phi(g)| \leq \varepsilon$, and $(1/t)[\mu_t(\phi \circ L_g) - \phi(g)]$ converges uniformly for $g \in G$ as $t \to 0$. Let U be a compact neighborhood of e such that $U^{-1} = U$ and $U^2 = UU \subset V$; let $f \in C_u(G)$ satisfy $0 \leq f \leq 1$, $f(e) = 0$, and $f(g) = 1$ for $g \in U^c$; let $0 < \varepsilon < \frac{1}{4}$; and let $\phi \in \mathcal{C}$ be chosen as before with $\eta = \inf\{\phi(g); g \in G\}$. Then $-\varepsilon \leq \eta \leq \varepsilon$, $f - \varepsilon \leq \phi \leq f + \varepsilon$, and $\exists z \in U$ with $\phi(z) = \eta$. Let $\psi = \phi - \eta$. Then $\psi \geq 0$ on G, $\psi \geq 1 - 2\varepsilon$ on U^c, and $\psi(z) = 0$. Let $\xi = \psi \circ L_z$. Then $\xi \geq 0$ on G, $\xi \geq 1 - 2\varepsilon$ on V^c, and $\xi(e) = 0$. Moreover, $\lim_{t \to 0} \frac{1}{t} \mu_t(\xi) = \lim_{t \to 0} \frac{1}{t}[\mu_t(\phi \circ L_z) - \phi(z)]$ exists and is finite. This implies the conclusion of the lemma. $\qquad \square$

Lemma 3.2. *Let E be a Banach space, E' be a dense subspace of E, F be a closed subspace of E of finite codimension, $y \in E$, and $M = F + y = \{x + y; x \in F\}$. Then $M \cap E'$ is dense in M.*

Proof. It suffices to prove the case when $\dim(E) = \infty$. Suppose the codimension of F in E is equal to k. Then there exist linearly independent $x_1, \ldots, x_k \in E'$ such that $E = F \oplus H$, where H is the k-dimensional subspace of E spanned by $x_1, \ldots, x_k$. Evidently, $H \subset E'$. Any $x \in E$ may be written uniquely as $x = f_x + h_x$ with $f_x \in F$ and $h_x \in H$. Let $p: E \to M$ be the map defined by $p(x) = f_x + h_y = f_x - f_y + y \in M$. Then p is continuous and fixes points in M. It now follows that $p(E')$ is dense in M. Since $H \subset E'$, it is easy to see that $p(E') \subset E'$. This proves that $E' \cap M$ is dense in M. $\qquad \square$

As in Section 1.2, let $X_1, \ldots, X_d$ be a basis of $\mathfrak{g}$ and let $x_1, \ldots, x_d \in C_c^\infty(G)$ be a set of associated coordinate functions. For any function f on G, let $\|f\| = \sup_{g \in G} |f(g)|$, and for $f \in C_0^{2,l}(G)$, let

$$\|f\|_2^l = \|f\| + \sum_{i=1}^{d} \|X_i^l f\| + \sum_{i,j=1}^{d} \|X_i^l X_j^l f\|.$$

Similarly, for $f \in C_0^{2,r}(G)$, $\|f\|_2^r$ is defined with the X_i^l replaced by X_i^r. Note that $C_u(G)$, $C_0^{2,l}(G)$, and $C_0^{2,r}$ are Banach spaces under the norms $\|f\|$, $\|f\|_2^l$, and $\|f\|_2^r$ respectively.

Lemma 3.3. $D(L) \cap C_0^{2,r}(G)$ *is dense in* $C_0^{2,r}(G)$.

Proof. It is easy to see that if $f \in C_0^{2,r}(G)$, then $P_t f \in C_0^{2,r}(G)$ and $\| P_t f \|_2^r \leq \| f \|_2^r$. By restriction, we may consider P_t as a semigroup of linear operators on the Banach space $C_0^{2,r}(G)$, and then by the Hille–Yosida Theorem (see, for example, theorem 1 in Yosida [62, section IX.3]), the domain of its generator is dense in $C_0^{2,r}(G)$. This implies that $D(L) \cap C_0^{2,r}(G)$ is dense in $C_0^{2,r}(G)$. $\qquad\square$

Lemma 3.4. *There exist* $y_1, \ldots, y_d, \psi \in D(L) \cap C_0^{2,r}(G)$ *such that*

(i) $y_i(e) = 0$ *and* $X_i^r y_j(e) = \delta_{ij}$ *for* $i, j = 1, 2, \ldots, d$;
(ii) $\psi(e) = X_i^r \psi(e) = 0$ *and* $X_i^r X_j^r \psi(e) = 2\delta_{ij}$ *for* $i, j = 1, 2, \ldots, d$;
(iii) $\exists$ *a neighborhood* V *of* e *and* $\delta > 0$ *such that* $\psi(g) \geq \delta \sum_{i=1}^d x_i(g)^2$ *for all* $g \in V$.

Proof. Part (i) follows from Lemma 3.2 by setting $E = C_0^{2,r}(G)$, $E' = D(L) \cap C_0^{2,r}(G)$,

$$F = \{ f \in C_0^{2,r}(G); \ f(e) = X_i^r f(e) = 0 \quad \text{for} \quad 1 \leq i \leq d \},$$

and $y \in C_0^{2,r}(G)$ satisfying $y(e) = 0$ and $X_i^r y(e) = \delta_{ij}, j = 1, 2, \ldots, d$. Similarly, (ii) can be proved with the same E and E', but with

$$F = \{ f \in C_0^{2,l}(G); \ f(e) = X_i^r f(e) = X_i^r X_j^r f(e) = 0 \text{ for } i, j = 1, 2, \ldots, d \}$$

and $y \in C_0^{2,r}(G)$ satisfying $y(e) = X_i^r y(e) = 0$ and $X_i^r X_j^r y(e) = 2\delta_{ij}$. To prove (iii), first assume the coordinate functions x_i satisfy $g = \exp[\sum_i x_i(g) X_i]$ for g in a neighborhood of e, then the Taylor expansion (1.9) holds. The expansion holds also when gh and X_i^l are replaced by hg and X_i^r, respectively; hence, if $\psi \in C^2(G)$ satisfies $\psi(e) = X_i^r \psi(e) = 0$, then, for g in a neighborhood of e,

$$\psi(g) = \frac{1}{2} \sum_{i,j=1}^d x_i(g) x_j(g) X_i^r X_j^r \psi(g'),$$

where $g' = \exp[t \sum_{i=1}^d x_i(g) X_i]$ for some $t \in [0, 1]$. Part (iii) follows from this and the fact that $X_i^r X_j^r \psi(e) = 2\delta_{ij}$. Note that this conclusion is independent of the choice of coordinate functions x_i because of (1.6). $\qquad\square$

Lemma 3.5. $\sup_{t>0} \frac{1}{t} \mu_t \left(\sum_{i=1}^d x_i^2 \right) < \infty$.

Proof. Let $\psi \in D(L) \cap C_0^{2,r}(G)$ be given in Lemma 3.4. Note that $\psi \in D(L) \implies \sup_{t>0} \frac{1}{t} \mu_t(\psi) < \infty$. However, by Lemma 3.1, $\sup_{t>0} \frac{1}{t} \mu_t(V^c) <$

∞ for any neighborhood V of e. This implies that $\sup_{t>0} \frac{1}{t} \int_V \psi d\mu_t < \infty$. From this and Lemma 3.4(iii), we obtain $\sup_{t>0} \frac{1}{t} \mu_t(\sum_{i=1}^d x_i^2 \cdot 1_V) < \infty$. By Lemma 3.1, $\sup_{t>0} \frac{1}{t} \mu_t(\sum_{i=1}^d x_i^2) < \infty$. $\qquad\square$

Lemma 3.6. *There is a constant $c > 0$ such that*

$$\forall f \in C_0^{2,r}(G) \quad \text{and} \quad t > 0, \quad \left| \frac{1}{t}[\mu_t(f) - f(e)] \right| \leq c\|f\|_2^r. \tag{3.1}$$

Proof. For $f \in C_0^{2,r}(G)$, let

$$\phi(g) = f(g) - f(e) - \sum_{i=1}^d y_i(g)X_i^r f(e)$$

for $g \in G$, where $y_1, \ldots, y_d \in D(L) \cap C_0^{2,r}(G)$ are given in Lemma 3.4. Then $\phi(e) = X_i^r \phi(e) = 0$. Let W be a neighborhood of e such that the exponential map $\exp$ is diffeomorphic on $\exp^{-1}(W)$. By the version of the Taylor expansion (1.9) using right invariant vector fields X_i^r, for a properly chosen W and all $g \in W$,

$$\phi(g) = \frac{1}{2} \sum_{i,j=1}^d x_i(g)x_j(g)X_i^r X_j^r \phi(g')$$

for some $g' \in W$. Hence, there is a constant $c_1 > 0$, independent of f and ϕ, such that

$$\forall g \in W, \quad |\phi(g)| \leq c_1 \|\phi\|_2^r \sum_{i=1}^d x_i(g)^2,$$

and therefore

$$\left| \frac{1}{t} \int_W \phi d\mu_t \right| \leq c_2 \|\phi\|_2^r, \tag{3.2}$$

where $c_2 = c_1 \sup_{t>0} \frac{1}{t} \mu_t(\sum_{i=1}^d x_i^2) < \infty$ by Lemma 3.5. However, for any neighborhood W of e,

$$\left| \frac{1}{t} \int_{W^c} \phi d\mu_t \right| \leq c_3 \|\phi\|, \tag{3.3}$$

where $c_3 = \sup_{t>0} \frac{1}{t} \mu_t(W^c) < \infty$ by Lemma 3.1. Note that $\|\phi\| \leq \|\phi\|_2^r \leq c_4 \|f\|_2^r$ for some constant $c_4 > 0$ independent of f and ϕ. Let

$c_5 = (c_2 + c_3)c_4$. Adding (3.2) and (3.3) yields

$$\left| \frac{1}{t}[\mu_t(f) - f(e)] - \frac{1}{t}\sum_{i=1}^{d} X_i^r f(e)\mu_t(y_i) \right| = \left| \frac{1}{t}\mu_t(\phi) \right| \le c_5 \|f\|_2^r.$$

Since $y_i \in D(L)$ and $y_i(e) = 0$, $\sup_{t>0} |\frac{1}{t}P_t y_i(e)| < \infty$. From this, (3.1) follows. □

We may regard P_t as a semigroup of operators on $C_u(G)$. Its generating functional A is defined by

$$Af = \lim_{t\to 0} \frac{1}{t}[\mu_t(f) - f(e)] \tag{3.4}$$

for $f \in C_u(G)$ if the limit exists and is finite. We may regard A as a linear functional on $C_u(G)$ with domain $D(A)$ being the set of $f \in C_u(G)$ for which this limit exists and is finite.

It is clear that $D(A)$ contains $D(L)$, the domain of the generator L, and for $f \in D(L)$, $Af = Lf(e)$. Moreover, by the left invariance of P_t, if $f \in D(L)$ and $g \in G$, then $f \circ L_g \in D(L) \subset D(A)$ and $Lf(g) = A(f \circ L_g)$.

Lemma 3.7. $C_u(G) \cap C^2(G) \subset D(A)$. *In particular, $D(A)$ contains both $C_0^{2,r}(G)$ and $C_0^{2,l}(G)$ and is dense in $C_u(G)$ under the norm $\|\cdot\|$.*

Proof. We first show that $D(A)$ contains $C_0^{2,r}(G)$. By Lemma 3.3, $C_0^{2,r}(G) \cap D(L)$ is dense in $C_0^{2,r}(G)$. Since, for $f \in D(L)$, $\lim_{t\to 0}(1/t)[\mu_t(f) - f(e)]$ exists and is finite, it follows from (3.1) that $C_0^{2,r}(G) \subset D(A)$.

Let $f \in C_0(G) \cap C^2(G)$. For any $\varepsilon > 0$, there is a relatively compact neighborhood V of e such that f may be written as $f = f_1 + f_2$, where $f_1 \in C_c^2(G) \subset C_0^{2,r}(G)$ and $f_2 \in C_0(G) \cap C^2(G)$ satisfying $f_2 = 0$ on V and $\|f_2\| < \varepsilon$. Then $f_1 \in D(A)$ and $|\sup_{t>0}(1/t)\mu_t(f_2)| \le c\varepsilon$, where $c = \sup_{t>0}(1/t)\mu_t(V^c) < \infty$. This implies $f \in D(A)$. Since any $f \in C_u(G) \cap C^2(G)$ can be written as $f = h + c$, where $h \in C_0(G) \cap C^2(G)$ and $c = \lim_{g\to\infty} f(g)$ is a constant, it follows that $C_u(G) \cap C^2(G) \subset D(A)$. □

Lemma 3.8. *There is a constant $c > 0$ such that*

$$\forall f \in C_0^{2,l}(G) \text{ and } t > 0, \quad \left\| \frac{1}{t}[P_t f - f] \right\| \le c\|f\|_2^l. \tag{3.5}$$

Proof. For $B \in \mathcal{B}(G)$, let $\hat{\mu}_t(B) = \mu_t(B^{-1})$. Then $\{\hat{\mu}_t; t \in \mathbb{R}_+\}$ is a continuous convolution semigroup of probability measures on G and (3.1) holds with

μ_t replaced by $\hat{\mu}_t$. For $f \in C_0^{2,l}(G)$, let $\hat{f}(g) = f(g^{-1})$. Then $\hat{f} \in C_0^{2,r}(G)$ with $\|f\|_2^l = \|\hat{f}\|_2^r$ and $\hat{\mu}_t(\hat{f}) = \mu_t(f)$. For $f \in C_0^{2,l}(G)$ and $g \in G$,

$$
\left| \frac{1}{t}[P_t f(g) - f(g)] \right| = \left| \frac{1}{t}[\mu_t(f \circ L_g) - (f \circ L_g)(e)] \right|
$$

$$
= \left| \frac{1}{t}\{\hat{\mu}_t[(f \circ L_g)\check{}] - (f \circ L_g)\check{}(e)\} \right|
$$

$$
\le c\|(f \circ L_g)\check{}\|_2^r = c\|f \circ L_g\|_2^l = c\|f\|_2^l. \qquad \square
$$

Let G_∞ denote the one-point compactification of G with ∞ being the point at infinity. Any function f in $C_u(G)$ will be regarded as a function in $C(G_\infty)$ by setting $f(\infty) = \lim_{g \to \infty} f(g)$. The function spaces $C_u(G)$ and $C(G_\infty)$ are thus naturally identified. The measures μ_t will be extended to G_∞ by setting $\mu_t(\{\infty\}) = 0$.

Let W be an open subset of G_∞ whose closure does not contain e. For $f \in D(A) \cap C_c(W)$, $Af = \lim_{t \to 0}(1/t)\mu_t(f)$; hence, $|Af| \le c_W \|f\|$, where $c_W = \sup_{t>0}(1/t)\mu_t(W)$ is finite by Lemma 3.1, and if $f \ge 0$, then $Af \ge 0$. Therefore, $f \mapsto Af$ is a positive bounded linear functional on $D(A) \cap C_c(W)$ under the norm $\|f\|$ and, because $D(A) \cap C_c(W)$ is dense in $C_c(W)$ by Lemma 3.7, it may be extended uniquely to be such a functional on $C_c(W)$. By the Riesz representation theorem on a locally compact Hausdorff space, there is a unique measure $\bar{\Pi}$ on W such that $Af = \bar{\Pi}(f)$ and $|\bar{\Pi}(f)| \le c_W \|f\|$ for $f \in D(A) \cap C_c(W)$. By letting $W \uparrow (G_\infty - \{e\})$, $\bar{\Pi}$ becomes a measure on $(G_\infty - \{e\})$ such that $\bar{\Pi}(U^c) < \infty$ for any neighborhood U of e. We can regard $\bar{\Pi}$ as a measure on G_∞ by setting $\bar{\Pi}(\{e\}) = 0$. Then

$$
Af = \bar{\Pi}(f) \quad \text{for any } f \in D(A) \text{ vanishing in a neighborhood of } e. \quad (3.6)
$$

Let $\Pi = \bar{\Pi}|_G$. Choose $\phi_n \in C_u(G) \cap C^\infty(G)$ such that $0 \le \phi_n \le 1$, the ϕ_n vanish in a neighborhood of e, $\phi_n = 1$ in a neighborhood of ∞, and $[\phi_n = 1] \uparrow (G - \{e\})$ as $n \uparrow \infty$, where $[\phi_n = 1] = \{g \in G; \phi_n(g) = 1\}$. Then

$$
\Pi\left(\sum_{i=1}^d x_i^2\right) = \lim_{n \to \infty} \Pi\left(\phi_n \sum_{i=1}^d x_i^2\right)
$$

$$
= \lim_{n \to \infty} A\left(\phi_n \sum_{i=1}^d x_i^2\right) \le \sup_{t>0} \frac{1}{t}\mu_t\left(\sum_{i=1}^d x_i^2\right) < \infty. \quad (3.7)
$$

Therefore, Π satisfies the condition (1.8) and hence is a Lévy measure on G.

Let

$$
a_{ij} = A(x_i x_j) - \Pi(x_i x_j) \quad \text{and} \quad c_i = A(x_i). \quad (3.8)
$$

It is clear that $a_{ij} = a_{ji}$. For $(\xi_1, \ldots, \xi_d) \in \mathbb{R}^d$, let $\psi = (\sum_{i=1}^d \xi_i x_i)^2$. Then

$$A(\psi) = \lim_{t \to 0} \frac{1}{t} \mu_t(\psi) \geq \lim_{t \to 0} \frac{1}{t} \mu_t(\phi_n \psi) = A(\phi_n \psi) = \Pi(\phi_n \psi) \uparrow \Pi(\psi)$$

as $n \uparrow \infty$; hence,

$$\sum_{i,j=1}^d a_{ij} \xi_i \xi_j = A(\psi) - \Pi(\psi) \geq 0.$$

This shows that the symmetric matrix $\{a_{ij}\}$ is nonnegative definite.

Let U be a neighborhood of e on which the coordinate functions $x_1, \ldots, x_d$ may be used as local coordinates. We will assume that these coordinate functions satisfy $g = \exp[\sum_{i=1}^d x_i(g)X_i]$ for $g \in U$. Because of (1.6), Lemma 3.9 below still holds without this additional assumption if the constants c_i are suitably changed.

Lemma 3.9. *For $f \in C_u(G) \cap C^2(G)$,*

$$Af = \sum_{i=1}^d c_i X_i^l f(e) + \frac{1}{2} \sum_{j,k=1}^d a_{jk} X_j^l X_k^l f(e)$$

$$+ \Pi \left[f - f(e) - \sum_{i=1}^d x_i X_i^l f(e) \right]. \tag{3.9}$$

Proof. Let $f \in C_u(G) \cap C^2(G)$. We may write

$$f(g) - f(e) = \sum_{i=1}^d x_i(g)X_i^l f(e) + \frac{1}{2} \sum_{j,k=1}^d x_j(g)x_k(g)X_j^l X_k^l f(e) + R_f(g)$$

$$\tag{3.10}$$

for $g \in G$ and some $R_f \in C_u(G) \cap C^2(G)$. By Taylor expansion (1.9), for $g \in U$,

$$R_f(g) = \frac{1}{2} \sum_{j,k=1}^d x_j(g)x_k(g)[X_j^l X_k^l f(g') - X_j^l X_k^l f(e)],$$

where $g' = \exp[t \sum_{i=1}^d x_i(g)X_i]$ for some $t \in [0, 1]$.

We now show $A(R_f) = \bar{\Pi}(R_f)$. Let ϕ_n be chosen as before. Then $A(\phi_n R_f) = \bar{\Pi}(\phi_n R_f) \to \bar{\Pi}(R_f)$ as $n \to \infty$. Because $|R_f| \leq \eta \sum_{i=1}^d x_i^2$ in a sufficiently small neighborhood of e for some $\eta \in C(G - \{e\})$ with $\eta(g) \to 0$ as $g \to e$, and because $[\phi_n < 1] \downarrow \{e\}$, it follows from Lemma 3.5 that

$$|A(R_f - \phi_n R_f)| \leq \sup_{t > 0} \frac{1}{t} \mu_t[(1 - \phi_n)|R_f|] \to 0$$

as $n \to \infty$. This proves $A(R_f) = \bar{\Pi}(R_f)$.

Applying $(1/t)\mu_t$ to (3.10) and then letting $t \to 0$, we obtain, for any $f \in C_u(G) \cap C^2(G)$,

$$
\begin{aligned}
Af &= \sum_{i=1}^{d} A(x_i)X_i^l f(e) + \frac{1}{2} \sum_{j,k=1}^{d} A(x_j x_k)X_j^l X_k^l f(e) + A(R_f) \\
&= \sum_{i=1}^{d} c_i X_i^l f(e) + \frac{1}{2} \sum_{j,k=1}^{d} a_{jk} X_j^l X_k^l f(e) \\
&\quad + \frac{1}{2} \sum_{j,k=1}^{d} \bar{\Pi}(x_j x_k)X_j^l X_k^l f(e) + \bar{\Pi}(R_f) \\
&= \sum_{i=1}^{d} c_i X_i^l f(e) + \frac{1}{2} \sum_{j,k=1}^{d} a_{jk} X_j^l X_k^l f(e) \\
&\quad + \bar{\Pi}\left[\frac{1}{2} \sum_{j,k=1}^{d} x_j x_k X_j^l X_k^l f(e) + R_f \right] \\
&= \sum_{i=1}^{d} c_i X_i^l f(e) + \frac{1}{2} \sum_{j,k=1}^{d} a_{jk} X_j^l X_k^l f(e) + \bar{\Pi}\left[f - f(e) - \sum_{i=1}^{d} x_i X_i^l f(e) \right] \\
&= \sum_{i=1}^{d} c_i X_i^l f(e) + \frac{1}{2} \sum_{j,k=1}^{d} a_{jk} X_j^l X_k^l f(e) + \Pi\left[f - f(e) - \sum_{i=1}^{d} x_i X_i^l f(e) \right] \\
&\quad + c[f(\infty) - f(e)], \tag{3.11}
\end{aligned}
$$

where $c = \bar{\Pi}(\{\infty\})$.

It remains to show $c = 0$. If $c > 0$, choose $f_n \in C_c(G) \cap C^2(G)$ such that $f_n = 1$ in a neighborhood of e, $0 \le f_n \le 1$, and $f_n \uparrow 1$ on G as $n \uparrow \infty$. By (3.11), $Af_n = \Pi(f_n - 1) - c \le -c$. It follows that, for sufficiently small $t > 0$, $\mu_t(f_n) \le 1 - (c/2)t$. This is impossible because $\mu_t(f_n) \uparrow 1$ as $n \uparrow \infty$. $\qquad\square$

We are now ready to prove the first half of Theorem 1.1 and the uniqueness in the second half, which may be stated as follows:

Theorem 3.1. *Let a_{ij}, c_i, and Π be defined as before. If $f \in C_0^{2,l}(G)$, then $f \in D(L)$ and, for $f \in C_0^{2,l}(G)$, (1.7) holds. Moreover, there is at most one left invariant Feller semigroup P_t on G whose generator L restricted to $C_c^\infty(G)$ is given by (1.7).*

Proof. Let $f \in C_0^{2,l}(G)$. Then $f \circ L_g \in C_0^{2,l}(G) \subset D(A)$ for any $g \in G$ and

$$
\lim_{t \to 0} \frac{1}{t}[P_t f(g) - f(g)] = \lim_{t \to 0} \frac{1}{t}[P_t(f \circ L_g)(e) - (f \circ L_g)(e)] = A(f \circ L_g).
$$

By (3.9), $A(f \circ L_g)$ is equal to the expression on the right-hand side of (1.7). To prove the first statement of the theorem, we need only to show $f \in D(L)$.

For $\lambda > 0$, let R_λ be the resolvent associated to the semigroup P_t defined by $R_\lambda f = \int_0^\infty e^{-\lambda t} P_t f \, dt$ for $f \in C_0(G)$. It is well known that the range of R_λ is equal to $D(L)$; that is, $D(L) = \{R_\lambda f; f \in C_0(G)\}$. Let $f \in C_0^{2,l}(G)$. By Lemma 3.7 and the invariance of $C_0^{2,l}(G)$ under left translations, $f \circ L_g \in D(A)$ for $g \in G$ and

$$\lim_{t \to 0} \frac{1}{t}[P_t f(g) - f(g)] = \lim_{t \to 0} \frac{1}{t}[\mu_t(f \circ L_g) - (f \circ L_g)(e)] = A(f \circ L_g).$$

By Lemma 3.8, $\|(1/t)(P_t f - f)\|$ is bounded in t. Therefore, writing $Bf(g) = A(f \circ L_g)$, we have

$$R_\lambda(\lambda f - Bf) = \lambda R_\lambda f - \lim_{t \to 0} \frac{1}{t} R_\lambda(P_t f - f)$$

$$= \lambda R_\lambda f - \lim_{t \to 0} \frac{1}{t}\left(e^{\lambda t} \int_t^\infty e^{-\lambda s} P_s f \, ds - R_\lambda f\right)$$

$$= \lambda R_\lambda f - \lim_{t \to 0} \frac{1}{t}(e^{\lambda t} - 1)R_\lambda f + \lim_{t \to 0} \frac{1}{t}e^{\lambda t} \int_0^t e^{-\lambda s} P_s f \, ds$$

$$= \lambda R_\lambda f - \lambda R_\lambda f + f = f.$$

This proves $f \in D(L)$.

We now prove the uniqueness. We will actually prove a slightly stronger result: that there is at most one left invariant Feller semigroup P_t of probability kernels on G if its generating functional restricted to $C_c^\infty(G)$ is given. The idea of this proof is taken from Hunt [32]. Suppose P_t and P_t' are two Feller semigroups on G whose generating functionals A and A' agree on $C_c^\infty(G)$. Since the restriction of the generating functional to $C_c^\infty(G)$ determines Π, c_i, and a_{ij} in (3.9), by Lemma 3.9, A and A' agree on $C_u(G) \cap C^2(G)$. For $f \in C_0^{2,r}(G)$ and $g \in G$, let $Nf(g) = A(f \circ L_g) = A'(f \circ L_g)$. Then $(d/dt)P_t f(g) = (d/ds)P_s P_t f(g) |_{s=0} = (d/ds)\mu_s[(P_t f) \circ L_g] |_{s=0} = A[(P_t f) \circ L_g] = N P_t f(g)$ and similarly $(d/dt)P_t' f(g) = N P_t' f(g)$. Let $f \in C_c^\infty(G)$ and define $h(t, g) = P_t f(g) - P_t' f(g)$ for $g \in G$. It suffices to show $h(t, g) = 0$ for any $(t, g) \in \mathbb{R}_+ \times G$. In fact, one just needs to show $h(t, g) \geq 0$ because replacing f by $-f$ will then lead to $h(t, g) \leq 0$.

Note that $h(t, \cdot) \in C_0^{2,r}(G)$ and that it may be regarded as a continuous function on $\mathbb{R}_+ \times G_\infty$ satisfying

$$h(0, \cdot) = 0, \quad h(\cdot, \infty) = 0, \quad \text{and} \quad \frac{d}{dt}h(t, g) = Nh(t, g).$$

By (3.9), Nf is given by the right-hand side of (1.7). From this it follows that if $h(s, u) = \min_{g \in G_\infty} h(s, g)$ for some $(s, u) \in (0, \infty) \times G$, then

$(d/dt)h(s, u) = Nh(s, u) \geq 0$. Let $h_c(t, g) = e^{-ct}h(t, g)$ for some fixed $c > 0$. If $h_c(t, g) < 0$ for some (t, g), then there are $0 < \delta < 1$, $s > 0$, and $u \in G$ such that $-\delta = h_c(s, u) = \min_{g \in G_\infty} h_c(s, g)$ and $h_c(t, g) > -\delta$ for all $(t, g) \in [0, s) \times G$. It follows that $(d/dt)h_c(s, u) \leq 0$. However, $h(s, u) = \min_{g \in G_\infty} h(s, g)$ and $(d/dt)h_c(s, u) = -ch_c(s, u) + e^{-cs}(d/dt)h(s, u) \geq c\delta > 0$, which is a contraction. This shows $h_c(t, g) \geq 0$ and hence $h(t, g) \geq 0$. $\qquad\square$

3.2. Existence and Uniqueness of the Solution to a Stochastic Integral Equation

In this section, we prove the existence and the uniqueness of the solution to the stochastic integral equation (1.16), that is, the second half of Theorem 1.2. From this result, it is easy to derive the existence stated in the second half of Theorem 1.1

Let $\{\mathcal{F}_t\}$ be a filtration, let B_t be a d-dimensional $\{\mathcal{F}_t\}$-Brownian motion with covariance matrix a_{ij}, let N be an $\{\mathcal{F}_t\}$-Poisson random measure on $\mathbb{R}_+ \times G$ with characteristic measure Π being a Lévy measure, such that $\{B_t\}$ and N are independent under $\{\mathcal{F}_t\}$, and let c_i be some constants. A càdlàg process g_t in G with $g_0 = e$, adapted to $\{\mathcal{F}_t\}$, is called a solution of the stochastic integral equation (1.16) if it satisfies (1.16) for any $f \in C_0^{2,l}(G)$. Equation (1.16) is said to have a unique solution if any two solutions agree almost surely for any time $t \in \mathbb{R}_+$.

Let $\{\mathcal{F}_t^{B,N}\}$ denote the filtration generated by $\{B_t\}$ and N as defined in Appendix B.3. It is clear that $\mathcal{F}_t^{B,N} \subset \mathcal{F}_t$.

In the rest of this chapter, a Lévy process in G will mean a left Lévy process unless explicitly stated otherwise.

Lemma 3.10. *Given* $\{B_t\}$, N, Π, *and* c_i *as before, let* g_t *be a Lévy process in* G *and suppose it is a solution of (1.16), let* N' *be a* $\{\mathcal{F}_t\}$*-Poisson random measure on* $\mathbb{R}_+ \times G$ *with a finite characteristic measure* Π' *and independent of* $\{B_t\}$ *and* N *under* $\{\mathcal{F}_t\}$, *and let* y_t *be the Lévy process in* G *obtained from* g_t *by adding the jumps determined by* N' *as described in the statement of Proposition 1.5. Then* y_t *is a solution of (1.16) with the same* $\{B_t\}$, *but with* N, Π, *and* c_i *replaced by* $N + N'$, $\Pi + \Pi'$, *and* $c_i + \int x_i d\Pi'$. *Moreover, if* g_t *is adapted to* $\{\mathcal{F}_t^{B,N}\}$, *then* y_t *is adapted to* $\{\mathcal{F}_t^{B,N+N'}\}$.

Note that the construction of y_t from g_t and the proof that y_t is a Lévy process, as stated in Proposition 1.5, do not depend on Theorem 1.1. Although

Propositions 1.4 and 1.5 are now available because their proofs depend only on the first half of Theorem 1.1, which has been proved, they will not be used in this section.

Proof. Fix B_t, N, Π, and c_i in (1.16). For $f \in C_0^{2,l}(G)$, a càdlàg process g_t in G adapted to $\{\mathcal{F}_t\}$, and $a < b$, let

$$S(f, g., a, b) = \sum_{i=1}^{d} \int_a^b X_i^l f(g_{s-}) \circ dB_s^i + \sum_{i=1}^{d} c_i \int_a^b X_i^l f(g_{s-}) \, ds$$
$$+ \int_a^b \int_G [f(g_{s-}h) - f(g_{s-})] \tilde{N}(ds\,dh)$$
$$+ \int_a^b \int_G \left[f(g_{s-}h) - f(g_{s-}) - \sum_{i=1}^{d} x_i(h) X_i^l f(g_{s-}) \right] ds\,\Pi(dh).$$

Equation (1.16) may be written as $f(g_t) = f(g_0) + S(f, g., 0, t)$. Note that, if g_t is a solution, then $S(f, g., a, b) = f(g_b) - f(g_a)$. Moreover, for $h \in G$, $S(f \circ L_h, g., a, b) = S(f, hg., a, b)$.

Let random times $T_n \uparrow \infty$ and G-valued random variables σ_n be the two sequences determined by N'. For simplicity, write $T = T_1$ and $\sigma = \sigma_1$. Then $y_t = g_t$ for $t < T$ and $y_t = g_{T-}\sigma g_T^{-1}g_t$ for $T \le t < T_2$.

Note that the process g_t only jumps when N does. Since N and N' are independent, g_t is almost surely continuous at $t = T$. Thus, $y_t = g_T\sigma g_T^{-1}g_t$ for $T \le t < T_2$ and, for $f \in C_0^{2,l}(G)$,

$$f(y_t) = f(g_T) + [f(g_T\sigma g_T^{-1}g_t) - f(g_T\sigma)] + [f(g_T\sigma) - f(g_T)]$$
$$= f(g_T) + [f(g_T\sigma g_T^{-1}g_t) - f(g_T\sigma)]$$
$$+ \int_0^t \int_G [f(y_{s-}h) - f(y_{s-})] N'(ds\,dh). \tag{3.12}$$

Let $B_t^{(T)}$ and $N^{(T)}$ be, respectively, the Brownian motion B_t and the Poisson random measure N shifted by the finite stopping time T; that is,

$$B_t^{(T)} = B_{T+t} - B_T \quad \text{and} \quad N^{(T)}([0, t] \times \cdot) = N([T, T+t] \times \cdot).$$

Then $B_t^{(T)}$ is a Brownian motion with same covariance matrix as B_t, and $N^{(T)}$ is a Poisson random measure on $\mathbb{R}_+ \times G$ with the same characteristic measure as N. Moreover, $\{B_t^{(T)}\}$, $N^{(T)}$, and $\mathcal{F}_T$ are independent.

Let $\tau = g_T\sigma g_T^{-1}$. Then τ is independent of $\{B_t^{(T)}\}$ and $N^{(T)}$. We have

$$f(g_T\sigma g_T^{-1}g_t) - f(g_T\sigma) = (f \circ L_\tau)(g_t) - (f \circ L_\tau)(g_T)$$
$$= S(f \circ L_\tau, g., T, t)$$
$$= S(f, \tau g., T, t) = S(f, y., T, t).$$

By (3.12), we obtain, for $T \le t < T_2$,

$$f(y_t) = f(g_T) + S(f, y_\cdot, T, t) + \int_0^t \int_G [f(y_{s-}h) - f(y_{s-})]N'(ds\,dh).$$

Since $f(g_T) = f(g_0) + S(f, g_\cdot, 0, T) = f(y_0) + S(f, y_\cdot, 0, T)$, it follows that

$$f(y_t) = f(y_0) + S(f, y_\cdot, 0, t) + \int_0^t \int_G [f(y_{s-}h) - f(y_{s-})]N'(ds\,dh).$$

This is just Equation (1.16) with g_t, N, Π, and c_i replaced by y_t, $N + N'$, $\Pi + \Pi'$, and $c_i + \int x_i\,d\Pi'$ respectively. We have proved that y_t satisfies Equation (1.16) for $0 \le t < T_2$. By a similar computation, we can prove successively that it satisfies (1.16) for $T_n \le t < T_{n+1}$ for $n = 2, 3, \ldots$, and hence for all $t \in \mathbb{R}_+$.

It is clear from the explicit construction of y_t from g_t that if g_t is adapted to $\{\mathcal{F}_t^{B,N}\}$, then y_t is adapted to $\{\mathcal{F}_t^{B,N+N'}\}$. $\qquad\square$

Let $U \subset V$ be two relatively compact neighborhoods of e such that the closure $\overline{UU}$ of $UU = \{gh; g, h \in U\}$ is contained in V and the coordinate functions x_i form a set of coordinates on V.

For the moment, we will assume supp$(\Pi) \subset U$. An adapted càdlàg process g_t in G with $g_0 = e$, adapted to $\{\mathcal{F}_t\}$, is called a local solution of (1.16) in U if for any $f \in C_0^{2,l}(G)$, (1.16) holds for $t \le \tau$, where $\tau = \inf\{t > 0; g_t \in U^c\}$. Equation (1.16) is said to have a unique local solution in U if two such solutions agree for $t \le \tau$ almost surely. Note that because Π is supported by U, $g_\tau \in \overline{UU} \subset V$. Therefore, when checking whether g_t is a local solution in U, its value for $t > \tau$ is not important. In fact, one may assume $g_t = g_\tau$ for $t > \tau$ and then the function space $C_0^{2,l}(G)$ can be replaced by $C^2(G)$.

Let σ be a finite $\{\mathcal{F}_t\}$-stopping time, let $B_t^{(\sigma)}$ and $N^{(\sigma)}$ be B_t and N shifted by σ as defined earlier, and let $\{\mathcal{F}_t^{(\sigma)}\}$ be the filtration generated by $\{B_t^{(\sigma)}\}$ and $N^{(\sigma)}$. Note that if (1.16) has a local solution in U, then it will still have a local solution in U when B_t, N, and $\mathcal{F}_t$ are replaced by $B_t^{(\sigma)}$, $N^{(\sigma)}$, and $\mathcal{F}_t^{(\sigma)}$.

The following lemma may be used to reduce the existence and uniqueness of the solution of (1.16) to that of a local solution.

Lemma 3.11. *Assume* supp$(\Pi) \subset U$. *If the stochastic integral equation (1.16) has a unique local solution x_t in U, then it has a unique solution g_t in G with $g_0 = e$. Moreover, if x_t is adapted to $\{\mathcal{F}_t^{B,N}\}$, then g_t is also adapted to $\{\mathcal{F}_t^{B,N}\}$ and is a Lévy process in G with generator L restricted to $C_0^{2,l}(G)$ given by (1.7).*

Proof. Suppose (1.16) has a unique local solution x_t in U. Then a solution g_t of (1.16) in G with $g_0 = e$ can be constructed as follows: Let $g_t^0 = x_t$, let $\tau_1 = \inf\{t > 0; g_t^0 \in U^c\} \wedge 1$, and let g_t^1, be the local solution of (1.16) in U with B_t, N, and $\mathcal{F}_t$ replaced by $B_t^{(\tau_1)}$, $N^{(\tau_1)}$, and $\mathcal{F}_t^{(\tau_1)}$. Inductively, let $\tau_n = \inf\{t > 0; g_t^{n-1} \in U^c\} \wedge 1$ and let g_t^n be the local solution of (1.16) in U with B_t, N, and $\mathcal{F}_t$ replaced by $B_t^{(\tau_n)}$, $N^{(\tau_n)}$, and $\mathcal{F}_t^{(\tau_n)}$. Then $\tau_1, \tau_2, \ldots$ are positive iid random variables and $T_n = \sum_{i=1}^n \tau_i \uparrow \infty$ almost surely as $n \uparrow \infty$. Let $g_t = g_t^0$ for $0 \le t \le T_1$ and inductively $g_t = g_{T_1} g_{t-T_n}^n$ for $T_n < t \le T_{n+1}$. Obviously, g_t is a solution of (1.16) for $t \le T_1$. Replacing g_t, f, B_t, and N in (1.16) by g_t^1, $f \circ L_g$ with $g = g_{T_1}^0$, $B_t^{(T_1)}$, and $N^{(T_1)}$, we obtain

$$f\left(g_{T_1}^0 g_t^1\right) = f\left(g_{T_1}^0\right) + \sum_{i=1}^d \int_0^t X_i^l f\left(g_{T_1}^0 g_{s-}^1\right) \circ d\left(B^{(T_1)}\right)_s^i$$

$$+ \sum_{i=1}^d c_i \int_0^t X_i^l f\left(g_{T_1}^0 g_{s-}^1\right) ds$$

$$+ \int_0^t \int_G [f\left(g_{T_1}^0 g_{s-}^1 h\right) - f\left(g_{T_1}^0 g_{s-}^1\right)] \tilde{N}^{(T_1)}(ds\, dh)$$

$$+ \int_0^t \int_G [f\left(g_{T_1}^0 g_{s-}^1 h\right) - f\left(g_{T_1}^0 g_{s-}^1\right)$$

$$- \sum_{i=1}^d x_i(h) X_i^l f\left(g_{T_1}^0 g_{s-}^1\right)] ds\, \Pi(dh)$$

$$= f(g_{T_1}) + \sum_{i=1}^d \int_{T_1}^{T_1+t} X_i^l f(g_{s-}) \circ dB_s^i + \sum_{i=1}^d c_i \int_{T_1}^{T_1+t} X_i^l f(g_{s-}) ds$$

$$+ \int_{T_1}^{T_1+t} \int_G [f(g_{s-}h) - f(g_{s-})] \tilde{N}(ds\, dh)$$

$$+ \int_{T_1}^{T_1+t} \int_G \left[f(g_{s-}h) - f(g_{s-}) - \sum_{i=1}^d x_i(h) X_i^l f(g_{s-}) \right] ds\, \Pi(dh).$$

Since $g_{T_1}^0 g_t^1 = g_{T_1+t}$ for $t \le T_2 - T_1$, it follows that g_t is a solution of (1.16) for $t \le T_2$. This argument can be continued to prove that g_t is a solution of (1.16) for any $t \in \mathbb{R}_+$.

Suppose g_t' is another solution of (1.16) with $g_0' = e$. By the uniqueness of the local solution in U, $g_t' = g_t$ for $t \le T_1$. Then $g_{T_1}^{-1} g_{T_1+t}'$ is the unique local solution of (1.16) in U with B_t replaced by $B_t^{(T_1)}$ and N by $N^{(T_1)}$. It follows that $g_t' = g_t$ for $t \le T_2$. This argument can be continued to prove $g_t' = g_t$ for any $t \in \mathbb{R}_+$. Therefore, g_t constructed in the last paragraph is the unique solution of (1.16) in G satisfying $g_0 = e$.

From the explicit construction of the solution g_t in G from the local solution x_t in U, it is clear that for any $b \in \mathbb{R}_+$, the process g_t for $t \in [0, b]$ depends only on x_t, B_t, and $N([0, t] \times \cdot)$ for $t \in [0, b]$. It follows that if the local solution x_t in U is adapted to $\{\mathcal{F}_t^{B,N}\}$, then so is the solution g_t in G. Fix $s \in \mathbb{R}_+$ and let $h = g_s^{-1}$. Using the notation in the proof of Lemma 3.10, we have, for $f \in C_0^{2,l}(G)$,

$$f(g_s^{-1}g_{s+t}) = f \circ L_h(g_{s+t}) = f \circ L_h(g_s) + S(f \circ L_h, g., s, s+t)$$
$$= f(e) + S(f, hg., s, s+t).$$

This shows that the process $g'_t = g_s^{-1}g_{s+t}^{-1}$ is a solution of (1.16) with B_t, N, and $\mathcal{F}_t$ replaced by $B_t^{(s)}$, $N^{(s)}$, and $\mathcal{F}_t^{(s)}$. It follows that $g_s^{-1}g_{s+t}$ is independent of $\mathcal{F}_s^{B,N}$ and, by the uniqueness, has the same distribution as g_t. This proves that g_t is a Lévy process in G. By (1.18), which is equivalent to (1.16), it is easy to show that the generator L of g_t restricted to $C_0^{2,l}(G)$ is given by (1.7). $\qquad\square$

We will prove that Equation (1.16) has a unique local solution in U that is adapted to $\{\mathcal{F}_t^{B,N}\}$.

We may identify V with an open subset of $\mathbb{R}^d$ and regard (1.16) as a stochastic integral equation on $\mathbb{R}^d$. More precisely, let $\phi = (x_1, \ldots, x_d)$: $V \to \mathbb{R}^d$ with $\phi(e) = 0$ be a diffeomorphism from V onto an open subset of $\mathbb{R}^d$ containing 0, let $W = \phi(U)$, and let $z: W \times W \to \phi(V)$ be the map that transfers the product structure on G to W—that is, $z(x, y) = \phi(\phi^{-1}(x)\phi^{-1}(y))$ for x, $y \in W$. Let $X'_i(x) = D\phi[X'_i(\phi^{-1}(x))]$ and $X'_0(x) = D\phi[\sum_{i=1}^d c_i X'_i(\phi^{-1}(x))]$ for $x \in W \subset \mathbb{R}^d$, $N' = (\mathrm{id}_{\mathbb{R}_+} \times \phi)N$, and $\Pi' = \phi\Pi$. Then $X'_0, X'_1, \ldots, X'_d$ are vector fields on W and N' is a Poisson random measure on $\mathbb{R}_+ \times W$ with characteristic measure Π'. Let $y_t = \phi(g_t)$. We may extend $X'_0, X'_1, \ldots, X'_d$ to be smooth vector fields on $\mathbb{R}^d$ and z to be a smooth function from $\mathbb{R}^d \times \mathbb{R}^d$ to $\mathbb{R}^d$, all with compact supports, and regard N' and Π' as measure on $\mathbb{R}_+ \times \mathbb{R}^d$ and $\mathbb{R}^d$ respectively. Then stochastic integral equation (1.16) on G for g_t with $g_0 = e$ becomes the following stochastic integral equation on $\mathbb{R}^d$ for $y_t = \phi(g_t)$: For $f \in C^2(\mathbb{R}^d)$,

$$f(y_t) = f(0) + \sum_{i=1}^d \int_0^t X'_i f(y_{s-}) \circ dB_s^i + \int_0^t X'_0 f(y_s)\, ds$$

$$+ \int_0^t \int_W [f(z(y_{s-}, x)) - f(y_{s-})]\tilde{N}'(ds\, dx)$$

$$+ \int_0^t \int_W \left[f(z(y_s, x)) - f(y_s) - \sum_{i=1}^d x_i X'_i f(y_s)\right] ds\, \Pi'(dx).$$

$$(3.13)$$

Let $\sigma_{ij}, b_j \in C_c^\infty(R^d)$ be defined by

$$X_i'(x) = \sum_{j=1}^d \sigma_{ij}(x) \frac{\partial}{\partial x_j} \quad \text{for } 1 \le i \le d \quad \text{and} \quad X_0'(x) = \sum_{j=1}^d b_j(x) \frac{\partial}{\partial x_j}.$$

Let $\sigma_{j\cdot} = (\sigma_{j1}, \ldots, \sigma_{jd})$ and $b = (b_1, \ldots, b_d)$. Applying (3.13) to the vector-valued function $f = (x_1, \ldots, x_d)$, we obtain the following integral equation for y_t in vector form:

$$\begin{aligned}
y_t = & \sum_{i=1}^d \int_0^t \sigma_{i\cdot}(y_{s-}) \circ dB_s^i + \int_0^t b(y_s)\,ds \\
& + \int_0^t \int_W [z(y_{s-}, x) - y_{s-}]\tilde{N}'(ds\,dx) \\
& + \int_0^t \int_W \left[z(y_s, x) - y_s - \sum_{i=1}^d x_i \sigma_{i\cdot}(y_s)\right] ds\,\Pi'(dx). \quad (3.14)
\end{aligned}$$

Lemma 3.12. *Assume* $\mathrm{supp}(\Pi) \subset U$. *If (3.14) has a unique solution y_t, then (1.16) has a unique local solution x_t in U. Moreover, if y_t is adapted to $\{\mathcal{F}_t^{B,N'}\}$, then x_t is adapted to $\{\mathcal{F}_t^{B,N}\}$.*

Proof. Different solutions of (1.16) will lead to different solutions of (3.13) and hence different solutions of (3.14). It suffices to show that, if y_t is a solution of (3.14), then it satisfies (3.13) for any $f \in C^2(\mathbb{R}^d)$.

We will write f_i for $(\partial/\partial x_i)f$ and $y_i(t)$ for the ith component of y_t. Let $h(t, x) = [z_i(y_t, x)) - y_i(t)]$ and $(h.\tilde{N}')_t = \int_0^t \int_W h(s-, x)\tilde{N}'(ds\,dx)$. By (B.10) and (B.11) in Appendix B.3, it can be shown that $[y_i(\cdot), (h.\tilde{N}').]_t^c = 0$, and then, by (B.3), $[f_i(y_\cdot), (h.\tilde{N}').]_t^c = 0$. It follows that

$$\begin{aligned}
\int_0^t f_i(y_{s-}) \circ d(h.\tilde{N}')_s &= \int_0^t f_i(y_{s-}) d(h.\tilde{N}')_s \\
&= \int_0^t \int_W f_i(y_{s-})[z_i(y_{s-}, x) - y_i(s-)]\tilde{N}'(ds\,dx).
\end{aligned}$$

By the Itô formula (B.4), for any $f \in C^2(\mathbb{R}^d)$,

$$\begin{aligned}
f(y_t) = & f(0) + \sum_{i=1}^d \int_0^t f_i(y_{s-}) \circ dy_i(s) \\
& + \sum_{0 < s \le t} \left[f(y_s) - f(y_{s-}) - \sum_{i=1}^d f_i(y_{s-})\Delta_s y_i(\cdot)\right].
\end{aligned}$$

Note that $\sum_{0<s\leq t}[f(y_s) - f(y_{s-}) - \sum_{i=1}^{d} f_i(y_{s-})\Delta_s y_i]$ is equal to

$$
\int_0^t \int_U \left\{ f \circ \phi(g_{s-\sigma}) - f \circ \phi(g_{s-}) \right.
$$
$$
\left. - \sum_{i=1}^{d} f_i \circ \phi(g_{s-})[x_i(g_{s-\sigma}) - x_i(g_{s-})] \right\} N(ds\, d\sigma)
$$
$$
= \int_0^t \int_W \left\{ f(z(y_{s-}, x)) - f(y_{s-}) \right.
$$
$$
\left. - \sum_{i=1}^{d} f_i(y_{s-})[z_i(y_{s-}, x) - y_i(s-)] \right\} N'(ds\, dx).
$$

Let $W_\varepsilon = \{x \in W; |x| > \varepsilon\}$. Then

$$
f(y_t) = f(0) + \sum_{i,j=1}^{d} \int_0^t f_i(y_{s-})\sigma_{ji}(y_{s-}) \circ dB_s^j + \sum_{i=1}^{d} \int_0^t f_i(y_s)b_i(y_s)\, ds
$$
$$
+ \sum_{i=1}^{d} \int_0^t \int_W f_i(y_{s-})[z_i(y_{s-}, x) - y_i(s-)]\tilde{N}'(ds\, dx)
$$
$$
+ \sum_{i=1}^{d} \int_0^t \int_W f_i(y_s)\left[z_i(y_s, x) - y_i(s) - \sum_{j=1}^{d} x_j\sigma_{ji}(y_s) \right] ds\, \Pi'(dx)
$$
$$
+ \int_0^t \int_W \left\{ f(z(y_{s-}, x)) - f(y_{s-}) \right.
$$
$$
\left. - \sum_{i=1}^{d} f_i(y_{s-})[z_i(y_{s-}, x) - y_i(s-)] \right\} N'(ds\, dx)
$$
$$
= f(0) + \sum_{j=1}^{d} \int_0^t X_j' f(y_{s-}) \circ dB_s^j + \int_0^t X_0' f(y_s)\, ds
$$
$$
+ \sum_{i=1}^{d} \lim_{\varepsilon\to 0} \left\{ \int_0^t \int_{W_\varepsilon} f_i(y_{s-})[z_i(y_{s-}, x) - y_i(s-)]N'(ds\, dx) \right.
$$
$$
\left. - \sum_{i=1}^{d} \int_0^t \int_{W_\varepsilon} f_i(y_{s-})[z_i(y_{s-}, x) - y_i(s-)]ds\, \Pi'(dx) \right\}
$$
$$
+ \sum_{i=1}^{d} \lim_{\varepsilon\to 0} \left\{ \int_0^t \int_{W_\varepsilon} f_i(y_s)[z_i(y_s, x) - y_i(s)]ds\, \Pi'(dx) \right.
$$
$$
\left. - \int_0^t \int_{W_\varepsilon} f_i(y_s) \sum_{j=1}^{d} x_j\sigma_{ij}(y_s)ds\, \Pi'(dx) \right\}
$$

$$+ \lim_{\varepsilon \to 0} \left\{ \int_0^t \int_{W_\varepsilon} [f(z(y_{s-}, x)) - f(y_{s-})] N'(ds\,dx) \right.$$

$$\left. - \sum_{i=1}^d \int_0^t \int_{W_\varepsilon} f_i(y_{s-})[z_i(y_{s-}, x) - y_i(s-)] N'(ds\,dx) \right\}$$

$$= f(0) + \sum_{j=1}^d \int_0^t X_j' f(y_{s-}) \circ dB_s^j + \int_0^t X_0' f(y_{s-})\,ds$$

$$+ \int_0^t \int_W [f(z(y_{s-}, x)) - f(y_{s-})] \tilde{N}'(ds\,dx)$$

$$+ \int_0^t \int_W \left[f(z(y_s, x)) - f(y_s) - \sum_{j=1}^d x_j X_j' f(y_s) \right] ds\,\Pi'(dx).$$

This is Equation (3.13) and the lemma is proved. $\qquad\square$

We will now prove that (3.14) has a unique solution that is adapted to $\{\mathcal{F}_t^{B,N'}\}$. This can be done by the usual successive approximation method. Set $y_t^0 = 0$. Let y_t^1 be the right-hand side of (3.14) with $y_s = y_s^0$ and inductively let y_t^{n+1} be the right-hand side of (3.14) with $y_s = y_s^n$. Recall that σ_{ij}, b_i, and z_i are smooth functions with compact supports. It is easy to show that there is a constant $C > 0$ such that

$$E\left[\sup_{0 \le s \le t} |y_s^{n+1} - y_s^n|^2 \right] \le C \int_0^t E\left[\sup_{0 \le u \le t_1} |y_u^n - y_u^{n-1}|^2 \right] dt_1$$

$$\le C^2 \int_0^t \int_0^{t_1} E\left[\sup_{0 \le u \le t_2} |y_u^{n-1} - y_u^{n-2}|^2 \right] dt_2 dt_1$$

$$\le \cdots \le C^n \int_0^t \int_0^{t_1} \cdots \int_0^{t_{n-1}} E\left[\sup_{0 \le u \le t_n} |y_u^1|^2 \right] dt_n dt_{n-1} \cdots dt_1$$

$$\le \frac{C^n t^n}{n!} E\left[\sup_{0 \le s \le t} |y_s^1|^2 \right] \le \frac{4C^n t^n}{n!} E[|y_t^1|^2],$$

where the last inequality follows from Doob's norm inequality (B.1). Because σ_{ij}, b_i, and z are smooth functions with compact support, it is easy to show that $E[|y_t^1|^2] < \infty$. It follows that a subsequence of y_t^n converges, uniformly on finite t-intervals, to some y_t that must be a solution of (3.14).

The uniqueness of the solution can be established by a similar computation. Suppose y_t and y_t' are two solutions of (3.14). Then

$$E\left[\sup_{0 \le s \le t} |y_s - y_s'|^2 \right] \le C \int_0^t E\left[\sup_{0 \le u \le s} |y_u - y_u'|^2 \right] ds$$

for some $C > 0$. Because σ_{ij}, b_i, and z are smooth functions with compact support, by (3.14), it is easy to show that $E[\sup_{0 \le s \le t} |y_s|^2]$ and $E[\sup_{0 \le s \le t} |y'_s|^2]$ are finite. Now Gronwall's inequality implies $y_t = y'_t$.

We have proved the following lemma.

Lemma 3.13. *Assume* $\text{supp}(\Pi) \subset U$. *The stochastic integral equation (3.14) has a unique solution* y_t. *Moreover,* y_t *is adapted to* $\{\mathcal{F}_t^{B,N'}\}$.

We can now easily prove the following result, which contains the second half of Theorem 1.2.

Theorem 3.2. *Given a filtration* $\{\mathcal{F}_t\}$, *a* G-*valued random variable u that is* $\mathcal{F}_0$-*measurable, a* d-*dimensional* $\{\mathcal{F}_t\}$-*Brownian motion* B_t *with covariance matrix* $\{a_{ij}\}$, *an* $\{\mathcal{F}_t\}$-*Poisson random measure* N *on* $\mathbb{R}_+ \times G$ *with characteristic measure* Π *being a Lévy measure, and constants* c_i, *such that* $\{B_t\}$ *and* N *are independent under* $\{\mathcal{F}_t\}$, *then there is a unique càdlàg process* g_t *in* G *with* $g_0 = u$, *adapted to* $\{\mathcal{F}_t\}$, *such that (1.16) holds for any* $f \in C_0^{2,l}(G)$. *The process* g_t *is a left Lévy process in* G, *its generator* L *restricted to* $C_0^{2,l}(G)$ *is given by (1.7), and it is adapted to the filtration generated by u,* $\{B_t\}$, *and* N.

Proof. By Lemmas 3.11, 3.12, and 3.13, we have proved that Equation (1.16) has a unique solution g_t with $g_0 = e$ that is a Lévy process in G, adapted to the filtration generated by $\{B_t\}$ and N, in the case when $\text{supp}(\Pi) \subset U$. The general case follows easily by applying Lemma 3.10 with N being $N|_{\mathbb{R}_+ \times U}$ and $N' = N|_{\mathbb{R}_+ \times U^c}$. To prove the existence of the unique solution g_t of (1.16) with $g_0 = u$, one just needs to replace f by $f \circ L_u$ in (1.16). $\square$

The existence stated in the second half of Theorem 1.1 follows directly from Theorem 3.2 because the transition semigroup P_t of the Lévy process g_t in Theorem 3.2 is the required Feller semigroup.

Corollary 3.1. *If* $\{a_{jk}\}$ *is a* $d \times d$ *nonnegative definite symmetric matrix,* c_i *are some constants, and* Π *is a Lévy measure on* G, *then there is a left invariant Feller semigroup* P_t *of probability kernels on* G *such that its generator* L *restricted to* $C_0^{2,l}(G)$ *is given by (1.7).*

Theorem 1.1 is now completely proved.

3.3. The Stochastic Integral Equation of a Lévy Process

Let g_t be a Lévy process in G. Assume its generator L restricted to $C_0^{2,l}(G)$ is given by (1.7) with coefficients a_{ij} and c_i and Lévy measure Π. Let N be the counting measure of right jumps of g_t defined by (1.12) and let $\{\mathcal{F}_t^e\}$ be the natural filtration of the process $g_t^e = g_0^{-1} g_t$.

We will prove the first half of Theorem 1.2; that there is a d-dimensional Brownian motion B_t with covariance matrix $\{a_{ij}\}$, independent of N and g_0, such that (1.16) holds for $f \in C_0^{2,l}(G)$. We will also show that $\{\mathcal{F}_t^e\}$ is the same as the filtration generated by $\{B_t\}$ and N.

By Proposition 1.6, which is now available because Theorem 1.1 has been completely proved, g_t can be obtained from a Lévy process x_t, whose Lévy measure is supported by an arbitrarily small neighborhood U of e, interlaced with jumps contained in U^c. By Lemma 3.10 with N being $N|_{\mathbb{R}_+ \times U}$ and $N' = N|_{\mathbb{R}_+ \times U^c}$, we see that if (1.16) holds for x_t, then it also holds for g_t. Therefore, without loss of generality, we may and will assume that the Lévy measure Π is supported by a relatively compact neighborhood U of e, which may be assumed to be as small as we want.

For $0 \le s < t$, let $g_{s,t} = g_s^{-1} g_t = (g_s^e)^{-1} g_t^e$, and, for any $f \in C_0^{2,l}(G)$ and $p \in G$, define

$$M_{s,t} f(p) = f(p g_{s,t}) - f(p) - \int_s^t L f(p g_{s,u}) \, du. \tag{3.15}$$

Then, for $v > t$,

$$
\begin{aligned}
E\left[M_{s,v} f(p) \mid \mathcal{F}_t^e \right] &= E\left[f(p g_{s,t} g_{t,v}) - f(p) \right.\\
&\quad \left. - \int_s^t L f(p g_{s,u}) \, du - \int_t^v L f(p g_{s,t} g_{t,u}) \, du \mid \mathcal{F}_t^e \right] \\
&= P_{v-t} f(p g_{s,t}) - f(p) - \int_s^t L f(p g_{s,u}) \, du - \int_t^v P_{u-t} L f(p g_{s,t}) \, du \\
&= P_{v-t} f(p g_{s,t}) - f(p) - \int_s^t L f(p g_{s,u}) \, du \\
&\quad - \int_t^v \frac{d}{du} P_{u-t} f(p g_{s,t}) \, du = M_{s,t} f(p).
\end{aligned}
$$

This shows that, for fixed s, $M_{s,t} f(p)$ is a L^2-martingale with respect to $\{\mathcal{F}_t^e\}$ for $t \ge s$. For any $f_1, f_2 \in C_0^{2,l}(G)$ and $p, q \in G$, $\langle M_{s,\cdot} f_1(p), M_{s,\cdot} f_2(q) \rangle_t$ and $[M_{s,\cdot} f_1(p), M_{s,\cdot} f_2(q)]_t$ are defined (see Appendix B.2). Although they are normally defined to vanish at $t = 0$, since both processes are defined only for $t \ge s$, we will now define them to vanish at $t = s$.

Let

$$B(f_1, f_2)(p, q) = \sum_{i,j=1}^{d} a_{ij} X_i^l f_1(p) X_j^l f_2(q)$$

$$+ \int_G [f_1(p\sigma) - f_1(p)][f_2(q\sigma) - f_2(q)]\Pi(d\sigma). \tag{3.16}$$

Lemma 3.14. $\langle M_{s,\cdot} f_1(p), M_{s,\cdot} f_2(q) \rangle_t = \int_s^t B(f_1, f_2)(pg_{s,u}, qg_{s,u}) \, du.$

Proof. We will first assume $f_1 = f_2 = f$ and $p = q = e$. Set $M_{s,t} f(e) = M_t$ and $g_{s,t}(e) = \phi_t$. Since $f(\phi_t) = f(e) + M_t + \int_s^t Lf(\phi_u) \, du$ is a semi-martingale, by the integration by parts formula and noting $\phi_s = e$,

$$f(\phi_t)^2 = f(e)^2 + 2\int_s^t f(\phi_{u-}) \cdot dM_u + 2\int_s^t f(\phi_u) Lf(\phi_u) du + [M, M]_t.$$

Therefore,

$$M_{s,t}(f^2)(e) = f(\phi_t)^2 - f(e)^2 - \int_s^t L(f^2)(\phi_u) \, du$$

$$= 2\int_s^t f(\phi_{u-}) \cdot dM_u + [M, M]_t - \langle M, M \rangle_t$$

$$+ \left\{ \langle M, M \rangle_t + 2\int_s^t f(\phi_u) Lf(\phi_u) \, du - \int_s^t L(f^2)(\phi_u) \, du \right\}.$$

Since $M_{s,t}(f^2)(e)$, $\int_s^t f(\phi_{u-}) \cdot dM_u$, and $[M, M]_t - \langle M, M \rangle_t$ are martingales, the process contained in the curly braces is a predictable martingale of finite variation. Such a process must be a constant. Since it vanishes at $t = s$, we obtain $\langle M, M \rangle_t = \int_s^t [L(f^2)(\phi_u) - 2f(\phi_u)Lf(\phi_u)] du$. A direct computation shows $B(f, f)(p, p) = L(f^2)(p) - 2f(p)Lf(p)$ for any $p \in G$. This proves the lemma in the case when $f_1 = f_2 = f$ and $p = q = e$. Setting $f = f_1 \pm f_2$ and using the polarization identity $\langle X, Y \rangle = \frac{1}{2}[\langle X + Y, X + Y \rangle - \langle X - Y, X - Y \rangle]$, we obtain the lemma for $f_1 \neq f_2$ and $p = q = e$. Now replacing f_1 and f_2 by $f_1 \circ L_p$ and $f_2 \circ L_q$, respectively, proves the lemma completely. $\square$

Using an inner product on $\mathfrak{g}$, one obtains a left invariant Riemannian metric on G. Let ρ be the associated Riemannian distance on G. For $s < t$, let $s = u_0 < u_1 < u_2 < \cdots < u_n = t$ be a partition of $[s, t]$.

Lemma 3.15. *For any $f \in C_0^{2,l}(G)$ and $p \in G$,*

$$E\left\{\left[M_{s,t}f(p) - \sum_{i=0}^{n-1} M_{u_i,u_{i+1}}f(p)\right]^2\right\}$$

$$\leq C(t-s)E\left[\sup_{u\in[0,t-s],\,v\in[0,\delta]} \rho(g_u^e g_v', g_v') \wedge 1\right]$$

for some constant $C > 0$ depending only on f, where $\delta = \max_{0\leq i\leq n-1}(u_{i+1} - u_i)$ and g_t' denotes a Lévy process that is independent of and identical in distribution to g_t^e.

Proof. For $f_1, f_2 \in C_0^{2,l}(G)$, $p, q \in G$, and $Y, Z \in \mathfrak{g}$,

$$\frac{d}{dt}B(f_1, f_2)(pe^{tY}, qe^{tZ})\,|_{t=0}$$

$$= \sum_{i,j=1}^{d} a_{ij}\left[Y^l X_i^l f_1(p) X_j^l f_2(q) + X_i^l f(p) Z^l X_j^l f_2(q)\right]$$

$$+ \int_G \left\{\left[(\mathrm{Ad}(\sigma^{-1})Y)^l f_1(p\sigma) - Y^l f_1(p)\right][f_2(q\sigma) - f_2(q)]\right.$$

$$\left. + [f_1(p\sigma) - f_1(p)]\left[(\mathrm{Ad}(\sigma^{-1})Z)^l f_2(q\sigma) - Z^l f_2(q)\right]\right\}\Pi(d\sigma).$$

Because Π is assumed to be supported by a relatively compact neighborhood U of e, this expression is bounded in (p, q); hence, $B(f_1, f_2)$ has a bounded first-order derivative taken with respect to any left invariant vector field on $G \times G$. Using Taylor expansion, it is now easy to show that, for $p, q, p', q' \in G$,

$$|B(f_1, f_2)(p, q) - B(f_1, f_2)(p', q')| \leq C_1\{[\rho(p, p') + \rho(q, q')] \wedge 1\}$$
(3.17)

for some constant $C_1 > 0$ depending only on f_1 and f_2.

Note that

$$M_{s,t}f(p) = \sum_{i=0}^{n-1} M_{u_i,u_{i+1}}f(pg_{s,u_i}).$$
(3.18)

If $i < j$, then $M_{u_i,u_{i+1}}f(pg_{s,u_i})$ and $M_{u_j,u_{j+1}}f(p)$ are independent; hence, their product has zero expectation. However, if $i > j$, then writing h for g_{s,u_i}, we have

$$E[M_{u_i,u_{i+1}}f(pg_{s,u_i})M_{u_j,u_{j+1}}f(p)] = E\left\{M_{u_j,u_{j+1}}f(p)E\left[M_{u_i,u_{i+1}}f(ph)\,|\,\mathcal{F}_{u_i}^e\right]\right\}$$

$$= E\{M_{u_j,u_{j+1}}f(p)\,E[M_{u_i,u_{i+1}}f(pz)]_{z=h}\} = 0.$$

Similarly, we can show

$$E[M_{u_i,u_{i+1}} f(pg_{s,u_i}) M_{u_j,u_{j+1}} f(pg_{s,u_j})]$$
$$= E\{M_{u_j,u_{j+1}} f(pg_{s,u_j}) E[M_{u_i,u_{i+1}} f(pz)]_{z=h}\} = 0$$

for $i > j$. It follows that

$$E\left\{\left[M_{s,t} f(p) - \sum_{i=0}^{n-1} M_{u_i,u_{i+1}} f(p)\right]^2\right\}$$

$$= E\left\{\left[\sum_{i=0}^{n-1} (M_{u_i,u_{i+1}} f(pg_{s,u_i}) - M_{u_i,u_{i+1}} f(p))\right]^2\right\}$$

$$= \sum_{i=0}^{n-1} E\left\{[M_{u_i,u_{i+1}} f(pg_{s,u_i}) - M_{u_i,u_{i+1}} f(p)]^2\right\}$$

$$= \sum_{i=0}^{n-1} E\left\{\int_{u_i}^{u_{i+1}} [B(f,f)(pg_{s,u}, pg_{s,u})\right.$$

$$\left. - 2B(f,f)(pg_{s,u}, pg_{u_i,u}) + B(f,f)(pg_{u_i,u}, pg_{u_i,u})]du\right\},$$

where the last equality is due to Lemma 3.14. Note that, for $u \in [u_i, u_{i+1}]$, $g_{s,u} = g_{s,u_i} g_{u_i,u}$ and the two factors on the right-hand side are independent, $g_{s,u_i} \stackrel{d}{=} g^e_{u_i-s}$, and $g_{u_i,u} \stackrel{d}{=} g^e_{u-u_i}$. Now the lemma follows from (3.17) and the left invariance of the distance ρ. □

The following theorem contains the first half of Theorem 1.2. Recall that N is the counting measure of the right jumps of the Lévy process g_t, and it is a Poisson random measure with characteristic measure equal to the Lévy measure Π and is associated to $\{\mathcal{F}^e_t\}$.

Theorem 3.3. *There exists an $\{\mathcal{F}^e_t\}$-adapted d-dimensional Brownian motion B_t with covariance matrix a_{ij} such that it is independent of N under $\{\mathcal{F}^e_t\}$ and, $\forall f \in C^{2,l}_0(G)$, (1.16) holds. Moreover, the pair of the Brownian motion $\{B_t\}$ and the Poisson random measure N on $\mathbb{R}_+ \times G$, which are independent and for which (1.16) holds for any $f \in C^{2,l}_0(G)$, is uniquely determined by the Lévy process g_t. Furthermore, $\{\mathcal{F}^e_t\}$ is the same as the filtration generated by $\{B_t\}$ and N.*

Proof. Let $\Delta: 0 = t_0 < t_1 < t_2 < \cdots < t_k < \cdots$ with $t_k \to \infty$ be a partition of $\mathbb{R}_+$ and assume its mesh $|\Delta| = \sup_i |t_{i+1} - t_i|$ is finite. For $f \in C^{2,l}_0(G)$

and $p \in G$, define

$$Y_t^\Delta f(p) = \sum_{n=0}^{\infty} M_{t_n \wedge t, \, t_{n+1} \wedge t} f(p). \tag{3.19}$$

Note that this sum contains only finitely many nonzero terms and that the process Y_t^Δ is an L^2-martingale. Let Δ_n be a sequence of partitions of $\mathbb{R}_+$ with $\Delta_n \subset \Delta_{n+1}$ and $|\Delta_n| \to 0$. By Lemma 3.15, and noting that $M_{s,t} f(p)$ and $M_{u,v} f(p)$ are independent for any two disjoint intervals $(s, \, t)$ and $(u, \, v)$, we can show that, for $n < m$,

$$E\left[|Y_t^{\Delta_n} f(p) - Y_t^{\Delta_m} f(p)|^2 \right] \le CtE\left[\sup_{0 \le u \le \delta_n, \, 0 \le v \le \delta_m} \rho\big(g_u^e g_v', \, g_v'\big) \wedge 1 \right],$$

where $\delta_n = |\Delta_n|$. It follows that $Y_t^{\Delta_n} f(p)$ is a Cauchy sequence in $L^2(P)$; therefore, $Y_t^{\Delta_n} f(p) \to Y_t f(p)$ in $L^2(P)$ as $n \to \infty$ for some L^2-martingale $Y_t f(p)$ with $Y_0 f(p) = 0$. By Lemma 3.14, it is easy to show that

$$\langle Y.f_1(p), Y.f_2(q) \rangle_t = t B(f_1, f_2)(p, q). \tag{3.20}$$

For any $\varepsilon > 0$, let $U_\varepsilon = \{g \in G; \rho(g, e) < \varepsilon\}$. Write the sum in (3.19), which defines $Y_t^\Delta f(p)$, as $\sum_t^1 + \sum_t^2$, where the second sum $\sum_t^2$ is taken over all the intervals $(t_i, \, t_{i+1}]$ during which the process g_t has a right jump contained in U_ε^c. Because such jumps are counted by the Poisson random measure N restricted to $\mathbb{R}_+ \times U_\varepsilon^c$, by (3.15), it can be shown that $\sum_t^2 \to \int_0^t \int_{U_\varepsilon^c} [f(p\sigma) - f(p)] N(du \, d\sigma)$ in $L^2(P)$ as $|\Delta| \to 0$. Let

$$H_t^\varepsilon = \int_0^t \int_{U_\varepsilon^c} [f(p\sigma) - f(p)] \tilde{N}(du \, d\sigma).$$

Then $J_t^\varepsilon = Y_t f(p) - H_t^\varepsilon$ is an L^2-martingale that does not have a jump of size exceeding $\sup\{|f(g) - f(e)|; \rho(g, e) \le \varepsilon\}$. Because H_t^ε depends only on the right jumps of g_t contained in U_ε^c, whereas J_t^ε depends on only the process g_t with those jumps removed, by Proposition 1.6, the two processes J_t^ε and H_t^ε are independent. As $\varepsilon \to 0$,

$$H_t^\varepsilon \to Y_t^d f(p) = \int_0^t \int_G [f(p\sigma) - f(p)] \tilde{N}(du \, d\sigma)$$

in $L^2(P)$, so J_t^ε converges in $L^2(P)$ to some continuous L^2-martingale $Y_t^c f(p)$ that is independent of $Y_t^d f(p)$. We obtain the decomposition $Y_t = Y_t^c + Y_t^d$.

By (B.9),

$$\langle Y_.^d f_1(p), Y_.^d f_2(q) \rangle_t = t \int_G [f_1(p\sigma) - f_1(p)][f_2(q\sigma) - f_2(q)] \Pi(d\sigma),$$

and by (3.16) and (3.20),

$$\langle Y_\cdot^c f_1(p), Y_\cdot^c f_2(q) \rangle_t = t \sum_{i,j=1}^{d} a_{ij} X_i^l f_1(p) X_j^l f_2(q). \qquad (3.21)$$

Let $B_t^i = Y_t^c(x_i)(e)$. Then $B_t = (B_t^1, \ldots, B_t^d)$ is a d-dimensional Brownian motion with covariance matrix a_{ij}, which is $\{\mathcal{F}_t^e\}$-adapted and is independent of N under $\{\mathcal{F}_t^e\}$. A direct computation using (3.21) shows that

$$\langle Y_\cdot^c f(p) - \sum_{i=1}^{d} X_i^l f(p) B^i, \ Y_\cdot^c f(p) - \sum_{i=1}^{d} X_i^l f(p) B^i \rangle_t = 0.$$

It follows that $Y_t^c f(p) = \sum_{i=1}^{d} X_i^l f(p) B_t^i$. Therefore,

$$Y_t f(p) = \sum_{i=1}^{d} X_i^l f(p) B_t^i + \int_0^t \int_G [f(p\sigma) - f(p)] \tilde{N}(du\, d\sigma). \qquad (3.22)$$

Let $0 = t_0 < t_1 < t_2 < \cdots < t_n = t$ be a partition of $[0, t]$ into n subintervals of equal length t/n, and for each $j = 0, 1, 2, \ldots, n-1$, let $t_j = u_0^j < u_1^j < \cdots < u_m^j = t_{j+1}$ be a partition of $[t_j, t_{j+1}]$ into m subintervals of equal length $t/(nm)$. Suppose $g_0 = p$. Note that $g_t = pg_t^e = pg_{0,t}$ and $M_{0,t} f(p) = \sum_{j=0}^{n-1} M_{t_j, t_{j+1}} f(g_{t_j})$. We have

$$M_{0,t} f(p) - \sum_{j=0}^{n-1} \sum_{k=0}^{m-1} M_{u_k^j, u_{k+1}^j} f(g_{t_j})$$

$$= \sum_{j=0}^{n-1} \left[M_{t_j, t_{j+1}} f(g_{t_j}) - \sum_{k=0}^{m-1} M_{u_k^j, u_{k+1}^j} f(g_{t_j}) \right].$$

Square this equation and then take the expectation. It is easy to see that the mixed terms coming from the right-hand side have zero expectation; hence, by Lemma 3.15,

$$E\left\{ \left[M_{0,t} f(p) - \sum_{j=0}^{n-1} \sum_{k=0}^{m-1} M_{u_k^j, u_{k+1}^j} f(g_{t_j}) \right]^2 \right\}$$

$$= \sum_{j=0}^{n-1} E\left\{ \left[M_{t_j, t_{j+1}} f(g_{t_j}) - \sum_{k=0}^{m-1} M_{u_k^j, u_{k+1}^j} f(g_{t_j}) \right]^2 \right\}$$

$$\leq CtE\left\{ \left[\sup_{0 \leq u \leq 1/n, \, 0 \leq v \leq 1/m} \rho(g_u^e g_v', g_v') \wedge 1 \right] \right\} \quad (\to 0 \text{ as } n \to \infty)$$

for some constant $C > 0$ and an independent copy g'_v of g^e_v. However, as $m \to \infty$,

$$\sum_{j=0}^{n-1}\sum_{k=0}^{m-1} M_{u^j_k, u^j_{k+1}} f(g_{t_j}) \overset{L^2}{\to} \sum_{j=0}^{n-1}[Y_{t_{j+1}} f(g_{t_j}) - Y_{t_j} f(g_{t_j})]$$

$$= \sum_{i=1}^{d}\sum_{j=0}^{n-1} X^l_i f(g_{t_j})\left[B^i_{t_{j+1}} - B^i_{t_j}\right]$$

$$+ \sum_{j=0}^{n-1}\int_{t_j}^{t_{j+1}}\int_G [f(g_{t_j}\sigma) - f(g_{t_j})]\tilde{N}(du\, d\sigma).$$

Because $g_t = g_{t-}$ almost surely, as $n \to \infty$, this expression will converge in $L^2(P)$ to

$$\Lambda = \sum_{i=1}^{d}\int_0^t X^l_i f(g_{s-})\, dB^i_s + \int_0^t\int_G [f(g_{s-}\sigma) - f(g_{s-})]\tilde{N}(ds\, d\sigma).$$

It follows that $M_{0,t}f(p) = \Lambda$. This is (1.18) and is equivalent to (1.16).

Now let $B'_t = (B'^1_t, \ldots, B'^d_t)$ be a Brownian motion with covariance matrix $\{a_{ij}\}$ and let N' be a Poisson random measure on $\mathbb{R}_+ \times G$ with characteristic measure Π such that $\{B'_t\}$ and N' are independent and (1.16) holds for any $f \in C^{2,l}_0(G)$ with B_t replaced by B'_t and N by N'. We want to show that $B'_t = B_t$ and $N' = N$. We may assume $p = g_0$ is arbitrary. Let $\Delta: 0 = t_0 < t_1 < \cdots < t_n$ be a partition of $[0, t]$. Then

$$Y^\Delta_t f(p) = \sum_{j=0}^{n-1} M_{t_j, t_{j+1}} f(p) = \sum_{j=0}^{n-1}\left[f(pg_{t_j, t_{j+1}}) \right.$$

$$\left. - f(p) - \int_{t_j}^{t_{j+1}} Lf(pg_{t_j, s})\, ds \right]$$

$$= \sum_{j=0}^{n-1}\left\{ \sum_{i=1}^{d}\int_{t_j}^{t_{j+1}} X^l_i f(pg_{t_j, s})\, dB'^i_s \right.$$

$$\left. + \int_{t_j}^{t_{j+1}}\int_G [f(pg_{t_j, s}\sigma) - f(pg_{t_j, s})]\tilde{N}'(ds d\sigma) \right\},$$

where the last equality follows from (1.18). It is easy to see that, as the mesh $|\Delta| \to 0$,

$$Y^\Delta_t f(p) \to \sum_{i=1}^{d} X^l_i f(p)B'^i_t + \int_0^t\int_G [f(p\sigma) - f(p)]\tilde{N}'(ds d\sigma).$$

However, it has been shown that $Y_t^\Delta f(p) \to Y_t f(p)$ in $L^2(P)$. Comparing the expression here with (3.22) and using the uniqueness of the decomposition of an L^2-martingale into continuous and purely discrete L^2-martingales (see, for example, [34, theorem 23.14]), we obtain

$$\sum_{i=1}^{d} X_i^l f(p) B_t^i = \sum_{i=1}^{d} X_i^l f(p) B_t^{\prime i}$$

and

$$\int_0^t \int_G [f(p\sigma) - f(p)] \tilde{N}(ds d\sigma) = \int_0^t \int_G [f(p\sigma) - f(p)] \tilde{N}'(ds d\sigma)$$

for any $f \in C_0^{2,l}(G)$ and $p \in G$. From this it follows that $B_t' = B_t$ and $N' = N$.

Since B_t is an $\{\mathcal{F}_t^e\}$-Brownian motion and N is an $\{\mathcal{F}_t^e\}$-Poisson random measure, the filtration $\{\mathcal{F}_t\}$ generated by $\{B_t\}$ and N is clearly contained in $\{\mathcal{F}_t^e\}$. However, by Theorem 3.2, as the unique solution of (1.16), g_t is adapted to the filtration generated by g_0, $\{B_t\}$, and N; hence, $\mathcal{F}_t^e \subset \mathcal{F}_t$. This proves $\mathcal{F}_t^e = \mathcal{F}_t$. $\qquad \square$

Theorem 1.2 is now completely proved.

3.4. Generator of an Invariant Markov Process

In this section, we prove Theorem 2.1. Let K be a compact subgroup of G and let Q_t be a G-invariant Feller semigroup of probability kernels on $M = G/K$ with generator $\tilde{L}$. As for the most part the arguments in the proof are similar to those in Section 3.1, we only outline the basic steps and indicate the necessary changes.

Let $\nu_t = Q_t(o, \cdot)$, the K-invariant convolution semigroup on M associated to Q_t, and let π be the natural projection $G \to G/K$. Recall that Q_t can be expressed in terms of ν_t via (2.13). By essentially the same proof of Lemma 3.1, changing G and P_t to M and Q_t, and letting U be a neighborhood of o such that $U = \pi(U')$ for some neighborhood U' of e satisfying $\pi(U'U') \subset V$ and $U'^{-1} = U'$, we can show that, for any neighborhood V of o,

$$\sup_{t>0} \frac{1}{t} \nu_t(V^c) < \infty. \tag{3.23}$$

As in Section 2.2, fix an Ad(K)-invariant inner product $\langle \cdot, \cdot \rangle$ on $\mathfrak{g}$ and let $\mathfrak{p}$ be the orthogonal complement of the Lie algebra $\mathfrak{k}$ of K. Let $\{X_1, \ldots, X_d\}$ be an orthonormal basis of $\mathfrak{g}$ such that $X_1, \ldots, X_n$ form a basis of $\mathfrak{p}$ and $X_{n+1}, \ldots, X_d$ form a basis of $\mathfrak{k}$. Let $y_1, \ldots, y_n$ be a set of canonical coordinate functions on M associated to this basis and let $|y|^2 = \sum_{i=1}^{n} y_i^2$. For $X \in \mathfrak{g}$, let X^* be the vector field on M defined by (2.3) and let $C_0^{2,*}(M)$ be the space of functions $f \in C_0(M) \cap C^2(M)$ such that $X^* f \in C_0(M)$ and $X^* Y^* f \in C_0(M)$ for any $X, Y \in \mathfrak{g}$. Then $C_0^{2,*}(M)$ is a Banach space under the norm

$$\| f \|_2^* = \| f \| + \sum_{i=1}^n \| X_i^* f \| + \sum_{i,j=1}^n \| X_i^* X_j^* f \|,$$

where $\| f \| = \sup_{x \in M} |f(x)|$. Choosing a local smooth section S in (2.13) and using the K-invariance of ν_t, we can easily to show that

$$\forall f \in C_0^{2,*}(M), \quad Q_t f \in C_0^{2,*}(M) \text{ and } \| Q_t f \|_2^* \le \| f \|_2^*.$$

The proofs of Lemmas 3.3–3.6 can be suitably adjusted, replacing $C_0^{2,r}(G)$ and $D(L)$ by $C_0^{2,*}(M)$ and $D(\tilde{L})$, to show that $D(\tilde{L}) \cap C_0^{2,*}(M)$ is dense in $D(\tilde{L})$,

$$\sup_{t>0} \frac{1}{t} \nu_t(|y|^2) < \infty, \tag{3.24}$$

and there is a constant $c > 0$ such that

$$\forall f \in C_0^{2,*}(M), \quad \sup_{t>0} \left| \frac{1}{t} [\nu_t(f) - f(o)] \right| \le c \| f \|_2^*. \tag{3.25}$$

We may regard Q_t as a semigroup of operators on the Banach space $C_u(M)$, the space of continuous functions f on M with finite $\lim_{x \to \infty} f(x)$, equipped with norm $\| f \|$. Its generating functional is defined by

$$\tilde{A} f = \lim_{t \to 0} \frac{1}{t} [\nu_t(f) - f(o)] \tag{3.26}$$

with domain $D(\tilde{A})$ being the space of $f \in C_u(M)$ for which this limit exists. By the proof of Lemma 3.7, suitably adjusted, it can be shown that $C_u(M) \cap C^2(M)$ is contained in $D(\tilde{A})$.

Let M_∞ be the one-point compactification of M with ∞ denoting the point at infinity. Identify the space $C_u(M)$ with $C(M_\infty)$ by setting $f(\infty) = \lim_{x \to \infty} f(x)$ for $f \in C_u(M)$. Regard ν_t as a probability measure on M_∞ by setting $\nu_t(\{\infty\}) = 0$. By essentially the same arguments as in Section 3.1 after introducing G_∞, we can show that there is a measure $\tilde{\Pi}$ on M_∞ such that $\tilde{A} f = \tilde{\Pi}(f)$ for any $f \in D(\tilde{A})$ vanishing in a neighborhood of o and that the restriction of $\tilde{\Pi}$ to M, denoted also by $\tilde{\Pi}$, satisfies the condition (2.23). Moreover, the $n \times n$ matrix $\{a_{ij}\}$, defined by $a_{ij} = \tilde{A}(y_i y_j) - \tilde{\Pi}(y_i y_j)$, is symmetric and nonnegative definite.

For $f \in C_u(M) \cap C^2(M)$, we may write

$$f - f(o) = \sum_{i=1}^{n} y_i X_i^l(f \circ \pi)(e) + \frac{1}{2} \sum_{i,j=1}^{n} y_i y_j X_i^l X_j^l(f \circ \pi)(e) + R_f$$

for some $R_f \in C_u(M) \cap C^2(M)$. By the Taylor expansion of $\phi(t) = (f \circ \pi)(\exp(t \sum_{i=1}^{n} y_i X_i))$, for x contained in a neighborhood of o,

$$R_f(x) = \frac{1}{2} \sum_{i,j=1}^{n} y_i(x) y_j(x) [X_i^l X_j^l(f \circ \pi)(g') - X_i^l X_j^l(f \circ \pi)(e)],$$

where $g' = \pi[\exp(t' \sum_{i=1}^{n} y_i(x) X_i)]$ for some $t' \in [0, 1]$. By essentially the same arguments as in the proof of Lemma 3.9, we can show that, for $f \in C_u(M) \cap C^2(M)$, $\tilde{\Pi}(R_f) = \tilde{A}(R_f)$ and

$$\tilde{A}f = \lim_{t \to 0} \frac{1}{t} \nu_t[f - f(o)]$$

$$= \sum_{i=1}^{n} c_i X_i^l(f \circ \pi)(e) + \frac{1}{2} \sum_{i,j=1}^{n} a_{ij} X_i^l X_j^l(f \circ \pi)(e)$$

$$+ \tilde{\Pi}\left[f - f(o) - \sum_{i=1}^{n} y_i X_i^l(f \circ \pi)(e)\right]$$

$$= Bf + \tilde{\Pi}\left[f - f(o) - \sum_{i=1}^{n} y_i X_i^l(f \circ \pi)(e)\right], \qquad (3.27)$$

where $c_i = \tilde{A}(y_i)$ and $Bf = \sum_{i=1}^{n} c_i X_i^l(f \circ \pi)(e) + (1/2) \sum_{i,j=1}^{n} a_{ij} X_i^l X_j^l$ $(f \circ \pi)(e)$. Moreover, $\tilde{\Pi}(\infty) = 0$.

The K-invariance of ν_t implies that $\tilde{A}$ is also K-invariant; that is, for any $f \in D(\tilde{A})$ and $k \in K$, $f \circ k \in D(\tilde{A})$ and $\tilde{A}(f \circ k) = \tilde{A}f$. Therefore, $\tilde{\Pi}$ is K-invariant. Because $X_i^l(f \circ k \circ \pi)(e) = [\mathrm{Ad}(k)X_i]^l(f \circ \pi)(e)$, by (2.16), the K-invariance of $\tilde{A}$ and $\tilde{\Pi}$, one sees that for any $k \in K$,

$$\tilde{\Pi}\left[f \circ k - f \circ k(o) - \sum_{i=1}^{n} y_i X_i^l(f \circ k \circ \pi)(e)\right]$$

$$= \tilde{\Pi}\left[f - f(o) - \sum_{i=1}^{n} y_i X_i^l(f \circ \pi)(e)\right].$$

It follows from (3.27) that B is K-invariant. Define $\tilde{T}f(x) = B(f \circ S_x)$ for $f \in C^2(M)$ and $x \in M$. It is easy to show that $\tilde{T}$ is a G-invariant diffusion generator on M and, by the K-invariance of B, its definition is independent of the choice of the section S. Noting $X_i^l(f \circ \pi)(e) = (\partial/\partial y_i)f(o)$, we have proved that, for $f \in C_u(M) \cap C^2(M)$,

$$\tilde{A}f = \tilde{T}f(o) + \tilde{\Pi}\left[f - f(o) - \sum_{i=1}^{n} y_i \frac{\partial}{\partial y_i}(f)(o)\right]. \qquad (3.28)$$

This expression coincides with the right-hand side of (2.22). If $f \in D(\tilde{L})$, then $f \in D(\tilde{A})$ and $\tilde{L}f(o) = \tilde{A}f$. To prove the restriction of $\tilde{L}$ to $C_c^\infty(M)$ at point o is given by (2.22), it remains to prove $C_c^\infty(M) \subset D(\tilde{L})$.

The uniqueness part in Theorem 2.1 can be proved by the same method that proves the uniqueness in Theorem 3.1. In fact, the proof shows that a G-invariant Feller semigroup Q_t on M is uniquely determined by its generating functional $\tilde{A}$ restricted to $C_c^\infty(M)$.

Given $\tilde{T}$ and $\tilde{\Pi}$ as in Theorem 2.1, the proof of Theorem 2.2 shows that there is a left Lévy process g_t in G with $g_0 = e$ such that g_t is right K-invariant and $x_t' = \pi(g_t)$ is a G-invariant Markov process in M whose generator, restricted to $C_c^\infty(M)$ at o, is given by (2.22). This proves the existence part in Theorem 2.1. It also proves $C_c^\infty(M) \subset D(\tilde{L})$ mentioned earlier because the generating functional of the transition semigroup Q_t' of x_t' agrees with $\tilde{A}$ on $C_c^\infty(M)$ and hence, by the uniqueness, $Q_t' = Q_t$.

4

Lévy Processes in Compact Lie Groups and Fourier Analysis

In this chapter, we apply Fourier analysis to study the distributions of Lévy processes g_t in compact Lie groups. After a brief review of the Fourier analysis on compact Lie groups based on the Peter–Weyl theorem, we discuss in Section 4.2 the Fourier expansion of the distribution density p_t of a Lévy process g_t in terms of matrix elements of irreducible unitary representations of G. It is shown that if g_t has an L^2 density p_t, then the Fourier series converges absolutely and uniformly on G, and the coefficients tend to 0 exponentially as time $t \to \infty$. In Section 4.3, for Lévy processes invariant under the inverse map, the L^2 distribution density is shown to exist, and the exponential bounds for the density as well as the exponential convergence of the distribution to the normalized Haar measure are obtained. The same results are proved in Section 4.4 for conjugate invariant Lévy processes. In this case, the Fourier expansion is given in terms of irreducible characters, a more manageable form of Fourier series. An example on the special unitary group $SU(2)$ is computed explicitly in the last section. The results of this chapter are taken from Liao [43].

4.1. Fourier Analysis on Compact Lie Groups

This section is devoted to a brief discussion of Fourier series of L^2 functions on a compact Lie group G based on the Peter–Weyl theorem. See Bröcker and Dieck [12] for more details on the representation theory of compact Lie groups and Helgason [27] for the related Fourier theory.

Let V be a complex vector space of complex dimension $\dim_{\mathbb{C}} V = n$. The set $GL(V)$ of all complex linear bijections: $V \to V$ is a Lie group. With a basis $\{v_1, \dots, v_n\}$ of V, $GL(V)$ may be identified with the complex general linear group $GL(n, \mathbb{C})$ via the Lie group isomorphism

$$ GL(V) \ni f \mapsto \{f_{ij}\} \in GL(n, \mathbb{C}), \quad \text{where} \quad f(v_j) = \sum_{i=1}^{n} f_{ij} v_i. $$

The unitary group $U(n)$ is the closed subgroup of $GL(n, \mathbb{C})$ consisting of the unitary matrices, that is, the set of $n \times n$ complex matrices X satisfying $X^{-1} = X^*$, where $X^* = \overline{X}'$, $\overline{X}$ is the complex conjugation of X, and the prime denotes the matrix transpose. Its Lie algebra $u(n)$ is the space of all $n \times n$ skew-Hermitian matrices, that is, the set of matrices X satisfying $X^* = -X$. The special unitary group $SU(n)$ is the closed subgroup of $U(n)$ consisting of unitary matrices of determinant 1, and its Lie algebra $su(n)$ is the space of traceless skew-Hermitian matrices.

Let G be a Lie group and let V be a complex vector space. A Lie group homomorphism $F: G \to GL(V)$ is called a representation of G on V. It is called faithful if F is injective. It is called irreducible if the only subspaces of V left invariant by F are $\{0\}$ and V. Two representations $F_1: G \to GL(V_1)$ and $F_2: G \to GL(V_2)$ are called equivalent if there is a linear bijection $f: V_1 \to V_2$ such that $f \circ [F_1(g)] = F_2(g) \circ f$ for any $g \in G$. Assume V is equipped with a Hermitian inner product $\langle \cdot, \cdot \rangle$. The representation F is called unitary if it leaves the Hermitian inner product invariant, that is, if $\langle F(v_1), F(v_2) \rangle = \langle v_1, v_2 \rangle$ for any $v_1, v_2 \in V$. Note that if G is compact, then any representation of G on a finite-dimensional complex space V may be regarded as unitary by properly choosing a Hermitian inner product on V.

Let U be a unitary representation of G on a complex vector space V of complex dimension $n = \dim_{\mathbb{C}}(V)$ equipped with a Hermitian inner product. Given an orthonormal basis $\{v_1, v_2, \ldots, v_n\}$ of V, U may be regarded as an unitary matrix-valued function $U(g) = \{U_{ij}(g)\}$ given by $U(g)v_j = \sum_{i=1}^{n} v_i U_{ij}(g)$ for $g \in G$. Let $\mathrm{Irr}(G, \mathbb{C})$ denote the set of all the equivalence classes of irreducible unitary complex representations. The compactness of G implies that $\mathrm{Irr}(G, \mathbb{C})$ is a countable set. For $\delta \in \mathrm{Irr}(G, \mathbb{C})$, let U^{δ} be a unitary representation belonging to the class δ and let d_{δ} be its dimension. We will denote by $\mathrm{Irr}(G, \mathbb{C})_+$ the set $\mathrm{Irr}(G, \mathbb{C})$ excluding the trivial one-dimensional representation given by $U^{\delta} = 1$.

For a compact Lie group G, the normalized Haar measure on G will be denoted either by ρ_G or by dg. Let $L^2(G)$ be the space of functions f on G with finite L^2-norm

$$\|f\|_2 = \left[\rho_G(|f|^2) \right]^{1/2} = \left[\int |f(g)|^2 dg \right]^{1/2},$$

identifying functions that are equal almost everywhere under ρ_G. We note that the normalized Haar measure is invariant under left and right translations, and the inverse map on G.

By the Peter–Weyl theorem (see section 4 in ch. II and section 3 in ch. III in Bröcker and Dieck [12]), the family

$$\{d_{\delta}^{1/2} U_{ij}^{\delta}; \ i, j = 1, 2, \ldots, d_{\delta} \text{ and } \delta \in \mathrm{Irr}(G, \mathbb{C})\}$$

is a complete orthonormal system on $L^2(G)$. The Fourier series of a function $f \in L^2(G)$ with respect to this orthonormal system may be written as

$$f = \rho_G(f) + \sum_{\delta \in \mathrm{Irr}(G,\mathbb{C})_+} d_\delta \, \mathrm{Trace}(A_\delta \, U^\delta) \quad \text{with} \quad A_\delta = \rho_G(f \, U^{\delta *}) \quad (4.1)$$

in L^2 sense; that is, the series converges to f in $L^2(G)$. The L^2-convergence of the series in (4.1) is equivalent to the convergence of the series of positive numbers in the following Parseval identity:

$$\|f\|_2^2 = |\rho_G(f)|^2 + \sum_{\delta \in \mathrm{Irr}(G,\mathbb{C})_+} d_\delta \mathrm{Trace}\,(A_\delta A_\delta^*). \quad (4.2)$$

The character of $\delta \in \mathrm{Irr}\,(G, \mathbb{C})$ is

$$\chi_\delta = \mathrm{Trace}\,(U^\delta), \quad (4.3)$$

which is independent of the choice of the unitary matrix U^δ in the class δ. By [12, chapter II, theorem (4.12)], a representation is uniquely determined by its character up to equivalence. The normalized character is

$$\psi_\delta = \chi_\delta/d_\delta. \quad (4.4)$$

Proposition 4.1. *The character χ_δ is positive definite in the sense that*

$$\sum_{i,j=1}^{k} \chi_\delta(g_i g_j^{-1}) \xi_i \overline{\xi_j} \geq 0$$

for any finite set of $g_i \in G$ and complex numbers ξ_i. In particular, $|\psi_\delta| \leq \psi_\delta(e) = 1$. Moreover, for any $u, v \in G$,

$$\int \psi_\delta(gug^{-1}v)\,dg = \psi_\delta(u)\psi_\delta(v). \quad (4.5)$$

Proof. Let U be a unitary representation in the class δ on a complex vector space V. Then

$$\sum_{i,j=1}^{k} \chi_\delta(g_i g_j^{-1}) \xi_i \overline{\xi_j} = \mathrm{Trace}\left[\sum_{i,j=1}^{k} U(g_i g_j^{-1}) \xi_i \overline{\xi_j} \right]$$

$$= \mathrm{Trace}\left\{ \left[\sum_i U(g_i)\xi_i \right] \left[\sum_j U(g_j)\xi_j \right]^* \right\} \geq 0.$$

The inequality involving ψ_δ follows easily from the positive definiteness of ψ_δ. To prove (4.5), let $A(u) = \int dg\, U(gug^{-1})$ for any $u \in G$. Then $A(u)$ is a linear map: $V \to V$. The invariance of dg implies that $A(u)$ commutes with $U(v)$ for any $v \in G$, that is,

$$U(v)A(u)U(v^{-1}) = \int dg\, U(vgug^{-1}v^{-1}) = \int dg\, U(gug^{-1}) = A(u).$$

By Schur's lemma (see [12, chapter I, theorem (1.10)]), this implies that $A(u)$ is a multiple of the identity map id_V on V. Taking the trace on both sides of $\int dg\, U(gug^{-1}) = A(u)$, we see that $\int dg\, U(gug^{-1}) = \psi_\delta(u)\,\mathrm{id}_V$. Multiplying by $U(v)$ on the right and taking the trace again proves (4.5). $\square$

˙A function f on G is called conjugate invariant if $f(hgh^{-1}) = f(g)$ for any $g, h \in G$. Such a function is also called a class function or a central function in the literature. Let $L^2_{ci}(G)$ denote the closed subspace of $L^2(G)$ consisting of conjugate invariant functions. The set of irreducible characters, $\{\chi_\delta;\ \delta \in \mathrm{Irr}(G, \mathbb{C})\}$, is an orthonormal basis of $L^2_{ci}(G)$ (see sections II.4 and III.3 in Bröcker and Dieck [12]). Therefore, for $f \in L^2_{ci}(G)$,

$$f = \rho_G(f) + \sum_{\delta \in \mathrm{Irr}(G,\mathbb{C})_+} d_\delta\, a_\delta\, \chi_\delta \quad \text{with} \quad a_\delta = \rho_G(f\overline{\psi_\delta}) \qquad (4.6)$$

in L^2 sense.

4.2. Lévy Processes in Compact Lie Groups

In this chapter, we consider exclusively left Lévy processes in compact Lie groups. Therefore, in the rest of this chapter, a Lévy process will be a left Lévy process and we will let g_t be such a process in a compact connected Lie group G with Lie algebra $\mathfrak{g}$ unless explicitly stated otherwise. Because G is compact, by Theorem 1.1, the domain $D(L)$ of the generator L of g_t contains $C^2(G)$ and, for $f \in C^2(G)$, Lf is given by (1.7), or by the simpler expression (1.11) when the Lévy measure Π has a finite first moment.

As in Section 1.1, the convolution of two probability measures μ and ν on G is a probability measure $\mu * \nu$ on G defined by $\mu * \nu(f) = \int f(gh)\mu(dg)\nu(dh)$ for $f \in \mathcal{B}(G)_+$.

The density of a measure on G will always mean the density function with respect to the normalized Haar measure dg unless explicitly stated otherwise. For any two functions p and q in $\mathcal{B}(G)_+$, their convolution is the function $p * q \in \mathcal{B}(G)_+$ on G defined by

$$p * q(g) = \int p(gh^{-1})q(h)dh = \int p(h)q(h^{-1}g)\, dh$$

for $g \in G$, where the second equality holds because of the invariance of the Haar measure dh. It is easy to show that if p is a density of μ and q is a density of ν, then $p * q$ is a density of $\mu * \nu$.

Lemma 4.1. *Let μ and ν be two probability measures on G such that one of them has a density p. Then $\mu * \nu$ has a density q with $\|q\|_2 \leq \|p\|_2$.*

Proof. We will only consider the case when p is the density of μ. The other case can be treated by a similar argument. For any $f \in C(G)$, by the translation invariance of dg,

$$\mu * \nu(f) = \iint f(gh)p(g)\, dg\, \nu(dh) = \iint f(g)p(gh^{-1})\, dg\, \nu(dh)$$

$$= \int f(g)\left[\int p(gh^{-1})\nu(dh)\right]dg.$$

Hence, $q(g) = \int p(gh^{-1})\nu(dh)$ is the density of $\mu * \nu$. It is easy to see, by the Schwartz inequality and the translation invariance of dg, that $\|q\|_2 \leq \|p\|_2$. $\qquad\square$

Remark. Using the Hölder inequality instead of the Schwartz inequality, we can prove the same conclusion in Lemma 4.1 with $\|\cdot\|_2$ replaced by the L^r-norm $\|\cdot\|_r$ for $1 \leq r \leq \infty$, where $\|f\|_\infty$ is the essential supremum of $|f|$ on G for any Borel function f on G.

We will assume the process g_t starts at the identity element e of G; that is, $g_0 = e$. Let μ_t be the distribution of g_t for $t \in \mathbb{R}_+$. Then $\mu_t * \mu_s = \mu_{t+s}$ for $s, t \in \mathbb{R}_+$. If μ_t has a density p_t for each $t > 0$, then $p_t * p_s = p_{t+s}$. By Lemma 4.1, if p_t is a density of μ_t for $t > 0$, then $\|p_t\|_2 \leq \|p_s\|_2$ for $0 < s < t$.

The Lévy process g_t will be called nondegenerate if the symmetric matrix $a = \{a_{ij}\}$ in (1.7) is positive definite. As in Section 1.4, let σ be an $m \times d$ matrix such that $a = \sigma'\sigma$ and let

$$Y_i = \sum_{j=1}^{d} \sigma_{ij} X_j \quad \text{for } 1 \leq i \leq m, \tag{4.7}$$

where $\{X_1, \ldots, X_d\}$ is the basis of the Lie algebra $\mathfrak{g}$ of G appearing in (1.7). Then the diffusion part of the generator L given by (1.7), $(1/2) \sum_{i,j=1}^{d} a_{ij} X_i^l X_j^l$, may be written as $(1/2) \sum_{i=1}^{m} Y_i^l Y_i^l$. The Lévy process g_t will be called hypo-elliptic if $\mathrm{Lie}(Y_1, Y_2, \ldots, Y_m)$, the Lie algebra generated by $\{Y_1, Y_2, \ldots, Y_m\}$, is equal to $\mathfrak{g}$. It can be shown (see Proposition 6.10 later) that this definition is independent of the choice of basis $\{X_1, \ldots, X_d\}$ and σ. It is easy to see that a nondegenerate Lévy process is hypo-elliptic.

A continuous hypo-elliptic Lévy process is an example of a hypo-elliptic diffusion process in G. It is well known (see, for example, [5, chapter 2]) that such a process has a smooth transition density function for $t > 0$. In this case, μ_t has a smooth density p_t for $t > 0$.

Theorem 4.1. *Let g_t be a nondegenerate Lévy process in a compact connected Lie group G starting at e with a finite Lévy measure. Then each distribution μ_t of g_t has a density $p_t \in L^2(G)$ for $t > 0$.*

Proof. Because the Lévy measure Π is finite, by the discussion in Section 1.4, the Lévy process g_t may be constructed from a continuous Lévy process x_t by interlacing jumps at exponentially spaced time intervals. More precisely, let x_t be a continuous Lévy process in G whose generator is given by (1.11) with $\Pi = 0$, let $\{\tau_n\}$ be a sequence of exponential random variables with a common rate $\lambda = \Pi(G)$, and let $\{\sigma_n\}$ be a sequence of G-valued random variables with a common distribution $\Pi(\cdot)/\Pi(G)$ such that all these objects are independent. Let $T_n = \tau_1 + \tau_2 + \cdots + \tau_n$ for $n \geq 1$ and set $T_0 = 0$. Let $g_t^0 = x_t$, $g_t^1 = g_t^0$ for $0 \leq t < T_1$, and $g_t^1 = g^0(T_1)\sigma_1 x(T_1)^{-1}x(t)$ for $t \geq T_1$, and, inductively, let $g_t^n = g_t^{n-1}$ for $t < T_n$ and $g_t^n = g^{n-1}(T_n)\sigma_n x(T_n)^{-1}x(t)$ for $t \geq T_n$. Define $g_t = g_t^n$ for $T_n \leq t < T_{n+1}$. Then g_t is a Lévy process in G with generator given by (1.11).

Note that T_n has a Gamma distribution with density $r_n(t) = \lambda^n t^{n-1} \times e^{-\lambda t}/(n-1)!$ with respect to the Lebesgue measure on $\mathbb{R}_+$. Let q_t denote the smooth density of the distribution of x_t for $t > 0$. For $f \in C(G)$ and $t > 0$, using the independence, we have

$$\mu_t(f) = E[f(x_t); \, t < T_1] + \sum_{n=1}^{\infty} E[f(g_t); \, T_n \leq t < T_n + \tau_{n+1}]$$

$$= E[f(x_t)]P(T_1 > t) + \sum_{n=1}^{\infty} \int_0^t r_n(s)\, ds \, E\left[f\left(g_s^{n-1}\sigma_n x_s^{-1} x_t\right)\right] \times$$

$$P(\tau_{n+1} > t - s). \tag{4.8}$$

We now show that, for $n \geq 1$ and $0 \leq s < t$,

$$E\left[f\left(g_s^{n-1}\sigma_n x_s^{-1}x_t\right)\right] = \int f(g)p_{s,t,n}(g)dg \quad \text{for some} \ \ p_{s,t,n} \ \ \text{with}$$

$$\|p_{s,t,n}\|_2 \leq \|q_{t/2^n}\|_2. \tag{4.9}$$

To prove (4.9) for $n = 1$, first assume $s \geq t/2$. We have $E[f(g_s^0\sigma_1 x_s^{-1}x_t)] = E[f(x_s\sigma_1 x_s^{-1}x_t)] = \mu * \nu(f)$, where μ and ν are respectively the distributions of x_s and $\sigma_1 x_s^{-1}x_t$. By Lemma 4.1, $\mu * \nu$ has a density $p_{s,t,1}$ with $\|p_{s,t,1}\|_2 \leq \|q_s\|_2 \leq \|q_{t/2}\|_2$. If $s \leq t/2$, then we may take μ and ν to be the distributions of $x_s\sigma_1$ and $x_s^{-1}x_t$, respectively, and still obtain a density $p_{s,t,1}$ of $\mu * \nu$ with $\|p_{s,t,1}\|_2 \leq \|q_{t/2}\|_2$. This proves (4.9) for $n = 1$. Now using induction, assume (4.9) is proved for $n = 1, 2, \ldots, k$ for some positive integer k. This implies in particular that the distribution of g_t^k has a density p_t^k with $\|p_t^k\|_2 \leq \|q_{t/2^k}\|_2$. Consider $E[f(g_s^k\sigma_{k+1}x_s^{-1}x_t)] = \mu * \nu(f)$, where μ are ν are taken to be the distributions of g_s^k and $\sigma_{k+1}x_s^{-1}x_t$, respectively, if $s \geq t/2$, and those of $g_s^k\sigma_{k+1}$ and $x_s^{-1}x_t$ if $s \leq t/2$. By Lemma 4.1, we can show that $\mu * \nu$ has a density whose L^2-norm is bounded by

$$\|p_s^k\|_2 \leq \|q_{s/2^k}\|_2 \leq \|q_{t/2^{k+1}}\|_2$$

if $s \geq t/2$ and bounded by $\|q_{t/2}\|_2$ if $s \leq t/2$. In either case, the L^2-norm of the density of $\mu * \nu$ is bounded by $\|q_{t/2^{k+1}}\|_2$. This proves (4.9) for any $n \geq 1$.

By (4.8) and the fact that $P(\tau_n > t) = e^{-\lambda t}$ for $t > 0$, we see that $\mu_t(f) = \int f(g)p_t(g)\,dg$ with

$$p_t = q_t e^{-\lambda t} + \sum_{n=1}^{\infty} \int_0^t r_n(s)\,ds\, e^{-\lambda(t-s)}\, p_{s,t,n}$$

and

$$\|p_t\|_2 \leq \|q_t\|_2 e^{-\lambda t} + \sum_{n=1}^{\infty} \int_0^t r_n(s)\,ds\, e^{-\lambda(t-s)}\|p_{s,t,n}\|_2$$

$$\leq \|q_t\|_2 e^{-\lambda t} + \sum_{n=1}^{\infty} \int_0^t r_n(s)\,ds\, e^{-\lambda(t-s)}\|q_{t/2^n}\|_2. \tag{4.10}$$

It is well known that the density of a nondegenerate diffusion process x_t on a d-dimensional compact manifold is bounded above by $Ct^{-d/2}$ for small $t > 0$, where C is a constant independent of t. See, for example, [5, chapter 9]. Therefore, $|q_t| \leq Ct^{-d/2}$ and $\|q_{t/2^n}\|_2 \leq C(2^n/t)^{d/2}$. Since $\int_0^t r_n(s)ds \leq (\lambda t)^n/n!$, it is easy to see that the series in (4.10) converges. This proves $p_t \in L^2(G)$. $\qquad \square$

For a square complex matrix A, let

$$A = Q \, \text{diag}[B_1(\lambda_1), \, B_2(\lambda_2), \, \ldots, \, B_r(\lambda_r)] \, Q^{-1} \qquad (4.11)$$

be the Jordan decomposition of A, where Q is an invertible matrix and $B_i(\lambda_i)$ is a Jordan block of the following form:

$$B(\lambda) = \begin{bmatrix} \lambda & 1 & 0 & 0 & \cdots & 0 \\ 0 & \lambda & 1 & 0 & \cdots & 0 \\ 0 & 0 & \lambda & 1 & \cdots & 0 \\ \vdots & \vdots & \vdots & \vdots & \cdots & \vdots \\ 0 & 0 & 0 & 0 & \cdots & \lambda \end{bmatrix}. \qquad (4.12)$$

We note that, if A is a Hermitian matrix (i.e., if $A^* = A$), then Q is unitary and all $B_i(\lambda_i) = \lambda_i$ are real.

Recall that the Euclidean norm of a matrix A is given by $|A| = \sqrt{\sum_{ij} A_{ij}^2} = \sqrt{\text{Trace}(AA^*)}$.

Proposition 4.2. *Let A be a square complex matrix. If all its eigenvalues λ_i have negative real parts $Re(\lambda_i)$, then $e^{tA} \to 0$ exponentially as $t \to \infty$ in the sense that, for any $\lambda > 0$ satisfying $\max_i Re(\lambda_i) < -\lambda < 0$, there is a constant $K > 0$ such that*

$$\forall t \in \mathbb{R}_+, \quad |e^{tA}| \leq K e^{-\lambda t}. \qquad (4.13)$$

Proof. Let the matrix A have the Jordan decomposition (4.11) with Jordan blocks $B_i(\lambda_i)$ given by (4.12). A direct computation shows that

$$e^{tB(\lambda)} = \begin{bmatrix} e^{\lambda t} & te^{\lambda t} & t^2 e^{\lambda t}/2! & t^3 e^{\lambda t}/3! & \cdots & t^{k-1} e^{\lambda t}/(k-1)! \\ 0 & e^{\lambda t} & te^{\lambda t} & t^2 e^{\lambda t}/2! & \cdots & t^{k-2} e^{\lambda t}/(k-2)! \\ 0 & 0 & e^{\lambda t} & te^{\lambda t} & \cdots & t^{k-3} e^{\lambda t}/(k-3)! \\ \vdots & \vdots & \vdots & \vdots & \cdots & \vdots \\ 0 & 0 & 0 & 0 & \cdots & e^{\lambda t} \end{bmatrix}.$$

Let $b_{ij}(t)$ be the element of the matrix $e^{tA} = Q\text{diag}[e^{t\,B_1(\lambda_1)}, \, e^{t\,B_2(\lambda_2)}, \ldots,$ $e^{t\,B_r(\lambda_r)}]Q^{-1}$ at place (i, j). From the expression for $e^{tB(\lambda)}$, it is easy to see that $b_{ij}(t) = \sum_{m=1}^r p_{ijm}(t)e^{\lambda_m t}$, where $p_{ijm}(t)$ are polynomials in t. Then $\text{Trace}[e^{tA}(e^{tA})^*] = \sum_{i,j} |\sum_m p_{ijm}(t)e^{\lambda_m t}|^2$ and from this (4.13) follows. $\square$

The following theorem is the main result of this section.

Theorem 4.2. *Let g_t be a Lévy process in a compact connected Lie group G with $g_0 = e$ and let L be its generator. Assume the distribution μ_t of g_t has a density $p_t \in L^2(G)$ for $t > 0$. Then the following statements hold:*

(a) For $t > 0$ and $g \in G$,

$$p_t(g) = 1 + \sum_{\delta \in \mathrm{Irr}(G, \mathbb{C})_+} d_\delta \, \mathrm{Trace}[A_\delta(t) \, U^\delta(g)], \qquad (4.14)$$

where

$$A_\delta(t) = \mu_t(U^{\delta *}) = \exp[t \, L(U^{\delta *})(e)], \qquad (4.15)$$

and the series converges absolutely on G and uniformly for $(t, g) \in [\eta, \infty) \times G$ for any fixed $\eta > 0$. Moreover, all the eigenvalues of $L(U^{\delta *})(e)$ have nonpositive real parts.

(b) If the Lévy process g_t is hypo-elliptic, then all the eigenvalues of $L(U^{\delta *})(e)$ have negative real parts. Consequently, $p_t \to 1$ uniformly on G as $t \to \infty$.

Remark. The uniform convergence of the series in (4.14) implies that the map $(t, g) \mapsto p_t(g)$ is continuous on $(0, \infty) \times G$.

Proof. For $f = p_t$, the series in (4.14) is just the Fourier series in (4.1) with $A_\delta = A_\delta(t) = \rho_G(p_t U^{\delta *}) = \mu_t(U^{\delta *})$. We have $\mu_0(U^{\delta *}) = I$, the $d_\delta \times d_\delta$ identity matrix, and

$$\mu_{t+s}(U^{\delta *}) = \int \mu_t(dg)\mu_s(dh)U^\delta(gh)^* = \int \mu_t(dg)\mu_s(dh)U^\delta(h)^* \, U^\delta(g)^*$$
$$= \mu_s(U^{\delta *})\mu_t(U^{\delta *}).$$

Therefore, $\mu_t(U^{\delta *}) = e^{tY}$ for some matrix Y. Because $(d/dt)\mu_t(U^{\delta *}) \mid_{t=0} = L(U^{\delta *})(e)$, we see that $Y = L(U^{\delta *})(e)$.

We now prove the absolute and uniform convergence of series in (4.14). Note that, by the Parseval identity, $\|p_t\|_2^2 = 1 + \sum_\delta d_\delta \mathrm{Trace}[A_\delta(t)A_\delta(t)^*]$, where the summation $\sum_\delta$ is taken over $\delta \in \mathrm{Irr}(G, \mathbb{C})_+$. For any $\eta > 0$ and $\varepsilon > 0$, there is a finite subset Γ of $\mathrm{Irr}(G, \mathbb{C})_+$ such that $\sum_{\delta \in \Gamma^c} d_\delta \mathrm{Trace}[A_\delta(\eta/2)A_\delta(\eta/2)^*] \leq \varepsilon^2$. By the Schwartz inequality and the fact that

U^δ is a unitary matrix, for any finite $\Gamma' \supset \Gamma$ and $t > \eta$,

$$
\begin{aligned}
\sum_{\delta \in \Gamma'-\Gamma} d_\delta \left| \text{Trace}\left[A_\delta(t) U^\delta\right]\right| &= \sum_{\delta \in \Gamma'-\Gamma} d_\delta \left|\text{Trace}\left[A_\delta(\eta/2) A_\delta(t-\eta/2) U^\delta\right]\right| \\
&\leq \sum_{\delta \in \Gamma'-\Gamma} d_\delta \left\{\text{Trace}[A_\delta(\eta/2) A_\delta(\eta/2)^*]\right\}^{1/2} \\
&\quad \times \left\{\text{Trace}[A_\delta(t-\eta/2) A_\delta(t-\eta/2)^*]\right\}^{1/2} \\
&\leq \left\{\sum_{\delta \in \Gamma'-\Gamma} d_\delta \, \text{Trace}[A_\delta(\eta/2) A_\delta(\eta/2)^*]\right\}^{1/2} \\
&\quad \times \left\{\sum_{\delta \in \Gamma'-\Gamma} d_\delta \, \text{Trace}[A_\delta(t-\eta/2) A_\delta(t-\eta/2)^*]\right\}^{1/2} \\
&\leq \varepsilon \, \|p_{t-\eta/2}\|_2 \leq \varepsilon \, \|p_{\eta/2}\|_2,
\end{aligned}
$$

where the last inequality follows from Lemma 4.1. This proves the absolute and uniform convergence stated in part (a).

To complete the proof, we will show that all the eigenvalues of the matrix $L(U^*)(e)$ have nonpositive real parts, and if g_t is hypo-elliptic, then all these real parts are negative. Note that this implies that $A_\delta(t) \to 0$ exponentially for $\delta \in \text{Irr}(G, \mathbb{C})_+$ and, combined with the uniform convergence of the series in (4.14), the uniform convergence of p_t to 1 as $t \to \infty$.

Write $U = U^\delta$ and $n = d_\delta$ for $\delta \in \text{Irr}(G, \mathbb{C})_+$. Consider the quadratic form $Q(z) = z^*[L(U^*)(e)]z$ for $z = (z_1, \ldots, z_n)'$, a column vector in $\mathbb{C}^n$. Since the eigenvalues of $L(U^*)(e)$ are the values of $Q(z)$ with $|z| = 1$, it suffices to show that $\text{Re}[Q(z)] \leq 0$ for all $z \in \mathbb{C}^n$ and that $\text{Re}[Q(z)] < 0$ for all nonzero $z \in \mathbb{C}^n$ if g_t is hypo-elliptic. For $X \in \mathfrak{g}$, let $\tilde{X} = X^l(U^*)(e)$. Then $\tilde{X}$ is a skew-Hermitian matrix (i.e., $\tilde{X}^* = -\tilde{X}$) and $U(e^{tX})^* = \exp(t\tilde{X})$. Moreover,

$$
X^l(U^*)(g) = \frac{d}{dt} U(ge^{tX})^* \mid_{t=0} = \frac{d}{dt} U(e^{tX})^* U(g)^* \mid_{t=0} = \tilde{X} U(g)^*.
$$

Therefore, $Y^l X^l(U^*)(e) = Y^l[\tilde{X}U^*](e) = \tilde{X}\tilde{Y}$ for $Y \in \mathfrak{g}$, and if $Z = [X, Y]$ (Lie bracket), then $\tilde{Z} = [\tilde{Y}, \tilde{X}]$. Let Y_i be defined in (4.7). Then $\sum_{i,j=1}^d a_{ij} X_i^l X_j^l U^*(e) = \sum_{i=1}^m \tilde{Y}_i \tilde{Y}_i = -\sum_{i=1}^m \tilde{Y}_i^* \tilde{Y}_i$ and, by (1.7),

$$
L(U^*)(e) = -\frac{1}{2} \sum_{i=1}^m \tilde{Y}_i^* \tilde{Y}_i + \tilde{Y}_V - \int_{V^c} [I - U(g)^*]\Pi(dg) + r_V, \quad (4.16)
$$

where V is a neighborhood of e, $\tilde{Y}_V = \sum_{i=1}^d c_i \tilde{X}_i - \int_{V^c} \sum_{i=1}^d x_i(g)\tilde{X}_i \Pi(dg)$, and

$$
r_V = \int_V \left[U(g)^* - I - \sum_{i=1}^d x_i(g)\tilde{X}_i\right] \Pi(dg) \to 0 \quad \text{as } V \downarrow \{e\}.
$$

Because $z^* W z = 0$ for any skew-Hermitian matrix W,

$$Q(z) = -\frac{1}{2} \sum_{i=1}^{m} |\tilde{Y}_i z|^2 - \int_{V^c} z^* [I - U(g)^*] z \, \Pi(dg) + z^* r_V z. \qquad (4.17)$$

Since $U(g)^*$ is unitary, $|z|^2 \geq |z^* U(g)^* z|$. It follows that $\mathrm{Re}[z^*(I - U(g)^*)z] \geq 0$. This shows that $\mathrm{Re}[Q(z)] \leq 0$. If $\mathrm{Re}[Q(z)] = 0$ for some nonzero $z \in \mathbb{C}^n$, then $\tilde{Y}_i z = 0$ for $1 \leq i \leq d$. For $Y = [Y_i, Y_j]$, we have $\tilde{Y} z = [\tilde{Y}_j, \tilde{Y}_i] z = \tilde{Y}_j \tilde{Y}_i z - \tilde{Y}_i \tilde{Y}_j z = 0$. If g_t is hypo-elliptic, then $\tilde{Y} z = 0$ for any $Y \in \mathfrak{g}$. Because $U(e^{tY})^* = \exp(t\tilde{Y})$, $U(g)^* z = z$ for all $g \in G$. This implies that $U(g)$ leaves the subspace of $\mathbb{C}^n$ that is orthogonal to z invariant for all $g \in G$. By the irreducibility of the representation U, this is impossible unless $n = 1$. When $n = 1$, $U(g)^* z = z$ would imply that U is the trivial representation, which contradicts the assumption that $\delta \in \mathrm{Irr}\,(G, \mathbb{C})_+$. Therefore, $\mathrm{Re}[Q(z)] < 0$ for nonzero $z \in \mathbb{C}^n$. □

The total variation norm of a signed measure ν on G is defined by $\|\nu\|_{tv} = \sup |\nu(f)|$ with f ranging over all Borel functions on G with $|f| \leq 1$. The following Corollary follows easily from the uniform convergence of p_t to 1 and the Schwartz inequality.

Corollary 4.1. *If the Lévy process g_t is hypo-elliptic in Theorem 4.2, then μ_t converges to the normalized Haar measure ρ_G under the total variation norm, that is,*

$$\|\mu_t - \rho_G\|_{tv} \to 0 \quad \text{as } t \to \infty.$$

4.3. Lévy Processes Invariant under the Inverse Map

Let g_t be a Lévy process in a Lie group G with distribution μ_t. It will be called invariant under a Borel measurable map $F \colon G \to G$, or F-invariant, if $F\mu_t = \mu_t$ for all $t \in \mathbb{R}_+$, where $F\mu_t$ is the probability measure on G given by $F\mu_t(f) = \mu_t(f \circ F)$ for $f \in \mathcal{B}(G)_+$ as defined in Section 2.2. This means that the process $F(g_t)$ has the same distribution as that of g_t.

In this section, we show that if g_t is a hypo-elliptic Lévy process in a compact Lie group G and is invariant under the inverse map

$$J \colon \ G \to G \quad \text{given by} \quad g \mapsto g^{-1},$$

on G, then its distribution μ_t has an L^2 density for $t > 0$ and converges exponentially to the normalized Haar measure ρ_G as $t \to \infty$. Some simple implications of the J-invariance of the Lévy process g_t are summarized in the following proposition.

Proposition 4.3. *Let g_t be a Lévy process in a compact connected Lie group G. Statements (a) and (b) that follow are equivalent. Moreover, they are also equivalent to statement (c) if the Lévy measure Π has a finite first moment.*

(a) g_t is invariant under the inverse map J on G.

(b) $L(U^{\delta*})(e)$ is a Hermitian matrix for all $\delta \in Irr(G, \mathbb{C})_+$.

(c) The Lévy measure Π is J-invariant and the generator L of g_t is given by

$$Lf(g) = \frac{1}{2} \sum_{i,j=1}^{d} a_{ij} X_i^l X_j^l f(g) + \int [f(gh) - f(g)]\Pi(dh) \quad (4.18)$$

for $g \in G$ and $f \in C^2(G)$.

Proof. We note that $L(U^{\delta*})(e)$ is a Hermitian matrix for all $\delta \in Irr(G, \mathbb{C})_+$ if and only if $A_\delta(t) = \exp[t \, L(U^{\delta*})(e)]$ is a Hermitian matrix for all $\delta \in Irr(G, \mathbb{C})_+$ and some (hence all) $t > 0$. Since $A_\delta(t)^* = \mu_t(U^\delta) = \mu_t(U^{\delta*} \circ J)$ and $\{d_\delta^{1/2} U_{ij}^\delta\}$ is a complete orthonormal system on $L^2(G)$, we see that this is also equivalent to the J-invariance of μ_t for all $t > 0$, that is, the invariance of the Lévy process g_t under the inverse map. This proves the equivalence of statements (a) and (b).

Suppose the Lévy measure Π of the Lévy process g_t has a finite first moment. Then its generator is given by (1.11). Assume the vector $X_0 = \sum_{i=1}^{d} b_i X_i$ in (1.11) vanishes. Then the generator L takes the form (4.18). Using the notation in the proof of Theorem 4.2, we have

$$L(U^*)(e) = -\frac{1}{2} \sum_{i=1}^{m} \tilde{Y}_i^* \tilde{Y}_i - \int (I - U^*) \, d\Pi$$

and $(1/2) \sum_{i=1}^{m} \tilde{Y}_i^* \tilde{Y}_i$ is a Hermitian matrix. It is easy to see that, if Π is J-invariant, then $\int (I - U^*) d\Pi$ is a Hermitian matrix and, hence, $L(U^*)(e)$ is a Hermitian matrix. This shows that the process g_t is J-invariant. Conversely, if g_t is J-invariant, then $L(f \circ J)(e) = Lf(e)$ for any $f \in C^2(G)$. Since $X^l(f \circ J)(e) = -X^l f(e)$ for any $X \in \mathfrak{g}$, by (1.11),

$$-X_0^l f(e) + \int [f(h) - f(e)]J\Pi(dh) = X_0^l f(e) + \int [f(h) - f(e)]\Pi(dh)$$

for any $f \in C^2(G)$. This implies that $J\Pi = \Pi$ and $X_0 = 0$. This proves the equivalence of (a) and (c). $\quad\square$

The main result of this section is the following theorem. For any function f on G, let $\|f\|_\infty = \sup_{g \in G} |f(g)|$.

Theorem 4.3. *Let G be a compact connected Lie group and let g_t be a Lévy process in G with $g_0 = e$ and generator L. Assume g_t is hypo-elliptic and is invariant under the inverse map on G.*

(a) *For $t > 0$, the distribution μ_t of g_t has a density $p_t \in L^2(G)$ and, for $g \in G$,*

$$p_t(g) = 1 + \sum_{\delta \in \mathrm{Irr}(G,\mathbb{C})_+} d_\delta \, \mathrm{Trace}\{Q_\delta \, \mathrm{diag}[\exp(\lambda_1^\delta t), \dots,$$

$$\exp(\lambda_{d_\delta}^\delta t)] \, Q_\delta^* \, U^\delta(g)\}, \tag{4.19}$$

where the series converges absolutely and uniformly for $(t, g) \in [\eta, \infty) \times G$ for any fixed $\eta > 0$, Q_δ is a unitary matrix, and $\lambda_1^\delta \leq \cdots \leq \lambda_{d_\delta}^\delta$ are the eigenvalues of the Hermitian matrix $L(U^{\delta*})(e)$, which are all negative.

(b) *There is a largest number $-\lambda < 0$ in the set of negative numbers λ_i^δ for $\delta \in \mathrm{Irr}(G, \mathbb{C})_+$ and $1 \leq i \leq d_\delta$, and for any $\eta > 0$, there are positive constants c and C such that, for $t > \eta$,*

$$\|p_t - 1\|_\infty \leq Ce^{-\lambda t}, \quad ce^{-\lambda t} \leq \|p_t - 1\|_2 \leq Ce^{-\lambda t}, \quad \text{and}$$

$$ce^{-\lambda t} \leq \|\mu_t - \rho_G\|_{tv} \leq Ce^{-\lambda t}.$$

Proof. Suppose first that μ_t has a density $p_t \in L^2(G)$ for $t > 0$. Since $L(U^{\delta*})(e)$ is a Hermitian matrix for all δ, $A_\delta(t) = Q_\delta \, \mathrm{diag}[\exp(\lambda_1^\delta), \dots, \exp(\lambda_{d_\delta}^\delta)] \, Q_\delta^*$, where Q_δ is a unitary matrix and $\lambda_1^\delta \leq \cdots \leq \lambda_{d_\delta}^\delta$ are the eigenvalues of $L(U^{\delta*})(e)$. It now follows from Theorem 4.2 that all $\lambda_i^\delta < 0$ and the series in (4.19) converges to $p_t(g)$ absolutely and uniformly.

The series in (4.19) also converges in $L^2(G)$. Because Q_δ is unitary, by the Parseval identity,

$$\|p_t - 1\|_2^2 = \sum_{\delta \in \mathrm{Irr}(G,\mathbb{C})_+} d_\delta \sum_{i=1}^{d_\delta} \exp(2\lambda_i^\delta t). \tag{4.20}$$

If g_t is continuous, then as a hypo-elliptic diffusion process, its density p_t is smooth and is given by (4.19). Using the notation in the proof of Theorem 4.2, we will write $U = U^\delta$, $n = d_\delta$, $Q(z) = z^* L(U^*)(e)z$, and $Q_0(z) = z^*[-(1/2)\sum_{i=1}^m \tilde{Y}_i^* \tilde{Y}_i]z = -(1/2)\sum_{i=1}^m |\tilde{Y}_i z|^2$ for $z \in \mathbb{C}^n$ regarded as a column vector. Note that $-(1/2)\sum_{i=1}^m \tilde{Y}_i^* \tilde{Y}_i$ is a Hermitian matrix. By assumption, so is $L(U^*)(e)$. Thus, $Q_1(z) = Q(z) - Q_0(z)$ is a Hermitian quadratic form. Letting $V \downarrow \{e\}$ in (4.17), we see that $Q_1(z) = -\int z^*(I - U^*)z \, d\Pi$, where the integral exists as the limit of $\int_{V^c} z^*(I - U^*)z \, d\Pi$ as $V \downarrow \{e\}$. Because $|z| \geq |z^* U z|$, $Q_1(z) \leq 0$ and hence $Q(z) \leq Q_0(z)$ for $z \in \mathbb{C}^n$.

It is known that the eigenvalues $\lambda_1 \leq \lambda_2 \leq \cdots \leq \lambda_n$ of an $n \times n$ Hermitian matrix A possess the following min–max representation:

$$\lambda_i = \min_{V_i} \max_{z \in V_i, |z|=1} z^* A z \quad \text{for } 1 \leq i \leq n, \tag{4.21}$$

where V_i ranges over all i-dimensional subspaces of $\mathbb{C}^n$ (see, for example, theorem 1.9.1 in Chatelin [14]). Let $\lambda_1 \leq \cdots \leq \lambda_n$ and $\lambda_1^0 \leq \cdots \leq \lambda_n^0$ be the eigenvalues of $L(U^*)(e)$ and $-(1/2) \sum_{i=1}^m \tilde{Y}_i^* \tilde{Y}_i$ respectively. Then $\lambda_i \leq \lambda_i^0$ for all i.

Now suppose g_t is not necessarily continuous. Then the series in (4.20) still converges because λ_i^δ can only become smaller and, hence, the series in (4.19) defines a function $p_t \in L^2(G)$, which may also be written as $p_t = 1 + \sum_\delta d_\delta \operatorname{Trace}[\mu_t(U^{\delta*}) U^\delta]$. Any $f \in L^2(G)$ has Fourier series $f = \rho_G(f) + \sum_\delta d_\delta \operatorname{Trace}[\rho_G(fU^{\delta*}) U^\delta]$. By the polarized Parseval identity,

$$\rho_G(f p_t) = \rho_G(f) \cdot 1 + \sum_\delta d_\delta \operatorname{Trace}[\rho_G(fU^{\delta*}) \mu_t(U^\delta)]$$

$$= \mu_t \left\{ \rho_G(f) + \sum_\delta d_\delta \operatorname{Trace}[\rho_G(fU^{\delta*}) U^\delta] \right\} = \mu_t(f).$$

This shows that p_t is the density of μ_t and proves (a).

From the convergence of the series in (4.20), it is easy to see that λ_i^δ should converge to $-\infty$ as δ leaves any finite subset of $\operatorname{Irr}(G, \mathbb{C})_+$. This implies that there is a largest number, denoted by $-\lambda$, in the set of negative numbers λ_i^δ for $\delta \in \operatorname{Irr}(G, \mathbb{C})_+$ and $1 \leq i \leq d_\delta$. By the computation proving the absolute and uniform convergence of the series in (4.14) in the proof of Theorem 4.2, replacing $\Gamma' - \Gamma$ and $\eta/2$ there by $\operatorname{Irr}(G, \mathbb{C})_+$ and η, respectively, we can show that, for $t > \eta > 0$,

$$|p_t - 1| \leq \sum_{\delta \in \operatorname{Irr}(G,\mathbb{C})_+} d_\delta |\operatorname{Trace}[A_\delta(t) U^\delta]|$$

$$\leq \|p_\eta\|_2 \left\{ \sum_{\delta \in \operatorname{Irr}(G,\mathbb{C})_+} d_\delta \operatorname{Trace}[A_\delta(t - \eta) A_\delta(t - \eta)^*] \right\}^{1/2}$$

$$= \|p_\eta\|_2 \left\{ \sum_{\delta \in \operatorname{Irr}(G,\mathbb{C})_+} d_\delta \sum_{i=1}^{d_\delta} \exp[2\lambda_i^\delta(t - \eta)] \right\}^{1/2}$$

$$\leq \|p_\eta\|_2 \left\{ e^{-2\lambda(t-2\eta)} \sum_{\delta \in \operatorname{Irr}(G,\mathbb{C})_+} d_\delta \sum_{i=1}^{d_\delta} \exp(2\lambda_i^\delta \eta) \right\}^{1/2}$$

$$\leq e^{-\lambda t} e^{2\lambda\eta} \|p_\eta\|_2 \|p_\eta - 1\|_2,$$

where the last inequality follows from (4.20). This proves the inequality for $\|p_t - 1\|_\infty$ in (b).

By this inequality, $\|p_t - 1\|_2 \le Ce^{-\lambda t}$ for $t > \eta$. However, by (4.20), $\|p_t - 1\|_2^2 \ge d_\delta \exp(2\lambda_i^\delta t)$ for any $\delta \in \mathrm{Irr}(G, \mathbb{C})_+$ and $1 \le i \le d_\delta$. This proves the inequalities for $\|p_t - 1\|_2$.

By $\|p_t - 1\|_2 \le Ce^{-\lambda t}$ and the Schwartz inequality, $\|\mu_t - \rho_G\|_{tv} \le Ce^{-\lambda t}$. However, since $|U_{ii}^\delta| \le 1$ and $\rho_G(U_{ii}^\delta) = 0$ for $\delta \in \mathrm{Irr}(G, \mathbb{C})_+$,

$$\|\mu_t - \rho_G\|_{tv} \ge |\mu_t(U_{ii}^\delta)| = |A_\delta(t)_{ii}| = \sum_{j=1}^{d_\delta} |(Q_\delta)_{ij}|^2 e^{\lambda_j^\delta t}.$$

For any j, $(Q_\delta)_{ij} \ne 0$ for some i; this completes the proof of (b). $\qquad\square$

4.4. Conjugate Invariant Lévy Processes

Recall that, for $h \in G$, $c_h \colon G \to G$ is the conjugation map defined by $c_h(g) = hgh^{-1}$. Its differential is the adjoint map $\mathrm{Ad}(h) = DL_h \circ DR_{h^{-1}}$: $\mathfrak{g} \to \mathfrak{g}$ and induces the adjoint action $\mathrm{Ad}(G)$ of G on $\mathfrak{g}$. A function f on G is called conjugate invariant if $f \circ c_h = f$ for any $h \in G$. A measure μ is called conjugate invariant if $c_h \mu = \mu$ for any $h \in G$. A Lévy process g_t in G with distributions μ_t is called conjugate invariant if each μ_t is conjugate invariant. This is equivalent to saying that, for any $h \in G$, the process $hg_t h^{-1}$ has the same distribution as g_t.

Let g_t be a conjugate invariant Lévy process in G. Then its generator L is also conjugate invariant. This means that, if $f \in D(L)$, the domain of L, then $f \circ c_h \in D(L)$ and $L(f \circ c_h) = (Lf) \circ c_h$ for any $h \in G$. In particular, this implies that, for any $f \in C^2(G)$ and $h \in G$, $[L(f \circ c_h)] \circ c_h^{-1} = Lf$.

Note that, for $g, h \in G$, $X \in \mathfrak{g}$ and $f \in C^1(G)$,

$$X^l(f \circ c_h)(c_h^{-1}(g)) = \frac{d}{dt}(f \circ c_h)(h^{-1}ghe^{tX})\,|_{t=0} = \frac{d}{dt} f(ge^{t\mathrm{Ad}(h)X})\,|_{t=0}$$
$$= [\mathrm{Ad}(h)X]^l f(g).$$

By (1.7), we can write $L(f \circ c_h)(c_h^{-1}(g))$ for $f \in C^2(G)$ explicitly as follows:

$$L(f \circ c_h)(h^{-1}gh) = \frac{1}{2}\sum_{i,j=1}^{d} a_{ij}[\mathrm{Ad}(h)X_i]^l[\mathrm{Ad}(h)X_j]^l f(g)$$

$$+ [\mathrm{Ad}(h)X_0]^l f(g)$$

$$+ \int \left\{ f(g\sigma) - f(g) - \sum_{i=1}^{d}[x_i \circ c_h^{-1}](\sigma)\,[\mathrm{Ad}(h)X_i]^l f(g) \right\}(c_h \Pi)(d\sigma),$$

$$\tag{4.22}$$

where $X_0 = \sum_{i=1}^{d} c_i X_i$. Note that $\{\mathrm{Ad}(h)X_1, \ldots, \mathrm{Ad}(h)X_d\}$ is a basis of $\mathfrak{g}$ and $x_i \circ c_h^{-1}$ are associated coordinate functions. By Proposition 1.3, the Lévy measure Π and the diffusion part of L, $(1/2)\sum_{i,j=1}^{d} a_{ij} X_i^l X_j^l$, are completely determined by the generator L and are independent of the choice of the basis $\{X_1, \ldots, X_d\}$ of $\mathfrak{g}$ and the associated coordinate functions x_i. It follows that if the Lévy process g_t is conjugate invariant, then $c_h \Pi = \Pi$ and

$$\sum_{i,j=1}^{d} a_{ij} X_i^l X_j^l = \sum_{i,j=1}^{d} a_{ij} [\mathrm{Ad}(h)X_i]^l [\mathrm{Ad}(h)X_j]^l \qquad (4.23)$$

for any $h \in G$. In particular, the Lévy measure Π is conjugate invariant.

Recall that $\psi_\delta = \chi_\delta / d_\delta$ is the normalized character. Because $\mathrm{Re}(\psi_\delta)$ takes the maximum value $\psi_\delta(e) = 1$ at e, all its first order derivatives vanish at e. It follows that

$$|\mathrm{Re}(\psi_\delta) - 1| = O(|x|^2).$$

Therefore, by (1.8), the integral $\int (1 - \mathrm{Re}\,\psi_\delta)d\Pi$ in the following theorem exists. Because $|\psi_\delta| \leq 1$, this integral is in fact nonnegative.

Theorem 4.4. *Let G be a compact connected Lie group and let g_t be a conjugate invariant hypo-elliptic Lévy process in G with $g_0 = e$ and generator L.*

(a) For $t > 0$, the distribution μ_t of g_t has a density $p_t \in L^2(G)$ and, for $g \in G$,

$$p_t(g) = 1 + \sum_{\delta \in \mathrm{Irr}(G,\mathbb{C})_+} d_\delta\, a_\delta(t)\, \chi_\delta(g) \quad \text{with } a_\delta(t) = \mu_t(\overline{\psi_\delta}) = e^{t\,L(\overline{\psi_\delta})(e)},$$

$$(4.24)$$

where the series converges absolutely and uniformly for $(t, g) \in [\eta, \infty) \times G$ for any fixed $\eta > 0$, and

$$|a_\delta(t)| = \exp\left\{ -\left[\lambda_\delta + \int (1 - \mathrm{Re}\,\psi_\delta)d\Pi \right] t \right\}$$

with $\lambda_\delta = -L_D\overline{\psi_\delta}(e) > 0$, where $L_D = (1/2)\sum_{i,j=1}^{d} a_{ij} X_i^l X_j^l$ is the diffusion part of L.

(b) Let

$$\lambda = \inf\left\{ \left[\lambda_\delta + \int (1 - \mathrm{Re}\,\psi_\delta)d\Pi \right];\ \delta \in \mathrm{Irr}(G, \mathbb{C})_+ \right\}.$$

Then $\lambda = [\lambda_\delta + \int (1 - \mathrm{Re}\,\psi_\delta)d\Pi] > 0$ for some $\delta \in \mathrm{Irr}(G, \mathbb{C}_+)$, and for any $\eta > 0$, there are positive constants c and C such that, for

$t > \eta$,

$$\|p_t - 1\|_\infty \le Ce^{-\lambda t}, \quad ce^{-\lambda t} \le \|p_t - 1\|_2 \le Ce^{-\lambda t}, \quad \text{and}$$
$$ce^{-\lambda t} \le \|\mu_t - \rho_G\|_{tv} \le Ce^{-\lambda t}.$$

Proof. Suppose that the distribution of g_t has an L^2 density p_t for $t > 0$. Then p_t is conjugate invariant and hence, by (4.6), may be expanded into a Fourier series in terms of irreducible characters as in (4.24) in L^2-sense with $a_\delta(t) = \int p_t(g)\overline{\psi_\delta}(g)\, dg = \mu_t(\overline{\psi_\delta})$. By the conjugate invariance of μ_t and (4.5),

$$a_\delta(t+s) = \mu_{t+s}(\overline{\psi_\delta}) = \int \overline{\psi_\delta(uv)}\mu_t(du)\mu_s(dv)$$

$$= \int \overline{\psi_\delta(gug^{-1}v)}\mu_t(du)\mu_s(dv)dg$$

$$= \int \overline{\psi_\delta(u)\psi_\delta(v)}\mu_t(du)\mu_s(dv) = a_\delta(t)a_\delta(s).$$

This combined with $\lim_{t \to 0} a_\delta(t) = \overline{\psi_\delta(e)} = 1$ implies that $a_\delta(t) = e^{ty}$ for some complex number y. We have $y = (d/dt)\mu_t(\overline{\psi_\delta})\,|_{t=0} = L\overline{\psi_\delta}(e)$ and, hence, $a_\delta(t) = \exp[t\, L(\overline{\psi_\delta})(e)]$.

As in the proof of Theorem 4.2, for fixed $\delta \in \mathrm{Irr}(G, \mathbb{C})_+$, write $U = U_\delta$ and $n = d_\delta$ and let $\tilde{X} = X(U^*)(e)$ for $X \in \mathfrak{g}$ and let Y_i be defined in (4.7). By (4.16),

$$L(\overline{\psi_\delta})(e) = \frac{1}{n}\mathrm{Trace}[L(U^*)(e)] = \frac{1}{n}\mathrm{Trace}\left[-\frac{1}{2}\sum_{i=1}^{n}\tilde{Y}_i^*\tilde{Y}_i + \tilde{Y}_V\right.$$

$$\left. - \int_{V^c}(I - U^*)d\Pi + r_V\right], \tag{4.25}$$

where $r_V \to 0$ as $V \downarrow \{e\}$. Since $\tilde{Y}_V$ is skew-Hermitian, $\mathrm{Trace}(\tilde{Y}_V)$ is purely imaginary. It follows that

$$|a_\delta(t)| = \exp\{t\, \mathrm{Re}[L(\overline{\psi_\delta})(e)]\} = \exp\left\{-\left[\frac{1}{2n}\sum_{i=1}^{n}\mathrm{Trace}(\tilde{Y}_i^*\tilde{Y}_i)\right.\right.$$

$$\left.\left. + \int(1 - \mathrm{Re}\,\psi_\delta)d\Pi\right]t\right\}$$

$$= \exp\left\{-\left[\lambda_\delta + \int(1 - \mathrm{Re}\,\psi_\delta)d\Pi\right]t\right\}, \tag{4.26}$$

where

$$\lambda_\delta = -\frac{1}{2}\sum_{i,j=1}^{d} a_{ij} X_i^l X_j^l \, \overline{\psi_\delta}(e) = \frac{1}{2n}\sum_{i=1}^{d} \mathrm{Trace}(\tilde{Y}_i^* \tilde{Y}_i)$$

is nonnegative and is zero only when $\tilde{Y}_i = 0$ for all i. Under the hypo-elliptic assumption and the irreducibility of $\delta \in \mathrm{Irr}(G, \mathbb{C})_+$, some $\tilde{Y}_i$ is nonzero. Therefore, $\lambda_\delta > 0$.

If g_t is a continuous hypo-elliptic Lévy process, then its distribution μ_t has a smooth density p_t for $t > 0$, for which (4.24) holds in L^2 sense. By the Parseval identity, $\|p_t\|_2^2 = 1 + \sum_\delta d_\delta^2 |a_\delta(t)|^2 = 1 + \sum_\delta d_\delta^2 |a_\delta(2t)|$. Since $\chi_\delta(e) = d_\delta$, the series in (4.24) evaluated at e is equal to $1 + \sum_\delta d_\delta^2 a_\delta(t)$, and we see that it actually converges absolutely at e. As a positive definite function on G, the character χ_δ satisfies $|\chi_\delta(g)| \le \chi_\delta(e)$ for any $g \in G$. It follows that the series in (4.24) converges absolutely and uniformly on G. In this case, the integral term in (4.26) does not appear because $\Pi = 0$.

Now assume that Π is not equal to zero. Because the hypo-elliptic assumption is still satisfied by the diffusion part of the generator L, we can still write down the series in (4.24) with $a_\delta(t) = \exp[t\, L(U^{\delta *})(e)]$. Because $\mathrm{Re}(1 - \psi_\delta) \ge 0$, we see that $|a_\delta(t)|$ becomes smaller than when $\Pi = 0$; hence, the series in (4.24) still converges absolutely and uniformly on G. Let p_t be its limit. As in the proof of Theorem 4.3, we can show that p_t is a density of μ_t using the polarized Parseval identity. By (4.26), it is easy to see that the series in (4.24) also converges uniformly in t for $t > \eta > 0$. We have proved (a).

The convergence of the series in (4.24) at e implies that $[\lambda_\delta + \int (1 - \mathrm{Re}\,\psi_\delta)d\Pi] \to \infty$ as δ leaves any finite subset of $\mathrm{Irr}(G, \mathbb{C})_+$. In particular, this implies that the set of positive numbers $[\lambda_\delta + \int (1 - \mathrm{Re}\,\psi_\delta)d\Pi]$, $\delta \in \mathrm{Irr}(G, \mathbb{C})_+$, has a smallest number $\lambda > 0$.

For $t > \eta > 0$, $|p_t - 1| \le e^{-\lambda(t-\eta)}\sum_\delta d_\delta |a_\delta(\eta)\chi_\delta| \le e^{-\lambda(t-\eta)}\sum_\delta d_\delta^2 |a_\delta(\eta)|$ $\le e^{-\lambda(t-\eta)}\|p_{\eta/2}\|_2^2$. This proves the inequality for $\|p_t - 1\|_\infty$ in (b), and from which the upper bounds for $\|p_t - 1\|_2$ and $\|\mu_t - \rho_G\|_{tv}$ follow. The lower bounds follow from $\|p_t - 1\|_2^2 \ge d_\delta^2 |a_\delta(t)|^2$ and $\|\mu_t - \rho_G\|_{tv} \ge |\mu_t(\psi_\delta)| = |a_\delta(t)|$. Part (b) is proved. $\qquad\square$

A Lie algebra $\mathfrak{g}$ with $\dim(\mathfrak{g}) > 1$ is called simple if it does not contain any ideal except $\{0\}$ and $\mathfrak{g}$. It is called semi-simple if it does not contain any abelian ideal except $\{0\}$. Here, an ideal $\mathfrak{i}$ of $\mathfrak{g}$ is called abelian if $[\mathfrak{i}, \mathfrak{i}] = \{0\}$. A Lie group G is called simple or semi-simple if its Lie algebra $\mathfrak{g}$ is so. Note that the center of $\mathfrak{g}$ is $\{0\}$ in the semi-simple case.

Because G is compact, there is an $\mathrm{Ad}(G)$-invariant inner product $\langle X, Y \rangle$ on $\mathfrak{g}$. This inner product induces a bi-invariant Riemannian metric on G, under which the Laplace–Beltrami operator is given by $\Delta = \sum_{i=1}^{d} X_i^l X_i^l$, where $\{X_1, \ldots, X_d\}$ is an orthonormal basis of $\mathfrak{g}$ (see Proposition 2.6).

Proposition 4.4. *Let G be a compact connected simple Lie group with Lie algebra $\mathfrak{g}$. Then up to a constant multiple, there is a unique $\mathrm{Ad}(G)$-invariant inner product $\langle \cdot, \cdot \rangle$ on $\mathfrak{g}$. Moreover, if g_t is a conjugate invariant Lévy process in G, then the diffusion part of its generator, $(1/2) \sum_{i,j=1}^{d} a_{ij} X_i^l X_j^l$, is equal to $c\Delta$ for some constant $c \geq 0$, where Δ is the Laplace–Beltrami operator on G under the bi-invariant Riemannian metric induced by $\langle \cdot, \cdot \rangle$.*

Proof. Fix an arbitrary $\mathrm{Ad}(G)$-invariant inner product on $\mathfrak{g}$. It suffices to prove the second assertion. We may assume the basis $\{X_1, \ldots, X_d\}$ is orthonormal. Then $\mathrm{Ad}(g)$ is an orthogonal transformation on $\mathfrak{g}$. By (4.23), the symmetric bilinear form $Q(x, y) = \sum_{i,j=1}^{d} a_{ij} x_i y_j$ on $\mathfrak{g} \equiv \mathbb{R}^d$ is $\mathrm{Ad}(G)$-invariant. Because G is simple, $\mathfrak{g}$ contains no proper $\mathrm{Ad}(G)$-invariant subspace; hence, this action is irreducible. By appendix 5 in Kobayashi and Nomizu [35], any symmetric bilinear form on $\mathbb{R}^d$ that is invariant under an irreducible action of a subgroup of the orthogonal group $O(d)$ is equal to a multiple of the standard Euclidean inner product on $\mathbb{R}^d$. It follows that the symmetric matrix $\{a_{ij}\}$ is equal to a multiple of the identity matrix I. This proves $L = c\Delta$ for some $c \geq 0$. $\qquad\square$

Proposition 4.5. *Let G be a compact connected semi-simple Lie group and let g_t be a conjugate invariant Lévy process in G such that its Lévy measure Π has a finite first moment. Then the generator L of g_t restricted to $C^2(G)$ is given by*

$$Lf = L_D f + \int_G (f \circ R_h - f)\Pi(dh), \tag{4.27}$$

where L_D is the diffusion part of L.

Proof. By (1.11), $Lf = L_D f + Y^l f + f_G(f \circ R_h - f)\Pi(dh)$ for some $Y \in \mathfrak{g}$. The conjugate invariance of L, L_D and Π implies that the operator Y^l is also conjugate invariant, hence, for any $g \in G$ and $f \in C^\infty(G)$, $Y^l(f \circ c_g) = (Y^l f) \circ c_g$. However, $Y^l(f \circ c_g) = \{[\mathrm{Ad}(g)Y]^l f\} \circ c_g$, which implies that $[\mathrm{Ad}(g)Y]^l f = Y^l f$ and hence $\mathrm{Ad}(g)Y = Y$ for any $g \in G$. The semi-simplicity of G implies that $Y = 0$. $\qquad\square$

4.5. An Example

In this section, we calculate explicitly the Fourier expansion of the distribution density of a conjugate invariant Lévy process g_t in $G = SU(2)$, the group of 2×2 unitary matrices of determinant 1. It is known that G is a simple Lie group with Lie algebra $\mathfrak{g} = \mathfrak{su}(2)$, the space of traceless 2×2 skew-Hermitian matrices. It is easy to see that $\langle X, Y \rangle = \text{Trace}(XY^*)$ is an $\text{Ad}(G)$-invariant inner product on $\mathfrak{g}$.

Let

$$T = \{\text{diag}(e^{\theta i}, e^{\theta i}); 0 \leq \theta \leq \pi\} \subset G,$$

where $i = \sqrt{-1}$. Any $g \in G$ is conjugate to an element in T, that is, $g = kak^{-1}$ for some $a \in T$ and $k \in G$. It follows that if f is a conjugate invariant function on G, then $f(g) = f(a)$, that is, the values of f on G are determined by the restriction of f to T.

Let $V_1 = \mathbb{C}$ and let U_1 be the trivial one-dimensional representation of $G = SU(2)$ on V_1. For $n \geq 2$, let V_n be the space of homogeneous polynomials of degree $n - 1$ in two variables z_1 and z_2 with complex coefficients. For $g = \{g_{ij}\} \in G$ and $P \in V_n$, define $gP \in V_n$ by setting

$$(gP)(z_1, z_2) = P(z_1 g_{11} + z_2 g_{21}, z_1 g_{12} + z_2 g_{22}).$$

This defines a representation U_n of G on V_n, given by $U_n(g)P = gP$ for $g \in G$ and $P \in V_n$, which is unitary with respect to a proper Hermitian inner product on V_n. By the discussion in [12, II.5], the set of all the finite dimensional irreducible representations of $SU(2)$, up to equivalence, is given by $\{U_n; n = 1, 2, 3, \ldots\}$. Note that $\dim_{\mathbb{C}}(V_n) = n$.

We now determine the character χ_n of U_n. Since characters are conjugate invariant, it suffices to calculate $\chi_n(a)$ for $a = \text{diag}(e^{\theta i}, e^{-\theta i}) \in T$. Because $P_k = z_1^k z_2^{n-1-k}$ for $k = 0, 1, 2, \ldots, n-1$ form a basis of V_n and $aP_k = (z_1 e^{\theta i})^k (z_2 e^{-\theta i})^{n-1-k} = e^{2k\theta i} e^{-(n-1)\theta i} P_k$,

$$\chi_n(a) = \text{Trace}[U(a)] = \sum_{k=0}^{n-1} e^{2k\theta i} e^{-(n-1)\theta i} = e^{-(n-1)\theta i} \frac{e^{2n\theta i} - 1}{e^{2\theta i} - 1}$$

$$= \frac{\sin(n\theta)}{\sin \theta}. \tag{4.28}$$

Let

$$X = \begin{bmatrix} i & 0 \\ 0 & -i \end{bmatrix}, \quad Y = \begin{bmatrix} 0 & 1 \\ -1 & 0 \end{bmatrix} \quad \text{and} \quad Z = \begin{bmatrix} 0 & i \\ i & 0 \end{bmatrix}.$$

Then $2^{-1/2}X$, $2^{-1/2}Y$, and $2^{-1/2}Z$ form an orthonormal basis of $\mathfrak{g}$. By Propositions 2.5 and 2.6, the Laplace-Beltrami operator Δ on G under the bi-invariant metric induced by the $\mathrm{Ad}(G)$-invariant inner product $\langle X, Y \rangle = \mathrm{Trace}\,(XY^*)$ is given by

$$\Delta = \frac{1}{2}[X^l X^l + Y^l Y^l + Z^l Z^l]. \tag{4.29}$$

We want to calculate $\Delta\psi_n(e) = \Delta\chi_n(e)/n$. For the sake of convenience, if f is a conjugate invariant function on G, we may write $f(\theta)$ for $f(g)$ with $g = kak^{-1}$, $a = \mathrm{diag}(e^{\theta i}, e^{-\theta i}) \in T$ and $k \in G$. Then

$$X^l X^l f(e) = \frac{d^2}{d\theta^2} f(\theta)|_{\theta=0}.$$

Since $Y = \mathrm{Ad}(k_1)X$ and $Z = \mathrm{Ad}(k_2)X$ for some $k_1, k_2 \in G$, it follows that $e^{tY} = k_1 e^{tX} k_1^{-1}$ and $e^{tZ} = k_2 e^{tX} k_2^{-1}$. Therefore, $Y^l Y^l f(e) = Z^l Z^l f(e) = X^l X^l f(e)$. We obtain

$$\Delta f(e) = \frac{1}{2}\left[X^l X^l f(e) + Y^l Y^l f(e) + Z^l Z^l f(e) \right] = \frac{3}{2} f''(\theta)|_{\theta=0}. \tag{4.30}$$

Let $f(\theta) = \chi_n(\theta) = \sin(n\theta)/\sin\theta$. We have

$$\chi_n'(\theta) = \frac{d}{d\theta} \frac{n[1 - (1/6)n^2\theta^2 + O(\theta^3)]}{1 - (1/6)\theta^2 + O(\theta^3)}$$

$$= -\frac{n^3}{3} \frac{\theta + O(\theta^2)}{1 - (1/6)\theta^2 + O(\theta^3)} + \frac{n}{3} \frac{[1 - (1/6)n^2\theta^2 + O(\theta^3)][\theta + O(\theta^2)]}{[1 - (1/6)\theta^2 + O(\theta^3)]^2}$$

and

$$\chi_n''(0) = -\frac{n^3}{3} + \frac{n}{3} = -\frac{n(n^2 - 1)}{3}. \tag{4.31}$$

By (4.30) and (4.31),

$$\Delta\psi_n(e) = \frac{1}{n}\Delta\chi_n(e) = -\frac{n^2 - 1}{2}. \tag{4.32}$$

Let g_t be a hypo-elliptic conjugate invariant Lévy process in $G = SU(2)$. By Proposition 4.4, the diffusion part of its generator is given by $c\Delta$ for some constant $c > 0$. Because any $g \in G$ is conjugate to $\mathrm{diag}(e^{\theta i}, e^{-\theta i})$ for a unique $\theta \in [0, \pi]$, $\xi(g) = \theta$ defines a map $\xi : G \to [0, \pi]$, which is in fact continuous. It follows from Theorem 4.4, (4.28), and (4.32), noting that $a_\delta(t)$ in Theorem 4.4 now takes positive real value, that the process g_t has a conjugate invariant distribution density p_t for $t > 0$ and

$$
p_t(\theta) = \sum_{n=1}^{\infty} n \exp \left\{ -t \left[\frac{c(n^2 - 1)}{2} + \int_0^\pi \right. \right.
$$

$$
\left. \left. \times \left(1 - \frac{\sin(nu)}{n \sin u} \right) \xi \Pi(du) \right] \right\} \frac{\sin(n\theta)}{\sin \theta}. \tag{4.33}
$$

We now derive an expression for the distribution μ_t of g_t. Any $g \in SU(2)$ is a matrix

$$
g = \begin{bmatrix} a & b \\ -\bar{b} & \bar{a} \end{bmatrix},
$$

where $a, b \in \mathbb{C}$ with $|a|^2 + |b|^2 = 1$. The map $g \to (a, b)$ provides an identification of $G = SU(2)$ with the three-dimensional unit sphere S^3 in $\mathbb{C}^2 \equiv \mathbb{R}^4$. Let x_0, x_1, x_2, x_3 be the standard coordinates on $\mathbb{R}^4$. The identification map from G to S^3 may be written as $g \to (x_0 + ix_1, x_2 + ix_3)$ with $\Sigma_{i=0}^3 x_i^2 = 1$. The spherical polar coordinates θ, ψ, ϕ on S^3 are defined by

$$
x_0 = \cos \theta, \ x_1 = \sin \theta \cos \psi, \ x_3 = \sin \theta \sin \psi \cos \phi \quad \text{and}
$$
$$
x_4 = \sin \theta \sin \psi \sin \phi,
$$

with $0 \le \theta \le \pi, 0 \le \psi \le \pi$ and $0 \le \phi \le 2\pi$. Under the identification of G and S^3, the normalized Haar measure on G is the uniform distribution on S^3 and may be expressed under the spherical polar coordinates as

$$
dg = \frac{1}{2\pi^2} \sin^2 \theta \sin \psi d\theta \, d\psi \, d\phi. \tag{4.34}
$$

The point on S^3 with polar coordinates (ϕ, ψ, ϕ) corresponds to the matrix

$$
g = \begin{bmatrix} \cos \theta + i \sin \theta \cos \psi & \sin \theta \sin \psi \cos \phi + i \sin \theta \sin \psi \sin \phi \\ -\sin \theta \sin \psi \cos \phi + i \sin \theta \sin \psi \sin \phi & \cos \theta - i \sin \theta \cos \psi \end{bmatrix}
$$

in $G = SU(2)$. It is easy to show that the eigenvalues of this matrix are equal to $e^{\pm \theta i}$, and hence this matrix is conjugate to $\mathrm{diag}(e^{\theta i}, e^{-\theta i}) \in T$. It follows that for any function f on $G \equiv S^3$ and $t > 0$,

$$
\mu_t(f) = \frac{1}{2\pi^2} \int_0^{2\pi} \int_0^\pi \int_0^\pi f(\theta, \psi, \phi) p_t(\theta) \sin^2 \theta \sin \psi \, d\theta \, d\psi \, d\phi. \tag{4.35}
$$

5

Semi-simple Lie Groups of Noncompact Type

In this chapter, we present an essentially self-contained introduction to semi-simple Lie groups of noncompact type. The first two sections deal with the basic definitions and the root system on the Lie algebra of such a group. The Cartan, Iwasawa, and Bruhat decompositions are introduced in Section 5.3. Two basic examples, the special linear group $SL(d, \mathbb{R})$ and the connected Lorentz group $SO(1, d)_+$, are discussed in some detail in Section 5.4. We give the complete definitions and results as well as all the short and simple proofs, but the longer and more complicated proofs are referred to standard references, mainly Helgason [26]. Examples in Section 5.4 may help to illustrate the general theory.

5.1. Basic Properties of Semi-simple Lie Groups

Let $\mathfrak{g}$ be a Lie algebra. Recall that, for $X \in \mathfrak{g}$, $\mathrm{ad}(X): \mathfrak{g} \to \mathfrak{g}$ is defined by $\mathrm{ad}(X)Y = [X, Y]$. The Killing form B of $\mathfrak{g}$ is defined by

$$\forall X, Y \in \mathfrak{g}, \quad B(X, Y) = \mathrm{Trace}[\mathrm{ad}(X)\,\mathrm{ad}(Y)]. \tag{5.1}$$

This is a symmetric bilinear form on $\mathfrak{g}$ and is invariant under any Lie algebra automorphism σ on $\mathfrak{g}$, in the sense that $B(\sigma X, \sigma Y) = B(X, Y)$, by the property of the trace operator. In particular, if $\mathfrak{g}$ is the Lie algebra of a Lie group G, then its Killing form B is $\mathrm{Ad}(G)$-invariant in the sense that $B(\mathrm{Ad}(g)X, \mathrm{Ad}(g)Y) = B(X, Y)$ for $X, Y \in \mathfrak{g}$ and $g \in G$, where $\mathrm{Ad}(g)$: $\mathfrak{g} \to \mathfrak{g}$ is the differential of the conjugation map $c_g: G \ni g' \mapsto gg'g^{-1} \in G$ defined before. Letting $\sigma = e^{t\,\mathrm{ad}(X)}$ and differentiating at $t = 0$, we obtain $B(\mathrm{ad}(X)Y, Z) + B(Y, \mathrm{ad}(X)Z) = 0$ for $X, Y, Z \in \mathfrak{g}$. Note that if $\mathfrak{h}$ is a subalgebra of $\mathfrak{g}$, its Killing form in general is not equal to the restriction of the Killing form of $\mathfrak{g}$ to $\mathfrak{h}$. However, this holds if $\mathfrak{h}$ is an ideal of $\mathfrak{g}$.

Recall that a Lie algebra $\mathfrak{g}$ is semi-simple if it does not contain any abelian ideal except $\{0\}$ and a Lie group is semi-simple if its Lie algebra is. By

103

Cartan's criterion (see theorem 1 in I.6 of Bourbaki [10]), a Lie algebra $\mathfrak{g}$ is semi-simple if and only if its Killing form B is nondegenerate, that is, if $B(X, \cdot)$ is not identically equal to zero for any $X \in \mathfrak{g}$.

Let G be a Lie group with Lie algebra $\mathfrak{g}$. A Cartan involution on G is a Lie group isomorphism $\Theta: G \to G$ such that $\Theta \neq \mathrm{id}_G$ and $\Theta^2 = \mathrm{id}_G$. Its differential $\theta = D\Theta: \mathfrak{g} \to \mathfrak{g}$ is a Lie algebra isomorphism such that $\theta \neq \mathrm{id}_\mathfrak{g}$ and $\theta^2 = \mathrm{id}_\mathfrak{g}$ and is called a Cartan involution on $\mathfrak{g}$. It is clear that θ has exactly two eigenvalues: 1 and -1. Let $\mathfrak{k}$ and $\mathfrak{p}$ be, respectively, the eigenspaces of θ corresponding to the eigenvalues 1 and -1. Then $\mathfrak{g} = \mathfrak{k} \oplus \mathfrak{p}$ is a direct sum and it is easy to show that

$$[\mathfrak{k}, \mathfrak{k}] \subset \mathfrak{k}, \quad [\mathfrak{k}, \mathfrak{p}] \subset \mathfrak{p}, \quad \text{and} \quad [\mathfrak{p}, \mathfrak{p}] \subset \mathfrak{k}. \tag{5.2}$$

The first relation implies that $\mathfrak{k}$ is a Lie subalgebra of $\mathfrak{g}$.

If the Killing form B is negative definite on $\mathfrak{g}$, then the Lie group G together with the Cartan involution Θ is said to be of compact type. In this case, G is compact (see [26, section II.6]). If B is negative definite on $\mathfrak{k}$ and positive definite on $\mathfrak{p}$, then G together with Θ is said to be of noncompact type. In this second case, G is noncompact and the direct sum $\mathfrak{g} = \mathfrak{k} \oplus \mathfrak{p}$ is called the Cartan decomposition of $\mathfrak{g}$. In both cases, $\mathfrak{g}$ is semi-simple because the Killing form B is nondegenerate.

In the rest of this book, we will always assume that G is a connected semi-simple Lie group of noncompact type with Lie algebra $\mathfrak{g}$ and Cartan involution Θ unless explicitly stated otherwise. The Killing form B induces an inner product $\langle \cdot, \cdot \rangle$ on $\mathfrak{g}$ given by

$$\langle X, Y \rangle = -B(X, \theta Y). \tag{5.3}$$

Under this inner product, $\mathfrak{g} = \mathfrak{k} \oplus \mathfrak{p}$ is an orthogonal decomposition.

Let $K = \{g \in G; \Theta(g) = g\}$ be the subset of G fixed by Θ. Then K is a closed subgroup of G with Lie algebra $\mathfrak{k}$. The homogeneous space G/K is called a symmetric space, and it is said to be of compact or noncompact type depending on whether G is of compact or noncompact type. The inner product $\langle \cdot, \cdot \rangle$ restricted to $\mathfrak{p}$ satisfies the condition (2.35) and, hence, it induces a G-invariant Riemannian metric on G/K under which the geodesics starting at the point $o = eK$ are given by $t \mapsto e^{tX} o$ for $X \in \mathfrak{p}$. The map $gK \mapsto \Theta(g)K$ on G/K induced by the Cartan involution Θ fixes o and maps any geodesic ray $\gamma(t)$ starting from o to the geodesic ray in the opposite direction, that is, the geodesic ray $\tilde{\gamma}(t)$ starting from o determined by $(d/dt)\tilde{\gamma}(0) = -(d/dt)\gamma(0)$. Such a map is called a geodesic symmetry at point o. Because G acts transitively and isometrically on G/K, there is a geodesic symmetry at every point of G/K. In fact, this property may be used to characterize symmetric spaces among Riemannian manifolds (see [26, section IV.3]). The semi-simple Lie groups and the associated symmetric spaces play very important roles in analysis and differential geometry. The reader is referred to Helgason [26] for a comprehensive treatment of this subject.

The following result is a direct consequence of [26, chapter VI, theorem 1.1].

Theorem 5.1. *(a) K is a connected closed subgroup of G containing the center Z of G.*

   (b) K is compact if and only if Z is finite. In this case, K is a maximal compact subgroup of G in the sense that there is no compact subgroup of G that properly contains K.

   (c) The map $(k, X) \mapsto ke^X$ is a diffeomorphism from $K \times \mathfrak{p}$ onto G.

By (c), the map $\mathfrak{p} \ni Y \mapsto e^Y o \in G/K$ is a diffeomorphism, which provides an identification of the symmetric space G/K with $\mathfrak{p} \equiv \mathbb{R}^d$ such that the point o corresponds to the origin in $\mathbb{R}^d$.

We note that the semi-simplicity of $\mathfrak{g}$ implies that the center $\mathfrak{z}$ of $\mathfrak{g}$ is trivial. However, the center Z of G may not be trivial, but as its Lie algebra $\mathfrak{z}$ is trivial, Z is at most countable. The following proposition can be easily proved.

Proposition 5.1. *(a) Both the inner product $\langle \cdot, \cdot \rangle$ and the space $\mathfrak{p}$ are $\mathrm{Ad}(K)$-invariant.*
(b) For $Y \in \mathfrak{p}$, $\mathrm{ad}(Y)$ is symmetric under $\langle \cdot, \cdot \rangle$ in the sense that

$$\forall X_1, X_2 \in \mathfrak{g}, \quad \langle \mathrm{ad}(Y)X_1, X_2 \rangle = \langle X_1, \mathrm{ad}(Y)X_2 \rangle.$$

   (c) For $Z \in \mathfrak{k}$, $\mathrm{ad}(Z)$ is skew-symmetric under $\langle \cdot, \cdot \rangle$ in the sense that

$$\forall X_1, X_2 \in \mathfrak{g}, \quad \langle \mathrm{ad}(Z)X_1, X_2 \rangle = -\langle X_1, \mathrm{ad}(Z)X_2 \rangle.$$

5.2. Roots and the Weyl Group

Let $\mathfrak{a}$ be a maximal abelian subspace of $\mathfrak{p}$ and let A be the Lie subgroup of G generated by $\mathfrak{a}$, that is, the connected Lie subgroup of G with Lie algebra $\mathfrak{a}$. This is an abelian Lie group. By Theorem 5.1(c), the exponential map exp on $\mathfrak{a}$ is a diffeomorphism: $\mathfrak{a} \to A$. Therefore, for $a \in A$, $X = \log a$ is well defined as the unique element $X \in \mathfrak{a}$ such that $\exp(X) = a$.

Let α be a linear functional on $\mathfrak{a}$. Define

$$\mathfrak{g}_\alpha = \{X \in \mathfrak{g}; \quad \mathrm{ad}(H)X = \alpha(H)X \text{ for } H \in \mathfrak{a}\}. \tag{5.4}$$

By the Jacobi identity of the Lie bracket, it is easy to verify that

$$[\mathfrak{g}_\alpha, \mathfrak{g}_\beta] \subset \mathfrak{g}_{\alpha+\beta} \tag{5.5}$$

for any two linear functionals α and β on $\mathfrak{a}$.

A nonzero α is called a root if $\mathfrak{g}_\alpha \neq \{0\}$. In this case, $\mathfrak{g}_\alpha$ is called the root space, its elements are called root vectors, and its dimension is called the multiplicity of the root α.

Let $\mathfrak{m}$ and M be, respectively, the centralizers of $\mathfrak{a}$ in $\mathfrak{k}$ and in K; that is,

$$\mathfrak{m} = \{X \in \mathfrak{k}; \ \mathrm{ad}(X)H = 0 \text{ for } H \in \mathfrak{a}\} \tag{5.6}$$

and

$$M = \{k \in K; \ \mathrm{Ad}(k)H = H \text{ for } H \in \mathfrak{a}\}. \tag{5.7}$$

Note that M is a closed Lie subgroup of K with Lie algebra $\mathfrak{m}$, and it is also the centralizer of A in K, that is, the set of elements in K that commute with every element of A. Moreover, by Theorem 5.1 (a), the center Z of G is contained in M.

An element $g \in G$ is said to normalize a subset $\mathfrak{t}$ of $\mathfrak{g}$ (resp. a subset T of G) if $\forall X \in \mathfrak{t}, \mathrm{Ad}(g)X \in \mathfrak{t}$ (resp. $\forall t \in T, gtg^{-1} \in T$). If every element of a Lie subgroup H of G normalizes $\mathfrak{t}$ (resp. T), then H is said to normalize $\mathfrak{t}$ (resp. T). Let S be a subgroup of G. Then the set of all the elements of S normalizing $\mathfrak{t}$ (resp. T) is a closed subgroup of S, called the normalizer of $\mathfrak{t}$ (resp. T) in S. When $S = G$, it will be called simply the normalizer of $\mathfrak{t}$ (resp. T). Note that if T is a connected Lie subgroup of G with Lie algebra $\mathfrak{t}$, then the normalizer of T in S is equal to the normalizer of $\mathfrak{t}$ in S.

Proposition 5.2. *Both A and M normalize $\mathfrak{g}_\alpha$, where α is a root or zero.*

Proof. Because $\mathrm{ad}(H)\mathfrak{g}_\alpha \subset \mathfrak{g}_\alpha$ for any $H \in \mathfrak{a}$, $\mathrm{Ad}(e^H)\mathfrak{g}_\alpha = e^{\mathrm{ad}(H)}\mathfrak{g}_\alpha \subset \mathfrak{g}_\alpha$. This shows $\mathrm{Ad}(a)\mathfrak{g}_\alpha = \mathfrak{g}_\alpha$. For $m \in M$, $X \in \mathfrak{g}_\alpha$, and $H \in \mathfrak{a}$, $\mathrm{Ad}(m)H = H$ and $[H, \mathrm{Ad}(m)X] = \mathrm{Ad}(m)[H, X] = \alpha(H)\mathrm{Ad}(m)X$. This shows $\mathrm{Ad}(m)X \in \mathfrak{g}_\alpha$. $\qquad \square$

Proposition 5.3. *(a) $\mathfrak{g} = \mathfrak{g}_0 \oplus \sum_\alpha \mathfrak{g}_\alpha$ is a direct sum whose components are mutually orthogonal under $\langle \cdot, \cdot \rangle$, where the summation $\sum_\alpha$ is taken over all the roots.*

(b) $\mathfrak{g}_0 = \mathfrak{a} \oplus \mathfrak{m}$.

(c) If α is a root, then so is $-\alpha$ with $\mathfrak{g}_{-\alpha} = \theta(\mathfrak{g}_\alpha)$.

Proof. Because $[\mathrm{ad}(X), \mathrm{ad}(Y)] = \mathrm{ad}([X, Y])$ by the Jacobi identity, it follows from Proposition 5.1(b) that $\{\mathrm{ad}(H); \ H \in \mathfrak{a}\}$ is a commutative family of linear operators on $\mathfrak{g}$ that are symmetric under $\langle \cdot, \cdot \rangle$ and hence can be simultaneous diagonalized over the real field. Therefore, $\mathfrak{g}$ can be expressed as an orthogonal direct sum of the common eigenspaces. Each of the common eigenspaces is $\mathfrak{g}_\alpha$, where α is either a root or zero. This proves (a). To show (b), first it is easy to see that $\mathfrak{a} \subset \mathfrak{g}_0$ and $\mathfrak{m} \subset \mathfrak{g}_0$. Let $X \in \mathfrak{g}_0$ and write $X = Z + Y$ with $Z \in \mathfrak{k}$ and $Y \in \mathfrak{p}$. Since $[H, X] = 0$ for any $H \in \mathfrak{a}$, by (5.2), $[H, Y] = 0$ and $[H, Z] = 0$. The latter implies $Z \in \mathfrak{m}$, and the former together with the fact that $\mathfrak{a}$ is a maximal abelian subspace of $\mathfrak{p}$ implies $Y \in \mathfrak{a}$. This proves $\mathfrak{g}_0 = \mathfrak{a} \oplus \mathfrak{m}$. To prove (c), note that $\exists$ nonzero $X \in \mathfrak{g}$ such that $\forall H \in \mathfrak{a}$, $[H, X] = \alpha(H)X$. Then $[H, \theta(X)] = -[\theta(H), \theta(X)] = -\theta([H, X]) = -\alpha(H)\theta(X)$. This shows that $-\alpha$ is also a root and $\theta(X) \in \mathfrak{g}_{-\alpha}$. $\qquad\square$

For each root α, the equation $\alpha = 0$ determines a subspace of $\mathfrak{a}$ of codimension 1. These subspaces divide $\mathfrak{a}$ into several open convex cones, called Weyl chambers. Fix a Weyl chamber $\mathfrak{a}_+$. A root α is called positive if it is positive on $\mathfrak{a}_+$. A root, if not positive, must be negative, that is, equal to $-\alpha$ for some positive root α, because it cannot vanish anywhere on $\mathfrak{a}_+$. Note that the definition of positive roots depends on the choice of the Weyl chamber $\mathfrak{a}_+$.

Proposition 5.4. *The set of positive roots span the dual space of* $\mathfrak{a}$.

Proof. It suffices to show that if $\alpha(H) = 0$ for some $H \in \mathfrak{a}$ and any positive root α, then $H = 0$. By Proposition 5.3, this implies $\mathrm{ad}(H)X = 0$ for any $X \in \mathfrak{g}$; hence, H must belong to the center of $\mathfrak{g}$, which is trivial by its semisimplicity. $\qquad\square$

Let α be a positive root and $X \in \mathfrak{g}_\alpha$. By (5.2), $X = Y + Z$ with $Y \in \mathfrak{p}$ and $Z \in \mathfrak{k}$ such that

$$\forall H \in \mathfrak{a}, \quad \mathrm{ad}(H)Y = \alpha(H)Z, \quad \text{and} \quad \mathrm{ad}(H)Z = \alpha(H)Y. \tag{5.8}$$

It follows that $\forall H \in \mathfrak{a}$, $\mathrm{ad}(H)^2 Y = \alpha(H)^2 Y$ and $\mathrm{ad}(H)^2 Z = \alpha(H)^2 Z$. Note that $\theta(X) = -Y + Z \in \mathfrak{g}_{-\alpha}$.

For each positive root α, let

$$\mathfrak{p}_\alpha = \{Y \in \mathfrak{p}; \ \mathrm{ad}(H)^2 Y = \alpha(H)^2 Y \text{ for } H \in \mathfrak{a}\} \tag{5.9}$$

and

$$\mathfrak{k}_\alpha = \{Z \in \mathfrak{k}; \ \mathrm{ad}(H)^2 Z = \alpha(H)^2 Z \text{ for } H \in \mathfrak{a}\}. \tag{5.10}$$

Proposition 5.5. $(a)\,\mathfrak{p} = \mathfrak{a} \oplus \sum_{\alpha>0} \mathfrak{p}_\alpha$ *and* $\mathfrak{k} = \mathfrak{m} \oplus \sum_{\alpha>0} \mathfrak{k}_\alpha$ *are direct sums whose components are mutually orthogonal under* $\langle \cdot, \cdot \rangle$, *where the summation* $\sum_{\alpha>0}$ *is taken over all the positive roots* α.

(b) $\mathfrak{g}_\alpha \oplus \mathfrak{g}_{-\alpha} = \mathfrak{p}_\alpha \oplus \mathfrak{k}_\alpha$ *for any positive root α.*

(c) For any positive root α and $H \in \mathfrak{a}_+$, the maps

$$\mathfrak{p}_\alpha \ni Y \mapsto \mathrm{ad}(H)Y \in \mathfrak{k}_\alpha \quad \text{and} \quad \mathfrak{k}_\alpha \ni Z \mapsto \mathrm{ad}(H)Z \in \mathfrak{p}_\alpha$$

are linear bijections.

Proof. Since $\mathfrak{p}_\alpha$ and $\mathfrak{k}_\alpha$ contain the orthogonal projections of $\mathfrak{g}_\alpha$ to $\mathfrak{p}$ and to $\mathfrak{k}$, respectively, by Proposition 5.3, $\mathfrak{p} = \mathfrak{a} + \sum_{\alpha>0} \mathfrak{p}_\alpha$ and $\mathfrak{k} = \mathfrak{m} + \sum_{\alpha>0} \mathfrak{k}_\alpha$, where the summations are not necessarily direct at the moment. To show that they are orthogonal direct sums, note that the summands are common eigenspaces of the commutative family of symmetric linear operators $\{\mathrm{ad}(H)^2; H \in \mathfrak{a}\}$ associated to distinct eigenvalues; hence, they are are mutually orthogonal. This proves (a). By (a), it is easy to see that $\mathfrak{p}_\alpha$ and $\mathfrak{k}_\alpha$ are, respectively, the orthogonal projections of $\mathfrak{g}_\alpha$ to $\mathfrak{p}$ and to $\mathfrak{k}$. Since $\mathfrak{g}_{-\alpha}$ has the same orthogonal projections, (b) follows. Part (c) can be easily proved using (5.8). $\qquad\square$

For any root α, let H_α be the element of $\mathfrak{a}$ representing α in the sense that $\alpha(H) = \langle H, H_\alpha \rangle$ for any $H \in \mathfrak{a}$. For $X \in \mathfrak{g}$, let $X_\mathfrak{a}$ be the orthogonal projection to $\mathfrak{a}$ of X under the inner product $\langle \cdot, \cdot \rangle$. The norm associated to this inner product will be denoted by $\| \cdot \|$.

Proposition 5.6. *If $X \in \mathfrak{g}_\alpha$ and $X = Y + Z$ with $Y \in \mathfrak{p}$ and $Z \in \mathfrak{k}$, then*

$$[X, \theta(X)] = -\|X\|^2 H_\alpha \quad \text{and} \quad [Z, Y] = \|Y\|^2 H_\alpha,$$

and $\|Y\| = \|Z\|$. Moreover, if $X' \in \mathfrak{g}_\beta$ and $X' = Y' + Z'$ with $Y' \in \mathfrak{p}$ and $Z' \in \mathfrak{k}$, then

$$[X, X']_\mathfrak{a} = 0 \text{ if } \alpha + \beta \neq 0 \quad \text{and} \quad [Z, Y']_\mathfrak{a} = \langle Z, Z' \rangle H_\beta.$$

Note that $\langle Z, Z' \rangle = 0$ if $\alpha \neq \pm\beta$.

Proof. $[X, \theta(X)] = [Y + Z, -Y + Z] = 2[Y, Z] \in \mathfrak{p}$. This combined with $[\mathfrak{g}_\alpha, \mathfrak{g}_{-\alpha}] \subset \mathfrak{g}_0$ implies that $[X, \theta(X)] \in \mathfrak{a}$. We have, for $H \in \mathfrak{a}$ and $X' \in \mathfrak{g}_\beta$,

$$\langle [X, X'], H \rangle = B([X, X'], H) = -B(X, [H, X'])$$
$$= -\beta(H)B(X, X') = \beta(H)\langle X, \theta(X') \rangle.$$

If $\beta = -\alpha$, then we may take $X' = \theta(X)$. This yields $\langle [X, \theta(X)], H \rangle = -\alpha(H)\langle X, X \rangle$, which implies $[X, \theta(X)] = -\langle X, X \rangle H_\alpha$. Since $\theta(X') \in \mathfrak{g}_{-\beta}$, $\langle X, \theta(X') \rangle = 0$ if $\beta \neq -\alpha$. This implies $[X, X']_\mathfrak{a} = 0$ if $\beta \neq -\alpha$. Choose

$H \in \mathfrak{a}$ with $\alpha(H) \neq 0$. Then

$$\langle Y, Y \rangle = B(Y, Y) = \frac{1}{\alpha(H)} B([H, Z], Y)$$

$$= -\frac{1}{\alpha(H)} B(Z, [H, Y]) = -B(Z, Z) = \langle Z, Z \rangle$$

and

$$[Z, Y] = \frac{1}{2}[Z - Y, Z + Y] = -\frac{1}{2}[X, \theta(X)] = \frac{1}{2}\|X\|^2 H_\alpha$$

$$= \frac{1}{2}(\|Y\|^2 + \|Z\|^2)H_\alpha = \|Y\|^2 H_\alpha.$$

For any $H \in \mathfrak{a}$,

$$\langle [Z, Y'], H \rangle = B([Z, Y'], H) = -B(Z, [H, Y'])$$

$$= -\beta(H)B(Z, Z') = \beta(H)\langle Z, Z' \rangle.$$

This proves $[Z, Y']_\mathfrak{a} = \langle Z, Z' \rangle H_\beta$. $\qquad\square$

Define

$$\mathfrak{n}^+ = \sum_{\alpha > 0} \mathfrak{g}_\alpha \quad \text{and} \quad \mathfrak{n}^- = \sum_{\alpha > 0} \mathfrak{g}_{-\alpha}. \qquad (5.11)$$

Because of (5.5), both $\mathfrak{n}^+$ and $\mathfrak{n}^-$ are Lie subalgebras of $\mathfrak{g}$. A Lie algebra $\mathfrak{n}$ is called nilpotent if $\exists$ integer $k > 0$ such that $\text{ad}(X)^k = 0$ for any $X \in \mathfrak{n}$. A Lie group is called nilpotent if its Lie algebra is nilpotent. It is easy to show that both $\mathfrak{n}^+$ and $\mathfrak{n}^-$ are nilpotent. Note that, to show that a subalgebra $\mathfrak{n}$ of $\mathfrak{g}$ is nilpotent, one should check $[\text{ad}_\mathfrak{n}(Y)]^k = 0$ for any $Y \in \mathfrak{n}$ and for some integer $k > 0$, instead of $\text{ad}(Y)^k = 0$, where $\text{ad}_\mathfrak{n}$ is the restriction of ad on $\mathfrak{n}$. However, by (5.5) and Proposition 5.3(a), the stronger condition $\text{ad}(Y)^k = 0$ holds for $\mathfrak{n} = \mathfrak{n}^+$ or $\mathfrak{n}^-$.

Let N^+ and N^- be, respectively, the (connected) Lie subgroups of G generated by $\mathfrak{n}^+$ and $\mathfrak{n}^-$.

Proposition 5.7. *The exponential maps* exp: $\mathfrak{n}^+ \to N^+$ *and* exp: $\mathfrak{n}^- \to N^-$ *are diffeomorphisms.*

Proof. By [26, chapter VI, corollary 4.4], the exponential map on a connected nilpotent Lie group is regular and onto. If $X, Y \in \mathfrak{n}^+$ with $e^X = e^Y$, then $e^{\text{ad}(X)} = \text{Ad}(e^X) = \text{Ad}(e^Y) = e^{\text{ad}(Y)}$. By [26, chapter VI, lemma 4.5], this implies that $\text{ad}(X) = \text{ad}(Y)$; hence, $X - Y$ belongs to the center of $\mathfrak{g}$, which is trivial, so we must have $X = Y$. This proves that exp is one-to-one on N^+. The same holds for exp on N^-.   □

By Proposition 5.7, for $n \in N^+$ (resp. $n \in N^-$), $\log n$ can be defined to be the unique element $Y \in \mathfrak{n}^+$ (resp. $Y \in \mathfrak{n}^-$) such that $n = e^Y$. By Proposition 5.3(c),

$$\theta(\mathfrak{n}^+) = \mathfrak{n}^- \quad \text{and} \quad \Theta(N^+) = N^-. \tag{5.12}$$

The following result follows directly from [26, chapter VI, corollary 4.4] mentioned earlier and Proposition 5.7 because a Lie subgroup of a nilpotent Lie group is also nilpotent.

Corollary 5.1. *Let N be a connected Lie subgroup of N^+ (resp. N^-) with Lie algebra $\mathfrak{n}$. Then $\exp: \mathfrak{n} \to N$ is a diffeomorphism. Consequently, N is a closed subgroup of N^+ (resp. N^-).*

Proposition 5.8. *Both A and M normalize $\mathfrak{n}^+$, $\mathfrak{n}^-$, N^+, and N^-.*

Proof. The conclusion follows easily from Proposition 5.2.   □

Proposition 5.9. *Let N be a connected Lie subgroup of N^+ or N^-, and let $\mathfrak{n}$ be its Lie algebra. Suppose N_1 and N_2 are two connected Lie subgroups of N with Lie algebras $\mathfrak{n}_1$ and $\mathfrak{n}_2$, respectively, and $\mathfrak{n} = \mathfrak{n}_1 \oplus \mathfrak{n}_2$ is a direct sum. Then the map $f: N_1 \times N_2 \ni (x, y) \mapsto xy \in N$ is a diffeomorphism under either of the following two conditions:*

    (i) either $\mathfrak{n}_1$ or $\mathfrak{n}_2$ is an ideal of $\mathfrak{n}$;
    (ii) both $\mathfrak{n}_1$ and $\mathfrak{n}_2$ are direct sums of some root spaces $\mathfrak{g}_\alpha$.

Proof. The conclusion can be derived by lemma 1.1.4.1 in Warner [61], but a direct proof is provided here. The Lie groups N, N_1, and N_2 are nilpotent; hence, their exponential maps are diffeomorphisms. The direct sum $\mathfrak{n} = \mathfrak{n}_1 \oplus \mathfrak{n}_2$ implies that f is regular at (e, e). Using the left translation on N_1 and the right translation on N_2, it is easy to show that f is regular on $N_1 \times N_2$. It is also one-to-one because by Corollary 5.1, $N_1 \cap N_2 = \exp(\mathfrak{n}_1) \cap \exp(\mathfrak{n}_2) = \exp(\mathfrak{n}_1 \cap \mathfrak{n}_2) = \{e\}$.

It remains to show that f is onto. Assume condition (i). Without loss of generality, we may assume that $\mathfrak{n}_1$ is an ideal of $\mathfrak{n}$. Then N_1 is a normal subgroup of N and $N_1 N_2 = f(N_1 \times N_2)$ is a connected subgroup of N. Using the regularity of the map f, it is easy to show that $N_1 N_2$ is a connected Lie subgroup of N. Since its Lie algebra is equal to $\mathfrak{n}_1 \oplus \mathfrak{n}_2 = \mathfrak{n}$, it follows that $N_1 N_2 = N$.

We now consider the condition (ii). We may assume that $\mathfrak{n} \subset \mathfrak{n}^+$. Let F, F_1, and F_2 be, respectively, the collections of positive roots such that

$$\mathfrak{n} = \sum_{\alpha \in F} \mathfrak{g}_\alpha, \quad \mathfrak{n}_1 = \sum_{\alpha \in F_1} \mathfrak{g}_\alpha, \quad \text{and} \quad \mathfrak{n}_2 = \sum_{\alpha \in F_2} \mathfrak{g}_\alpha.$$

For any two distinct positive roots α and β, $\alpha = \beta$ determines a proper subspace of $\mathfrak{a}$; hence, there is $H \in \mathfrak{a}_+$ such that $\alpha(H) \neq \beta(H)$ for any two distinct positive roots. We now introduce an order on the set of all the positive roots by setting $\alpha < \beta$ if $\alpha(H) < \beta(H)$, and let $\alpha_1 < \alpha_2 < \cdots < \alpha_f$ be the ordered set of all the roots in F. By [26, chapter VII, corollary 2.17], the only multiple of a positive root α that is also a positive root is either 2α or $(1/2)\alpha$. Note that $\mathfrak{n} \ominus \mathfrak{g}_{\alpha_1}$, the sum of $\mathfrak{g}_{\alpha_k}$ for $2 \leq k \leq f$, is an ideal of $\mathfrak{n}$. If $[\mathfrak{g}_{\alpha_1}, \mathfrak{g}_{\alpha_1}] = 0$, then $\mathfrak{g}_{\alpha_1}$ is an abelian Lie algebra. Otherwise, if $[\mathfrak{g}_{\alpha_1}, \mathfrak{g}_{\alpha_1}] \neq 0$, then $2\alpha_2$ must be contained in F and $\mathfrak{g}_{\alpha_1} \oplus \mathfrak{g}_{2\alpha_1}$ is a Lie algebra. Moreover, if β and γ are any two roots in F, not both equal to α_1, then $\beta + \gamma > 2\alpha_1$; hence, $\mathfrak{n} \ominus (\mathfrak{g}_{\alpha_1} \oplus \mathfrak{g}_{2\alpha_1})$ is an ideal of $\mathfrak{n}$.

Suppose $\alpha_1 \in F_1$. Let $\mathfrak{n}_{\alpha_1} = \mathfrak{g}_{\alpha_1}$ if $[\mathfrak{g}_{\alpha_1}, \mathfrak{g}_{\alpha_1}] = 0$. Otherwise, let $\mathfrak{n}_{\alpha_1} = \mathfrak{g}_{\alpha_1} \oplus \mathfrak{g}_{2\alpha_1}$. Note that, in the latter case, $2\alpha_1 \in F_1$. In either case, let $\mathfrak{n}' = \mathfrak{n} \ominus \mathfrak{n}_{\alpha_1}$. Then $\mathfrak{n}'$ is an ideal of $\mathfrak{n}$ and $\mathfrak{n} = \mathfrak{n}_{\alpha_1} \oplus \mathfrak{n}'$ is a direct sum. Applying the conclusion proved under the condition (i), we obtain $N = N_{\alpha_1} N'$, where N_{α_1} and N' are, respectively, connected Lie subgroups of N with Lie algebras $\mathfrak{n}_{\alpha_1}$ and $\mathfrak{n}'$. If $\alpha_1 \in F_2$, then we have that $N = N' N_{\alpha_1}$.

Repeating the same argument with N' in place of N and continuing in this fashion, we can show that

$$N = N^{(1)} N^{(2)} \cdots N^{(p)} N^{(p+1)} \cdots N^{(q)},$$

where $N^{(1)}, \ldots, N^{(p)}$ are Lie subgroups of N_1 and $N^{(p+1)}, \ldots, N^{(q)}$ are Lie subgroups of N_2. This proves that $N \subset N_1 N_2$ and hence the map f is onto. $\qquad\square$

Recall that M given by (5.7) is the centralizer of $\mathfrak{a}$ in K with Lie algebra $\mathfrak{m}$. Let M' be the normalizer of $\mathfrak{a}$ in K; that is,

$$M' = \{k \in K; \quad \mathrm{Ad}(k)\mathfrak{a} \subset \mathfrak{a}\}. \tag{5.13}$$

It is clear that M' is a closed subgroup of K and is also the normalizer of A in K.

Proposition 5.10. *(a) M and M' have the same Lie algebra* $\mathfrak{m}$.

    (b) M is a closed normal subgroup of M' and the quotient group M'/M is finite.

    (c) If the center Z of G is finite, then M and M' are compact.

Proof. Let Y belong to the Lie algebra $\mathfrak{m}'$ of M'. Then for any $H \in \mathfrak{a}$, $[Y, H] = (d/dt)\mathrm{Ad}(e^{tY})H \mid_{t=0} \in \mathfrak{a}$ and hence $\mathrm{ad}(H)^2 Y = 0$. It follows that $B(\mathrm{ad}(H)Y, \mathrm{ad}(H)Y) = -B(\mathrm{ad}(H)^2 Y, Y) = 0$, $[Y, H] = 0$ and $Y \in \mathfrak{m}$. This proves (a). For $m \in M$, $m' \in M'$, and $H \in \mathfrak{a}$,

$$\mathrm{Ad}(m'mm'^{-1})H = \mathrm{Ad}(m')\mathrm{Ad}(m)\mathrm{Ad}(m'^{-1})H = \mathrm{Ad}(m')\mathrm{Ad}(m'^{-1})H = H.$$

It follows that $m'mm'^{-1} \in M$ and M is a normal subgroup of M'. The center Z of G is a closed normal subgroup of G. The quotient group $\tilde{G} = G/Z$ has a trivial center and is also a semi-simple Lie group of noncompact type. If G is replaced by $\tilde{G}$, then K, M, and M' should be replaced by $\tilde{K} = K/Z$, $\tilde{M} = M/Z$, and $\tilde{M}' = M'/Z$, respectively. By Theorem 5.1, $\tilde{K}$ is compact, so as its closed subgroups, both $\tilde{M}$ and $\tilde{M}'$ are compact. Because $\tilde{M}$ and $\tilde{M}'$ have the same Lie algebra, they have the same identity component $\tilde{M}_0$, which is both open and closed in $\tilde{M}'$. The compactness of $\tilde{M}'$ implies that $\tilde{M}'/\tilde{M}_0$ is finite, and so is $\tilde{M}'/\tilde{M}$. Since $M'/M = \tilde{M}'/\tilde{M}$, (b) is proved. Part (c) follows directly from Theorem 5.1 (b). $\qquad\square$

The finite group $W = M'/M$ is called the Weyl group. For $s = m_s M \in W$, $\mathrm{Ad}(m_s)$: $\mathfrak{a} \to \mathfrak{a}$ is a linear map that does not depend on the choice of $m_s \in M'$ to represent s; therefore, $W \ni s \mapsto \mathrm{Ad}(m_s) \in GL(\mathfrak{a})$ is a faithful representation of W on $\mathfrak{a}$, where $GL(\mathfrak{a})$ is the group of the linear automorphisms on $\mathfrak{a}$. We may regard $s \in W$ as the linear map $\mathrm{Ad}(m_s)$: $\mathfrak{a} \to \mathfrak{a}$ and W as a group of linear transformations on $\mathfrak{a}$.

Let Δ be the set of all roots and let e_W be the identity element of W.

Proposition 5.11.

    (a) W permutes the Weyl chambers and is simply transitive on the set of Weyl chambers in the sense that $\forall$ two Weyl chambers C_1 and C_2, $\exists s \in W$ such that $s(C_1) = C_2$, and if $s \neq e_W$ (identity element of W), then $\forall$ Weyl chamber C, $s(C) \neq C$.

    (b) For any $H \in \overline{\mathfrak{a}_+}$ (the closure of $\mathfrak{a}_+$), the orbit $\{sH; s \in W\}$ intersects $\overline{\mathfrak{a}_+}$ only at H.

    (c) For $s \in W$ and $\alpha \in \Delta$, $\alpha \circ s \in \Delta$ and if $s \neq e_W$, then, for some $\alpha \in \Delta$, $\alpha \circ s \neq \alpha$.

    (d) For $s \in W$ and $\alpha \in \Delta$, $\mathrm{Ad}(m_s)\mathfrak{g}_\alpha = \mathfrak{g}_{\alpha \circ s^{-1}}$.

    (e) Let $s \in W$. If $s(\mathfrak{a}_+) = -\mathfrak{a}_+$, then $\mathrm{Ad}(m_s)\mathfrak{n}^+ = \mathfrak{n}^-$.

Proof. The reader is referred to [26, chapter VII, theorems 2.12 and 2.22] for the proofs of (a) and (b). For $H \in \mathfrak{a}$ and $X \in \mathfrak{g}_\alpha$,

$$[H, \mathrm{Ad}(m_s)X] = \mathrm{Ad}(m_s)[\mathrm{Ad}(m_s^{-1})H, X] = \alpha(\mathrm{Ad}(m_s^{-1})H)\mathrm{Ad}(m_s)X$$
$$= \alpha(s^{-1}(H))\mathrm{Ad}(m_s)X.$$

It follows that $\alpha \circ s^{-1} \in \Delta$ and $\mathrm{Ad}(m_s)\mathfrak{g}_\alpha = \mathfrak{g}_{\alpha \circ s^{-1}}$. If $\alpha \circ s = \alpha$ for any $\alpha \in \Delta$, then $\alpha(\mathrm{Ad}(m_s)(H) - H) = 0$ for any $H \in \mathfrak{a}$ and $\alpha \in \Delta$. This implies that $\exp[\mathrm{Ad}(m_s)(H) - H]$ belongs to the center Z of G. Because $Z \subset K$ and $\mathrm{Ad}(m_s)(H) - H \in \mathfrak{a}$, it follows that $\mathrm{Ad}(m_s)H = H$ for any $H \in \mathfrak{a}$; hence, $m_s \in M$ and $s = e_W$. This proves (c) and (d). If s maps $\mathfrak{a}_+$ into $-\mathfrak{a}_+$, then so does s^{-1}. In this case, if α is a positive root, then $\alpha \circ s^{-1}$ is a negative root. By (d), $\mathrm{Ad}(m_s)\mathfrak{n}^+ = \sum_{\alpha>0}\mathrm{Ad}(m_s)\mathfrak{g}_\alpha = \sum_{\alpha>0}\mathfrak{g}_{-\alpha} = \mathfrak{n}^-$. This proves (e). $\qquad \square$

A positive root α is called simple if it is not the sum of two positive roots. Let $\Sigma = \{\beta_1, \beta_2, \ldots, \beta_l\}$ be the set of all the simple roots. The following result is a direct consequence theorem 2.19 of Helgason [26, chapter VII].

Proposition 5.12. *The number l of simple roots is equal to* $\dim(\mathfrak{a})$ *and any positive root can be written as* $\alpha = \sum_{i=1}^{l} c_i \beta_i$, *where the coefficients c_i are nonnegative integers.*

We may identify $\mathfrak{a}$ with $\mathbb{R}^l$. Recall that the Weyl group W may be regarded as a group of linear transformations on $\mathfrak{a}$. The following result is a direct consequence of corollary 2.13 and lemma 2.21 in Helgason [26, chapter VII]. Let Δ_+ denote the set of positive roots. For a root α, the reflection s_α about the hyperplane $\alpha = 0$ in $\mathfrak{a}$, with respect to the inner product $\langle \cdot, \cdot \rangle$, is a linear map $\mathfrak{a} \to \mathfrak{a}$ given by

$$s_\alpha(H) = H - 2\frac{\alpha(H)}{\alpha(H_\alpha)}H_\alpha, \quad H \in \mathfrak{a},$$

where H_α is the element of $\mathfrak{a}$ representing α; that is, $\alpha(H) = \langle H, H_\alpha \rangle$ for $H \in \mathfrak{a}$.

Proposition 5.13. *(a) The Weyl group W is generated by* $\{s_\alpha; \alpha \in \Delta_+\}$.

(b) Let s_i be the reflection in $\mathfrak{a}$ about the hyperplane $\beta_i = 0$, where β_i is a simple root. Then s_i permutes all the roots in Δ_+ that are not proportional to β_i; that is, the map $\alpha \mapsto \alpha \circ s_i$ permutes all the roots $\alpha \in \Delta_+$ not proportional to β_i.

5.3. Three Decompositions

We will continue to use the notation introduced in the previous section. Let $A_+ = \exp(\mathfrak{a}_+)$. Because the exponential map is a diffeomorphism from $\mathfrak{a}$ onto $A = \exp(\mathfrak{a})$, $\overline{A_+} = \exp(\overline{\mathfrak{a}_+})$, where the overline denotes the closure.

Theorem 5.2. *Any $g \in G$ can be written as $g = \xi a^+ \eta$, where $\xi, \eta \in K$ and $a^+ \in \overline{A_+}$. Moreover, a^+ is uniquely determined by g and when $a^+ \in A_+$, all the possible choices for (ξ, η) are given by $(\xi m, m^{-1}\eta)$ for $m \in M$.*

The decomposition given in Theorem 5.2 is called the Cartan decomposition of G and may be written as $G = K \overline{A_+} K$. We note that it does not exactly correspond to the Cartan decomposition $\mathfrak{g} = \mathfrak{k} \oplus \mathfrak{p}$ of $\mathfrak{g}$.

See [26, chapter IX, theorem 1.1] for the proof of the decomposition $g = \xi a^+ \eta$ and the uniqueness of a^+. To show that all the possible choices for (ξ, η) are given by $(\xi m, m^{-1}\eta)$ when $a^+ \in A_+$, it suffices to prove that if $\xi a^+ \eta = a^+$, then $\xi = \eta^{-1} \in M$. Since $a^+ = \xi a^+ \eta = (\xi\eta)\eta^{-1}a^+\eta = (\xi\eta)e^{\mathrm{Ad}(\eta^{-1})\log a^+}$, by Theorem 5.1 (c), $\xi\eta = e$ and $\mathrm{Ad}(\eta^{-1})\log a^+ = \log a^+$. By the following proposition, $\eta^{-1} \in M$. The claim is proved.

Proposition 5.14. *Let $H \in \mathfrak{a}_+$ and $k \in K$. If $\mathrm{Ad}(k)H = H$ (resp. $\mathrm{Ad}(k)H \in \mathfrak{a}$), then $k \in M$ (resp. $k \in M'$).*

Proof. Suppose $\mathrm{Ad}(k)H \in \mathfrak{a}$. Let $H' = \mathrm{Ad}(k)H$. For $X \in \mathfrak{g}_0$, let $\mathrm{Ad}(k^{-1})X = X_0 + \sum_{\beta \neq 0} X_\beta$ with $X_0 \in \mathfrak{g}_0$ and $X_\beta \in \mathfrak{g}_\beta$. Because

$$X_0 = \mathrm{Ad}(k^{-1})[H', X] = [H, \mathrm{Ad}(k^{-1})X] = \sum_\beta \beta(H)X_\beta$$

and $\beta(H) \neq 0$, we obtain $\mathrm{Ad}(k^{-1})\mathfrak{g}_0 \subset \mathfrak{g}_0$. Since $\mathrm{Ad}(k^{-1})$ is a linear automorphism on $\mathfrak{g}$, we may conclude that $\mathrm{Ad}(k)$ leaves $\mathfrak{g}_0$ and hence $\mathfrak{a}$ invariant. Therefore, $k \in M'$. If $H' = H$, by Proposition 5.11(a), $k \in M$. $\square$

A subset of a manifold will be said to have a lower dimension if it is contained in the union of finitely many submanifolds of lower dimension.

Proposition 5.15. *We have $\mathfrak{p} = \mathrm{Ad}(K)\overline{\mathfrak{a}_+}$. In particular, $\mathfrak{p} = \mathrm{Ad}(K)\mathfrak{a}$. Moreover, $\mathrm{Ad}(K)\mathfrak{a}_+$ is an open subset of $\mathfrak{p}$ whose complement has a lower dimension.*

Proof. For $Y \in \mathfrak{p}$, let $e^Y = ke^H h$ be the Cartan decomposition with $k, h \in K$ and $H \in \overline{\mathfrak{a}}_+$. Then $e^Y = khe^{\mathrm{Ad}(h^{-1})H}$. By Theorem 5.1(c), $Y = \mathrm{Ad}(h^{-1})H$. This proves $\mathfrak{p} = \mathrm{Ad}(K)\overline{\mathfrak{a}_+}$.

Consider the map $\Phi \colon K \times \mathfrak{a}_+ \to \mathfrak{p}$ given by $\Phi(k, H) = \mathrm{Ad}(k)H$. For any positive root α and $Z \in \mathfrak{k}_\alpha$, $(d/dt)\mathrm{Ad}(e^{tZ})H \mid_{t=0} = [Z, H] = -\mathrm{ad}(H)Z$. Because $\mathrm{ad}(H) \colon \mathfrak{k}_\alpha \to \mathfrak{p}_\alpha$ is a linear isomorphism and $\mathfrak{p} = \alpha \oplus \sum_{\alpha>0} \mathfrak{p}_\alpha$, it follows that $D\Phi \colon T_{e,H}(K \times \mathfrak{a}_+) \to T_H \mathfrak{p}$ is a surjection for any $H \in \mathfrak{a}_+$. Using the left translation on K, we see that $D\Phi \colon T_{k,H}(K \times \mathfrak{a}_+) \to T_{\mathrm{Ad}(k)H}\mathfrak{p}$ is a surjection at every point $(k, H) \in K \times \mathfrak{a}_+$. Therefore, Φ is a submersion from $K \times \mathfrak{a}_+$ into $\mathfrak{p}$. It follows that its image, $\mathrm{Ad}(K)\mathfrak{a}_+$, is an open subset of $\mathfrak{p}$.

For a nonempty set E of positive roots, let $\mathfrak{a}_1 = \{H \in \mathfrak{a}; \alpha(H) = 0 \text{ for } \alpha \in E\}$ and let $\mathfrak{a}_1^+ = \{H \in \mathfrak{a}_1; \beta(H) > 0 \text{ for } \beta \in E^c\}$, where E^c is the set of the positive roots not contained in E. Since the boundary of $\overline{\mathfrak{a}_+}$ is contained in the union of finitely many sets of the form $\mathfrak{a}_1^+$, it suffices to show that $\mathrm{Ad}(K)\mathfrak{a}_1^+$ is a submanifold of $\mathfrak{p}$ of lower dimension. Let M_1 be the centralizer of $\mathfrak{a}_1$ in K. It is easy to see that its Lie algebra is equal to $\mathfrak{m}_1 = \mathfrak{m} \oplus \sum_{\alpha \in E} \mathfrak{k}_\alpha$. Consider the map $\Phi_1 \colon (K/M_1) \times \mathfrak{a}_1^+ \to \mathfrak{p}$ given by $\Phi_1(\bar{k}, H) = \mathrm{Ad}(k)H$, where $\bar{k} = kM_1$. The tangent space $T_{\bar{e}}(K/M_1)$ may be naturally identified with $\sum_{\beta \in E^c} \mathfrak{k}_\beta$. Since for any $Z \in \mathfrak{k}_\beta$ with $\beta \in E^c$ and $H \in \mathfrak{a}_1^+$, $(d/dt)\mathrm{Ad}(e^{tZ})H \mid_{t=0} = [Z, H] \neq 0$, it follows that $D\Phi_1$ is injective at $(\bar{e}, H)$ and its image has a dimension equal to $\dim(\sum_{\beta \in E^c} \mathfrak{k}_\beta) + \dim(\mathfrak{a}_1)$, which is less than $\dim(\mathfrak{p})$. Using the left translation of K, the same holds at every point of $(K/M_1) \times \mathfrak{a}_1^+$. This proves that the image of Φ_1, which is $\mathrm{Ad}(K)\mathfrak{a}_1^+$, is a submanifold of $\mathfrak{p}$ of a lower dimension. $\square$

Recall that G/K is a symmetric space of noncompact type.

Proposition 5.16. *Let* $\pi \colon G \to G/K$ *be the natural projection. Then* $\pi(KA_+)$ *is an open subset of* G/K *with a lower dimensional complement and the map* $(K/M) \times \mathfrak{a}_+ \to \pi(KA_+)$, *given by* $(\xi M, H) \mapsto \pi(\xi e^H)$, *is a diffeomorphism.*

Proof. Any element $k_1 a k_2$ in $K A_+ K$ may be written as ke^Y for $k = k_1 k_2 \in K$ and $Y = \mathrm{Ad}(k_2^{-1})\log a \in \mathfrak{p}$. By Theorem 5.1(c) and Proposition 5.15, KA_+K is an open subset of G with a lower dimensional complement. It follows that $\pi(KA_+)$ is such a subset of G/K. We want to show that the map $(K/M) \times A_+ \ni (\xi M, a^+) \mapsto \pi(\xi a^+) \in \pi(KA_+)$ is a diffeomorphism. This map is obviously onto. By Theorem 5.2, it is also one-to-one. To show that it is diffeomorphic, first verify that it is regular at (eM, a) for any $a \in A_+$, then using the left translation on K to establish the regularity at every point in $(K/M) \times A_+$. $\square$

The identification of a point $x = \pi(\xi e^H)$ in $\pi(KA_+)$ with $(\xi M, H)$ in $(K/M) \times \mathfrak{a}_+$ via the diffeomorphism in Proposition 5.16 is called the polar decomposition on G/K with ξM and H being called respectively "angular" and "radial" components of x.

Theorem 5.3. *The map* $(k, a, n) \mapsto g = kan$ *is a diffeomorphism from* $K \times A \times N^+$ *onto* G.

The decomposition of G given in Theorem 5.3 is called the Iwasawa decomposition and will be denoted simply as $G = KAN^+$. The corresponding direct sum decomposition at the Lie algebra level, namely, $\mathfrak{g} = \mathfrak{k} \oplus \mathfrak{a} \oplus \mathfrak{n}^+$, can be easily verified. It is clear that the map $K \times A \times N^+ \ni (k, a, n) \mapsto kan \in G$ is smooth and is regular at (e, e, e). Using the left and right translations, and the fact that N^+ is normalized by A, it is easy to show that the map is also regular at (k, a, n) for any $k \in K$, $a \in A$, and $n \in N^+$. To prove Theorem 5.3, it remains to show that the map is a bijection. The reader is referred to [26, chapter VI, theorem 5.1] for the complete proof.

There are other versions of the Iwasawa decomposition, for example, $G = KAN^-$, $G = N^+AK$, and $G = N^-AK$. These different versions of the Iwasawa decomposition can be proved easily by applying either the Cartan involution Θ or the inverse operation to the decomposition $G = KAN^+$. Because A normalizes N^+ and N^-, we also have $G = KN^+A = KN^-A = AN^+K = AN^-K$. Note that, as a consequence of the Iwasawa decomposition, A, N^+, and N^- are closed subgroups of G.

Let $H_1, H_2, \ldots, H_n$ be the Lie subgroups of a Lie group H. If $H_1 H_2 \cdots H_n$ is an embedded submanifold of H and if the map

$$H_1 \times H_2 \times \cdots \times H_n \ni (h_1, h_2, \ldots, h_n) \mapsto h_1 h_2 \cdots h_n \in H_1 H_2 \cdots H_n$$

is a diffeomorphism, then we will say that $H_1 H_2 \cdots H_n$ is a diffeomorphic product. The various versions of the Iwasawa decomposition are all diffeomorphic products.

Note that Lie subgroups do not in general form diffeomorphic products even if their Lie algebras form a direct sum, and a diffeomorphic product is not necessarily a group. The following proposition provides some special examples when these are true. It can be proved easily by Proposition 5.8 and the Iwasawa decompositions.

Proposition 5.17. $AN^+ = N^+A$, $AN^- = N^-A$,

$$MAN^+ = MN^+A = AMN^+ = AN^+M = N^+AM = N^+MA,$$

and

$$MAN^- = MN^-A = AMN^- = AN^-M = N^-AM = N^-MA$$

are diffeomorphic products and are closed subgroups of G.

The following result, which will be needed later, is a direct consequence of Proposition 5.9 and Proposition 5.11 (d).

Proposition 5.18. *Let $s \in W$. Then $m_s N^- m_s^{-1} = N_1 N_2$ is a diffeomorphic product, where N_1 and N_2 are respectively the Lie subgroups of N^- and N^+ generated by the following Lie algebras:*

$$\mathfrak{n}_1 = \sum_{\alpha>0,\, \alpha \circ s^{-1}>0} \mathfrak{g}_{-\alpha \circ s^{-1}} \quad \text{and} \quad \mathfrak{n}_2 = \sum_{\alpha>0,\, \alpha \circ s^{-1}<0} \mathfrak{g}_{-\alpha \circ s^{-1}}. \qquad (5.14)$$

The same holds for $m_s N^+ m_s^{-1}$ with $-\alpha \circ s^{-1}$ replaced by $\alpha \circ s^{-1}$.

Theorem 5.4.

$$G = \bigcup_{s \in W} N^- m_s MAN^+ \qquad (5.15)$$

is a disjoint union. Moreover, $N^- MAN^+$ is a diffeomorphic product and is an open subset of G and, for $s \neq e_W$, $N^- m_s MAN^+$ is a lower dimensional submanifold of G. Consequently, $N^- MAN^+$ has a lower dimensional complement in G.

The disjoint union $G = \bigcup_{s \in W} N^- m_s MAN^+$ given in (5.15) is called the Bruhat decomposition. Note that the set $N^- m_s MAN^+$ is determined by $s \in W$ and does not depend on the choice of $m_s \in M'$ to represent s. The Bruhat decomposition can be expressed in different forms. For example, the factors in the product MAN^+ can be permuted arbitrarily. Applying Θ to (5.15), we obtain the following version of Bruhat decomposition:

$$G = \bigcup_{s \in W} N^+ m_s N^- AM \quad \text{(disjoint union)}. \qquad (5.16)$$

For the proof of Theorem 5.4, note that, by Proposition 5.11 (a), there is $s^* \in W$ such that $s^*(\mathfrak{a}_+) = -\mathfrak{a}_+$. Then $s^* = (s^*)^{-1}$, $\mathrm{Ad}(m_{s^*}^{-1})\mathfrak{n}^\pm = \mathfrak{n}^\mp$, and $m_{s^*}^{-1} N^\pm m_{s^*} = N^\mp$. Because A and M normalize N^-, and they are normalized by m_s for any $s \in W$, the collection of subsets $N^- m_s M A N^+$ for $s \in W$ is identical to the collection of subsets $N^- m_{s^*} m_s M A N^+ = m_{s^*} N^+ m_s M A N^+$ for $s \in W$. Let $B = M A N^+$. Since $G = m_{s^*} G$, except for the statement that $N^- M A N^+$ is a diffeomorphic product, Theorem 5.4 is equivalent to saying that $G = \cup_{s \in W} B m_s B$ is a disjoint union, $B m_{s^*} B = m_{s^*} N^- N^+ M A$ is an open subset of G, and $B m_s B = m_{s^*} N^- m_{s^*}^{-1} m_s M A N^+$ is a lower dimensional submanifold of G for $s \neq s^*$. This is precisely corollary 18 in [26, chapter IX]. The claim about $N^- M A N^+$ being a diffeomorphic product follows from the Iwasawa decomposition $G = K A N^+$ and corollary 19 in [26, chapter IX], which says that the map $N^- \to G/(M A N^+)$ given by the natural projection $G \to G/(M A N^+)$ is a diffeomorphism.

Let Q be a closed subgroup of G and let $Q \backslash G$ be the space of right cosets Qg, $g \in G$. As for G/Q, there is also a unique manifold structure on $Q \backslash G$ under which the natural right action of G on $Q \backslash G$, defined by $(Q \backslash G) \times G \ni (Qx, g) \mapsto Qxg \in Q \backslash G$, is smooth. Moreover, the natural projection $G \ni g \mapsto Qg \in Q \backslash G$ is an open and smooth map.

Proposition 5.19. *Let Q be a closed subgroup of G containing $A N^+$ (resp. $A N^-$) and let $\pi_{G/Q} \colon G \to G/Q$ (resp. $\pi_{Q \backslash G} \colon G \to Q \backslash G$) be the natural projection. Then $\pi_{G/Q}(N^- M)$ (resp. $\pi_{Q \backslash G}(N^+ M)$) is an open subset of G/Q (resp. $Q \backslash G$) with a lower dimensional complement.*

Proof. We will only prove the conclusion for $\pi_{G/Q}$. The claim for $\pi_{Q \backslash G}$ can be proved by a similar argument. Note that $\pi_{G/Q}(N^- M) = \pi_{G/Q}(N^- M A N^+)$ is an open subset of G/Q because $\pi_{G/Q}$ is an open map. Note also that the complement C of $\pi_{G/Q}(N^- M)$ in G/Q has no interior point because

$$\pi_{G/Q}^{-1}(C) \subset \bigcup_{s \in [W - \{e_W\}]} N^- m_s M A N^+$$

is a lower dimensional subset of G. Suppose $\pi_{G/Q}(N^- m_s M)$ intersects $\pi_{G/Q}(N^- M)$ for some $s \in W$. Then $n_1 m_s Q = n_2 m Q$ for some $n_1, n_2 \in N^-$ and $m \in M$. This implies that $m_s Q = n_1^{-1} n_2 m Q$ and hence $\pi_{G/Q}(N^- m_s M) \subset \pi_{G/Q}(N^- M)$. It follows that C is equal to the union of the submanifolds $\pi_{G/Q}(N^- m_s M) = \pi_{G/Q}(m_s N_s^- M)$ that do not intersect $\pi_{G/Q}(N^- M)$. Such a submanifold of G/Q is necessarily lower dimensional because C has no interior point. $\qquad\square$

Let Q be a closed subgroup of G containing $A N^+$ and let $L = Q \cap K$. Then L is a closed subgroup of K. It follows from the Iwasawa decomposition

$G = KAN^+$ that $Q = LAN^+$. Moreover, $G/Q = (KAN^+)/(LAN^+)$ can be identified with K/L via the map $G/Q \ni gQ \mapsto kL \in K/L$ with k being the K-component of g in the Iwasawa decomposition $G = KAN^+$. Note that K/L is compact if G has a finite center. The natural action of G on G/Q given by $(G/Q) \ni g'Q \mapsto gg'Q \in (G/Q)$ for $g \in G$ induces an action of G on K/L, which may be denoted by $g(kL)$. However, this notation can be confusing, especially when $Q = AN^+$ and $K/L = K$. Therefore, the action of G on K/L will also be denoted by $g * (kL)$ and will be called the $*$-action of G on K/L. We have $g * (kL) = k'L$, where k' is the K-component of gk in the Iwasawa decomposition $G = KAN^+$. Via the identification of G/Q with K/L, the open subset $\pi_{G/Q}(N^-M)$ of G/Q corresponds to

$$(N^-M) * (eL) = \{(n'm) * (eL); \ n' \in N^- \text{ and } m \in M\},$$

which is an open subset of K/L with a lower dimensional complement.

If $Q = AN^+$, then $L = \{e\}$ and the $*$-action of G on $K/\{e\}$ becomes an action of G on K. For $g \in G$ and $k \in K$, $g * k$ is the K-component of gk in the Iwasawa decomposition $G = KAN^+$. Note that if $g \in K$, then $g * k$ has the same meaning as the product gk. The set $(N^-M) * e$ is an open subset of K with a lower dimensional complement.

In the literature, the space G/Q, where Q is a closed subgroup of G containing MAN^+, is called a Furstenberg boundary of G because such spaces arise in the compactification of the symmetric space G/K (see Furstenberg [19] and Moore [45]). In particular, $G/(MAN^+)$ is called the maximal Furstenberg boundary. We will see that $G/(MAN^+)$ plays an important role in the study of the limiting properties of Lévy processes in G.

5.4. Examples

Let d be an integer ≥ 2 in this section.

Special Linear Groups

Let $SL(d, \mathbb{R})$ be the group of all the $d \times d$ real matrices of determinant 1. This is a closed subgroup of the general linear group $GL(d, \mathbb{R})$, called the special linear group on $\mathbb{R}^d$. Its Lie algebra is the space $\mathfrak{sl}(d, \mathbb{R})$ of all the $d \times d$ real matrices of trace 0. The identity component $GL(d, \mathbb{R})_+ = \{g \in GL(d, \mathbb{R}); \det(g) > 0\}$ of $GL(d, \mathbb{R})$ can be identified with the product group $\mathbb{R}_+ \times SL(d, \mathbb{R})$ via the map $\mathbb{R}_+ \times SL(d, \mathbb{R}) \ni (c, g) \mapsto cg \in GL(d, \mathbb{R})_+$, where $\mathbb{R}_+$ is the multiplicative group of positive numbers. Therefore, $SL(d, \mathbb{R})$ is a normal subgroup of $GL(d, \mathbb{R})_+$ and $\mathfrak{sl}(d, R)$ is an ideal of $\mathfrak{gl}(d, \mathbb{R})$. Let

$G = SL(d, \mathbb{R})$ and $\mathfrak{g} = \mathfrak{sl}(d, \mathbb{R})$. We will compute the Killing form $\hat{B}$ of $\mathfrak{gl}(d, \mathbb{R})$ and then obtain the Killing form B of $\mathfrak{g} = \mathfrak{sl}(d, \mathbb{R})$ as the restriction of $\hat{B}$ to $\mathfrak{g}$.

Let E_{ij} denote the matrix in $\mathfrak{gl}(d, \mathbb{R})$ that has 1 at place (i, j) and 0 elsewhere, that is, $(E_{ij})_{pq} = \delta_{ip}\delta_{jq}$. Then $\{E_{ij}; i, j = 1, 2, \ldots, d\}$ is a basis of $\mathfrak{gl}(d, \mathbb{R})$. For $X, Y \in \mathfrak{gl}(d, \mathbb{R})$, $\mathrm{ad}(X)\mathrm{ad}(Y)E_{ij} = [X, [Y, E_{ij}]] = XYE_{ij} - XE_{ij}Y - YE_{ij}X + E_{ij}YX$ and $[\mathrm{ad}(X)\mathrm{ad}(Y)E_{ij}]_{ij} = \sum_k X_{ik}Y_{ki} - X_{ii}Y_{jj} - Y_{ii}X_{jj} + \sum_k Y_{jk}X_{kj}$. Summing over (i, j), we obtain

$$\forall X, Y \in \mathfrak{gl}(d, \mathbb{R}), \quad \hat{B}(X, Y) = 2d\,\mathrm{Trace}(XY) - 2(\mathrm{Trace}\,X)(\mathrm{Trace}\,Y).$$
(5.17)

Since $\mathrm{Trace}(X) = 0$ for $X \in \mathfrak{g} = \mathfrak{sl}(d, R)$, we have

$$\forall X, Y \in \mathfrak{g}, \quad B(X, Y) = 2d\,\mathrm{Trace}(XY).$$
(5.18)

The map $\Theta\colon G \to G$ defined by $\Theta(g) = (g')^{-1}$ is a nontrivial automorphism on $G = SL(d, \mathbb{R})$ with $\Theta^2 = \mathrm{id}_G$; hence, it is a Cartan involution on G. Its differential $\theta = D\Theta$ is given by $\theta(X) = -X'$. The eigenspaces of θ associated to the eigenvalues 1 and -1 are, respectively, $\mathfrak{k} = \{X \in \mathfrak{g}; X' = -X\}$, the space of skew-symmetric matrices in $\mathfrak{g}$ that is usually denoted by $o(d)$, and $\mathfrak{p} = \{X \in \mathfrak{g}; X' = X\}$, the space of symmetric matrices in $\mathfrak{g}$. By (5.18), it is easy to see that B is negative definite on $\mathfrak{k}$ and positive definite on $\mathfrak{p}$. Therefore, $G = SL(d, \mathbb{R})$ is a semi-simple Lie group of noncompact type.

A $d \times d$ real matrix g satisfying $g' = g^{-1}$ is called orthogonal. The set of such matrices form a compact group $O(d)$, called the orthogonal group on $\mathbb{R}^d$. The subset K of $G = SL(d, \mathbb{R})$ fixed by Θ is the group of $d \times d$ orthogonal matrices of determinant 1, called the special orthogonal group or the rotation group on $\mathbb{R}^d$, and is denoted by $SO(d)$. Note that $SO(d)$ is the identity component of $O(d)$, and they have the same Lie algebra $\mathfrak{k} = o(d)$.

Let $\mathfrak{a}$ be the subspace of $\mathfrak{p}$ consisting of all the diagonal matrices of trace 0. This is a maximal abelian subspace of $\mathfrak{p}$. The Lie subgroup A of G generated by $\mathfrak{a}$ consists of all the diagonal matrices with positive diagonal elements and of determinant one. The roots are given by α_{ij} for distinct $i, j \in \{1, \ldots, d\}$, defined by $\alpha_{ij}(H) = H_i - H_j$ for $H = \mathrm{diag}\{H_1, \ldots, H_d\} \in \mathfrak{a}$. The root space $\mathfrak{g}_{\alpha_{ij}}$ is one dimensional and is spanned by E_{ij}. We may let

$$\mathfrak{a}_+ = \{\mathrm{diag}(H_1, \ldots, H_d) \in \mathfrak{a};\ H_1 > H_2 > \cdots > H_d\}$$

be the chosen Weyl chamber. Then the positive roots are given by α_{ij} for $1 \le i < j \le d$. The nilpotent Lie algebra $\mathfrak{n}^+$ (resp. $\mathfrak{n}^-$) is the subspace of $\mathfrak{g}$ consisting of all the upper (resp. lower) triangular matrices of zero diagonal and the associated nilpotent group N^+ (resp. N^-) is the subgroup of G formed by all the upper (resp. lower) triangular matrices with all diagonal elements equal to 1. The centralizer M of A in K is the finite group formed by all the diagonal matrices whose diagonal elements equal ± 1 with an even number of -1s. Its Lie algebra is trivial, that is, $\mathfrak{m} = \{0\}$.

For $G = SL(d, \mathbb{R})$, the Cartan decomposition $G = K\overline{A_+}K$ takes the following form:

$$\forall g \in G = SL(d, \mathbb{R}), \; \exists \xi, \eta \in K = SO(d) \text{ such that}$$
$$g = \xi \operatorname{diag}\{\mu_1, \dots, \mu_d\} \eta,$$

where $\mu_1 \geq \cdots \geq \mu_d > 0$ are uniquely determined by g. We now provide a direct proof of this fact. Since gg' is a positive definite symmetric matrix, there is $\xi \in K = SO(d)$ such that $gg' = \xi \operatorname{diag}\{b_1, \dots, b_d\}\xi'$, where $b_1 \geq b_2 \geq \cdots \geq b_d > 0$. Let $\mu_i = \sqrt{b_i}$, $a = \operatorname{diag}\{\mu_1, \dots, \mu_d\}$, and $\eta = (\xi a)^{-1}g$. Then $gg' = \xi a^2 \xi' = \xi a(\xi a)'$ and $g = \xi a \eta$, and hence $\eta \eta' = [(\xi a)^{-1}g][(\xi a)^{-1}g]' = e$ and $\eta \in K$. Note that the positive numbers $\mu_1, \dots, \mu_d$ are the eigenvalues of gg', so are uniquely determined by g. They are called the singular values of the matrix g. The Cartan decomposition for $g \in G = SL(d, \mathbb{R})$ is also called the singular value decomposition. The set of $g \in G$ with distinct singular values is KA_+K with $A_+ = \exp(\mathfrak{a}_+)$.

The polar decomposition $(K/M) \times A_+$ on the symmetric space G/K can be described as follows. Let S_1 be the space of all the $d \times d$ real positive definite symmetric matrices of determinant 1. For $s \in S_1$, the equation $\sum_{i,j=1}^{d} s_{ij}x_i x_j = 1$ describes an ellipsoid of unit volume in $\mathbb{R}^d$ center at the origin; hence, S_1 may be regarded as the space of such ellipsoids. The map $G \ni g \mapsto gg' \in S_1$ is onto and its kernel is $K = SO(d)$; hence, G/K may be identified with S_1. Under this map, the image of KA_+K is the set of symmetric matrices $s \in S_1$ with distinct eigenvalues, or the set of ellipsoids whose axes have different lengths. Such a matrix takes the form $x = \xi(a^+)^2 \xi^{-1}$ with $a^+ \in A_+$ and $\xi \in K$. Its "angular" component ξM determines the orientation of the ellipsoid in $\mathbb{R}^d$, whereas the "radial" component $a^+ = \operatorname{diag}(a_1, \dots, a_d)$ determines the lengths of its axes. (More precisely, the half lengths of its axes are given by $1/a_1, \dots, 1/a_d$.)

We now provide a direct proof of the Iwasawa decomposition $G = KAN^+$ for $G = SL(d, \mathbb{R})$. By the discussion after the statement of Theorem 5.3, the map $K \times A \times N^+ \ni (k, a, n) \mapsto kan \in G$ is a regular. For $g \in SL(d, \mathbb{R})$, we may apply the Gram–Schmidt orthogonalization procedure to the column vectors of the matrix g to obtain an orthogonal matrix k in $SO(d)$. The procedure amounts to multiplying g on the right by an upper triangular matrix u whose diagonal elements are positive. Such a u is unique. Then $g = ku^{-1}$.

Since u^{-1} is also an upper triangular matrix with positive diagonal elements, letting a be its diagonal and putting $u = an$, we obtain $g = kan$, where n is an upper triangular matrix whose diagonal elements are equal to 1. This proves the Iwasawa decomposition $G = KAN^+$ for $G = SL(d, \mathbb{R})$.

For $G = SL(d, \mathbb{R})$, a matrix $k \in K = SO(d)$ belongs to M' if and only if $\sum_{p=1}^{d} b_p k_{ip} k_{jp} = 0$ for $i \neq j$ and any $H = \text{diag}\{b_1, \ldots, b_d\} \in \mathfrak{a}$. We claim that each column of k can have only one nonzero element. Otherwise, there exist p and $i \neq j$ such that k_{ip} and k_{jp} are both nonzero. Let $b_p = 1$ and $b_q = -1/(d-1)$ for $q \neq p$. Then $H = \text{diag}\{b_1, \ldots, b_d\} \in \mathfrak{a}$ and $\frac{1}{d-1} \sum_{q \neq p} k_{iq} k_{jq} = k_{ip} k_{jp}$. However, since $k \in SO(d)$, $\sum_{q \neq p} k_{iq} k_{jq} = -k_{ip} k_{jp}$. This implies $k_{ip} k_{jp} = 0$, which is a contradiction. The claim is proved. It is easy to show that if $k \in SO(d)$ has the stated property, then $k \in M'$. We have proved

$$M' = \{k \in SO(d); \ k_{pq} = \pm \delta_{p\,\sigma(q)} \text{ for some } \sigma \in S_d\}, \tag{5.19}$$

where S_d is the permutation group on the set $\{1, 2, \ldots, d\}$. To any $k \in M'$, we associate the permutation σ specified in (5.19). Then for any matrix $X \in \mathfrak{gl}(d, \mathbb{R})$, kX (resp. Xk) is obtained from X by permuting its rows by σ (resp. its columns by σ^{-1}). Such a matrix k will be called a permutation matrix. Note that if $k \in M'$ is associated to σ, then the set of matrices in M' associated to the same permutation σ is given by kM. It follows that the Weyl group $W = M'/M$ can be identified with the permutation group S_d in the sense that the map $W \ni w = kM \mapsto \sigma_w \in S_d$, where σ_w is associated to k, is a group isomorphism. Note that if $H = \text{diag}\{b_1, \ldots, b_d\} \in \mathfrak{a}$, then $wH = \text{diag}\{b_{\sigma_w^{-1}(1)}, \ldots, b_{\sigma_w^{-1}(d)}\}$.

For $G = SL(d, \mathbb{R})$, the simple roots are given by $\alpha_{12}, \alpha_{23}, \ldots, \alpha_{(d-1)d}$. Propositions 5.12 and 5.13 can be easily verified in this case.

For any $d \times d$ matrix g and $1 \leq i \leq d$, let $g[i]$ be the submatrix of g formed by the first i rows and i columns of g. Recall that, for $g \in G$ and $k \in K$, $g * k$ is the K-component of gk in the Iwasawa decomposition $G = KAN^+$.

Proposition 5.20. *Let* $G = SL(d, \mathbb{R})$. *Then*

$$N^- MAN^+ = \{g \in SL(d, \mathbb{R}); \ \det(g[i]) \neq 0 \text{ for } 1 \leq i \leq d\},$$

$$N^- AN^+ = \{g \in SL(d, \mathbb{R}); \ \det(g[i]) > 0 \text{ for } 1 \leq i \leq d\},$$

$$(N^- M) * e = \{k \in SO(d); \ \det(k[i]) \neq 0 \text{ for } 1 \leq i \leq d\}, \text{ and}$$

$$N^- * e = \{k \in SO(d); \ \det(k[i]) > 0 \text{ for } 1 \leq i \leq d\}.$$

Proof. Note that MAN^+ is the set of all the upper triangular matrices in $SL(d, \mathbb{R})$. Any $g \in SL(d, \mathbb{R})$ belongs to N^-MAN^+ if and only if there is an $n' \in N^-$, a lower triangular matrix with all the diagonal elements equal to 1, such that $n'g \in MAN^+$; that is, $g_{ij} + \sum_{p<i} n'_{ip}g_{pj} = 0$ for $i > j$. Fix $i \in \{2, \ldots, d\}$. This system of linear equations can be solved for $n'_{i1}, n'_{i2}, \ldots, n'_{i\,i-1}$ if $\det(g[i-1]) \neq 0$. Note that $\det(g[d]) = 1$. This shows that if all $\det(g[i]) \neq 0$, then $g \in N^-MAN^+$. However, it is easy to show that, for any square matrix g and upper triangular matrix u (resp. lower triangular matrix v),

$$(gu)[i] = g[i] \cdot u[i] \quad (\text{resp.} \quad (vg)[i] = v[i] \cdot g[i]). \tag{5.20}$$

It follows that if $g = n'u$ for $n' \in N^-$ and $u \in MAN^+$, then $g[i] = n'[i]u[i]$ and $\det(g[i]) = \det(n'[i])\det(u[i]) = \det(u[i]) \neq 0$. This proves the expression for N^-MAN^+. An upper triangular matrix $u \in SL(d, \mathbb{R})$ belongs to AN^+ if and only if all its diagonal elements are positive, hence, if and only if all $\det(u[i]) > 0$. By (5.20), this proves the expression for N^-AN^+ given in the second formula. Let $g = kan$ be the Iwasawa decomposition $G = KAN^+$. Then $g[i] = k[i](an)[i]$. From this the last two formulas can be derived from the first two and (5.20).

A shorter proof of Proposition 5.20 can be obtained from (5.20) and the Bruhat decomposition for $G = SL(d, \mathbb{R})$ by noting that if $n \in N^-$ and $m' \in (M' - M)$, then $(nm')[i] = 0$ for some i with $1 \leq i < d$. $\qquad\square$

General Linear Groups

Although $GL(d, \mathbb{R})_+$, the group of $d \times d$ invertible real matrices of positive determinants, is not semi-simple, because of the product structure $GL(d, \mathbb{R})_+ = \mathbb{R}_+ \times SL(d, \mathbb{R})$, almost all results stated for $SL(d, \mathbb{R})$ hold also for $GL(d, \mathbb{R})_+$ if $\mathbb{R}_+$, which corresponds to the subgroup $\{cI_d; c \in \mathbb{R}_+\}$ of $GL(d, \mathbb{R})_+$, is absorbed into A. More precisely, let $K = SO(d)$, N^+, N^-, M, and M' be the same subgroups of $G = SL(d, \mathbb{R})$ defined before, but let A be the subgroup of $GL(d, \mathbb{R})_+$ consisting of all the diagonal matrices with arbitrary positive diagonal elements (not necessarily of unit determinant) and let A_+ be the subset of A consisting of all the diagonal matrices with strictly descending diagonal elements. Then the Cartan, Iwasawa, and Bruhat decompositions, namely, Theorems 5.2, 5.3, and 5.4 (including the other versions of the Iwasawa and Bruhat decomposition) hold for $G = GL(d, \mathbb{R})_+$. In fact, Theorem 5.1 holds for $G = GL(d, \mathbb{R})_+$ as well except that K no longer contains the center of $GL(d, \mathbb{R})_+$, which is equal to $\{cI_d; c \in \mathbb{R} - \{0\}\}$ when d is even and $\{cI_d; c \in \mathbb{R}_+\}$ when d is odd.

Lorentz Groups

Let $O(1, d)$ be the group of all the $(d + 1) \times (d + 1)$ real matrices g satisfying $g'Jg = J$, where $J = \text{diag}(1, -I_d)$. This is a closed Lie subgroup of $GL(d + 1, \mathbb{R})$, called the Lorentz group on $\mathbb{R}^{d+1}$. Its Lie algebra is $o(1, d) = \{X \in \mathfrak{gl}(d + 1, \mathbb{R}); X'J + JX = 0\}$. Any $X \in o(1, d)$ can be written in block form as

$$X = \begin{bmatrix} 0 & y' \\ y & B \end{bmatrix}, \tag{5.21}$$

where $y \in \mathbb{R}^d$ is a column vector and $B \in o(d)$.

Any $(d + 1) \times (d + 1)$ real matrix g may be written as

$$g = \begin{bmatrix} u & y' \\ x & B \end{bmatrix},$$

where $u \in \mathbb{R}$, and $x, y \in \mathbb{R}^d$ are regarded as column vectors, and B is a $d \times d$ real matrix. Then the condition $g'Jg = J$ is equivalent to the following three equations:

$$u^2 = 1 + x'x, \quad uy - B'x = 0, \quad \text{and} \quad B'B = I_d + yy'. \tag{5.22}$$

It is clear that $u \neq 0$ and $\det(B) \neq 0$. It follows that $O(1, d)$ has four connected components determined by the signs of u and $\det(B)$. In particular, the identity component, denoted by $SO(1, d)_+$, is the set of all $g \in O(1, d)$ with positive u and $\det(B)$. It follows from $g'Jg = J$ that $\det(g)^2 = 1$ for $g \in O(1, d)$. In particular, $\det(g) = 1$ for $g \in SO(1, d)_+$; hence, $SO(1, d)_+$ is a closed subgroup of $SL(d + 1, \mathbb{R})$.

Let $G = SO(1, d)_+$ and $\mathfrak{g} = o(1, d)$. For $y \in \mathbb{R}^d$ regarded as a column vector, let

$$\xi_y = \begin{bmatrix} 0 & y' \\ y & 0 \end{bmatrix}, \tag{5.23}$$

and for $B \in o(d)$, let $\eta_B = \text{diag}(0, B)$. Then

$$[\xi_x, \xi_y] = \eta_{(xy'-yx')}, \quad [\eta_B, \xi_y] = \xi_{By}, \quad \text{and} \quad [\eta_B, \eta_C] = \eta_{[B,C]}. \tag{5.24}$$

Let $y^i \in \mathbb{R}^d$ be the column vector whose ith component is 1 and with all other components 0. By (5.21), it is easy to see that a basis of $\mathfrak{g}$ is given by $\{\xi_i = \xi_{y^i}$ and $\eta_{jk} = \eta_{(E_{jk}-E_{kj})}; 1 \leq i \leq d$ and $1 \leq j < k \leq d\}$.

As for $SL(d, \mathbb{R})$, the map $\Theta: G \ni g \mapsto (g')^{-1} \in G$ is a Cartan involution with $\theta = D\Theta: \mathfrak{g} \ni X \mapsto -X' \in \mathfrak{g}$. The eigenspaces of θ of eigenvalues ± 1 are, respectively, $\mathfrak{k} = \text{diag}\{0, o(d)\} = \{\eta_A; A \in o(d)\}$ and $\mathfrak{p} = \{\xi_y; y \in \mathbb{R}^d\}$. A matrix $g \in G$ is fixed by Θ if and only if $gg' = I_{d+1}$. Using (5.22), it can be shown that the subset of G fixed by Θ is the compact Lie group $K = \text{diag}\{1, SO(d)\}$ with Lie algebra $\mathfrak{k} = \text{diag}\{0, o(d)\}$.

If we index the rows and columns of a matrix in $\mathfrak{g} = o(1, d)$ by $0, 1, \ldots, d$, then, for $X, Y \in \mathfrak{g}$, the trace of $\text{ad}(X)\text{ad}(Y)$ is given by

$$\sum_{i=1}^{d} [\text{ad}(X)\text{ad}(Y)\xi_i]_{i0} + \sum_{1 \leq i < j \leq d} [\text{ad}(X)\text{ad}(Y)\eta_{ij}]_{ij}.$$

With the help of (5.24), this can be computed to get the Killing form $B(X, Y)$ on $\mathfrak{g} = o(1, d)$. We obtain, for $x, y \in \mathbb{R}^d$ and $C, D \in o(d)$,

$$\begin{cases} B(\xi_x, \xi_y) = 2(d-1)(x \cdot y), \\ B(\eta_C, \eta_D) = -2(d-1)\sum_{1 \leq i < j \leq d} C_{ij} D_{ij}, \\ B(\xi_x, \eta_C) = 0, \end{cases} \tag{5.25}$$

where $(x \cdot y) = \sum_{i=1}^{d} x_i y_i$ is the Euclidean inner product on $\mathbb{R}^d$. From this, it follows that B is negative definite on $\mathfrak{k}$ and positive definite on $\mathfrak{p}$, so $G = SO(1, d)_+$ is a semi-simple Lie group of noncompact type.

Any nontrivial abelian subspace of $\mathfrak{p}$ is one dimensional and is spanned by ξ_y for some nonzero $y \in \mathbb{R}^d$. Fix a vector $v \in \mathbb{R}^d$ with $|v| = 1$, where $|v| = (v \cdot v)^{1/2}$ is the Euclidean norm, and let $\mathfrak{a}$ be the linear span of ξ_v. We may take $\mathfrak{a}_+ = \{c\xi_v; c > 0\}$ to be our Weyl chamber. There are only two roots $\pm\alpha$ given by $\alpha(c\xi_v) = c$. The root spaces $\mathfrak{g}_{\pm\alpha} = \mathfrak{n}^\pm$ are given by

$$\mathfrak{n}^\pm = \left\{ \begin{bmatrix} 0 & y' \\ y & \pm(vy' - yv') \end{bmatrix}; \ y \in \mathbb{R}^d \text{ with } (y \cdot v) = 0 \right\}. \tag{5.26}$$

Let $\zeta_C = \text{diag}(1, C)$ for $C \in SO(d)$. Then $\text{Ad}(\zeta_C)\xi_y = \xi_{Cy}$. The centralizer M of $\mathfrak{a}$ in K consists of the matrices $\text{diag}(1, C)$, where $C \in SO(d)$ satisfying $Cv = v$.

Let $v = (1, 0, \ldots, 0)' \in \mathbb{R}^d$, where the prime denotes the transpose as usual and thus v is a column vector. Then

$$A = \{\exp(t\xi_v); \ t \in \mathbb{R}\} = \left\{ \begin{bmatrix} \cosh t & \sinh t & 0 \\ \sinh t & \cosh t & 0 \\ 0 & 0 & I_{d-1} \end{bmatrix}; \ t \in \mathbb{R} \right\}, \tag{5.27}$$

$$M = \text{diag}\{1, 1, SO(d-1)\}, \quad \text{and} \quad M' = M \cup \text{diag}\{1, -1, h\, SO(d-1)\}, \tag{5.28}$$

where $h \in O(d-1)$ with $\det(h) = -1$. The Weyl group $W = M'/M$ has only two elements; these correspond to the identity map and the reflection about 0 in $\mathfrak{a} \equiv \mathbb{R}$.

The symmetric space G/K is called a hyperbolic space. Let o denote the point eK in G/K. Under the identification of G/K with $\mathfrak{p} \equiv \mathbb{R}^d$, via the map $\mathfrak{p} \ni Y \mapsto e^Y o \in G/K$, the polar decomposition on G/K given by Proposition 5.16 is just the usual spherical polar decomposition on $\mathbb{R}^d$.

For $z = (z_1, \ldots, z_{d-1})' \in \mathbb{R}^{d-1}$, let $n^{\pm}(z)$ be the $(d+1) \times (d+1)$ matrix defined by

$$n^{\pm}(z) = \exp\left\{\begin{bmatrix} 0 & y' \\ y & \pm(vy' - yv') \end{bmatrix}\right\}$$

$$= \exp\left\{\begin{bmatrix} 0 & 0 & z' \\ 0 & 0 & \pm z' \\ z & \mp z & 0 \end{bmatrix}\right\} = \begin{bmatrix} 1 + \frac{b}{2} & \mp\frac{b}{2} & z' \\ \pm\frac{b}{2} & 1 - \frac{b}{2} & \pm z' \\ z & \mp z & I_{d-1} \end{bmatrix}, \quad (5.29)$$

where $y = (0, z_1, \ldots, z_{d-1})' \in \mathbb{R}^d$ and $b = |y|^2 = |z|^2$. By (5.26) and the bijectivity of the exponential maps on $n^{\pm}$, the two nilpotent groups are given by $N^{\pm} = \{n^{\pm}(z); z \in \mathbb{R}^{d-1}\}$.

Let Σ_2 be the set of all the oriented two-dimensional subspaces of $\mathbb{R}^d$. For $\sigma \in \Sigma_2$, let $R(\sigma, \alpha)$ be the element of $SO(d)$ that rotates σ by an angle α and fixes the orthogonal complement of σ in $\mathbb{R}^d$. Let $\{e_1, \ldots, e_d\}$ be the standard basis of $\mathbb{R}^d$. Recall that, for $g \in G$ and $k \in K$, $g * k$ is the K-component of gk in the Iwasawa decomposition $G = KAN^+$.

Proposition 5.21. *For $G = SO(1, d)_+$ and $K = \mathrm{diag}\{1, SO(d)\}$, $N^- * e$ consists of all the matrices $\mathrm{diag}(1, R(\sigma, \alpha))$, where $\sigma \in \Sigma_2$ contains e_1 and $0 \leq \alpha < \pi$.*

Proof. The Iwasawa decomposition of $n^-(z)$ has the form $n^-(z) = k \exp(s\xi_v)n^+(w)$, where $k = \mathrm{diag}\{1, C\}$ for some $C \in SO(d)$, $s \in \mathbb{R}$, and $w \in \mathbb{R}^{d-1}$. One verifies directly using (5.23) and (5.29) that

$$C = \begin{bmatrix} \frac{1-b}{1+b} & -\frac{2}{1+b}z' \\ \frac{2}{1+b}z & I_{d-1} - \frac{2}{1+b}zz' \end{bmatrix},$$

for $b = |z|^2$, $s = \log(1 + b)$, and $w = \frac{1}{1+b}z$. Let

$$Z = \begin{bmatrix} 0 & -z' \\ z & 0 \end{bmatrix}.$$

Then

$$e^{tZ} = \begin{bmatrix} \cos\sqrt{b}t & -\frac{\sin\sqrt{b}t}{\sqrt{b}}z' \\ \frac{\sin\sqrt{b}t}{\sqrt{b}}z & I_{d-1} - \frac{1-\cos\sqrt{b}t}{b}zz' \end{bmatrix}.$$

However, $e^{tZ} = R(\sigma, \sqrt{b}t)$, where $\sigma \in \Sigma_2$ contains e_1 and $(0, z')' \in \mathbb{R}^d$. If t is chosen such that $\cos\sqrt{b}t = \frac{1-b}{1+b}$, then $C = e^{tZ}$ and $k = \mathrm{diag}\{1, e^{tZ}\}$. $\square$

6

Limiting Properties of Lévy Processes

In this chapter, we present basic limiting properties of a Lévy process in a semi-simple Lie group G of noncompact type. In Section 6.1, contracting sequences are introduced. These are a type of sequences in G convergent to infinity and they will play an important role in the discussion of limiting properties of Lévy processes in G. In Section 6.2, we obtain the limiting properties of a special type of continuous Lévy process. Because the stochastic differential equations satisfied by these processes possess a certain symmetry, a rather elementary computation may be carried out to obtain the desired results. As an application, a continuous Lévy process in the general linear group $GL(d, \mathbb{R})$ is discussed in Section 6.3. In Section 6.4, the concepts of invariant and irreducible measures are introduced, and an important result connecting these notions is established as a necessary preparation for our main results. The basic limiting properties of a general Lévy process, and some variations, are stated and proved in Section 6.5. It is shown that if g_t is a left (resp. right) Lévy process in G satisfying a certain nondegeneracy condition with the Cartan decomposition $g_t = \xi_t a_t^+ \eta_t$ and the Iwasawa decomposition $g_t = n_t a_t k_t$ of $G = N^- A K$ (resp. $g_t = k_t a_t n_t$ of $G = K A N^+$), then, almost surely, both a_t^+ and a_t are positive contracting, and $\xi_t M$ (resp. $M \eta_t$) and n_t converge. Some sufficient conditions that guarantee these limiting properties are discussed in Section 6.6. In Sections 6.1, 6.4, and 6.5, we follow closely the ideas in Guivarc'h and Raugi [24], but much more detail is provided here with considerable modifications. The materials in Section 6.2 are taken from Liao [38].

6.1. Contracting Properties

Let G be a connected semi-simple Lie group of noncompact type with Lie algebra $\mathfrak{g}$. The notations and definitions introduced in Chapter 5 will be used in the rest of this work. For simplicity, we will assume from now on that G has a finite center Z. Then the subgroups K, M, and M' are compact. Moreover,

G/Q (resp. $Q\backslash G$) is a compact homogeneous space for any closed subgroup Q of G containing AN^+ (resp. AN^-). Note that this additional assumption is really not essential because if Z is infinite, then the quotient group $\tilde{G} = G/Z$ has a trivial center and is a semi-simple Lie group of noncompact type. Our theory can be applied to $\tilde{G}$. The subgroups K, M, and M' of G become the subgroups $\tilde{K} = K/Z$, $\tilde{M} = M/Z$, and $\tilde{M}' = M'/Z$ of $\tilde{G}$; respectively, and the subgroups A, N^+, and N^- remain unchanged. Because Z is always discrete, all the Lie subalgebras remain the same.

Let Δ_+ be the set of positive roots as before. A sequence $\{a_j\}$ in A will be called positive contracting if $\forall \alpha \in \Delta_+, \alpha(\log a_j) \to \infty$.

Lemma 6.1. *Let a_j be a positive contracting sequence in A.*

(a) *If F is a compact subset of N^+, then $a_j^{-1} n a_j \to e$ uniformly for $n \in F$.*
(b) *If F' is a compact subset of N^-, then $a_j n a_j^{-1} \to e$ uniformly for $n \in F'$.*
(c) *If $n_j \in N^+$ (resp. $n_j \in N^-$) and $a_j n_j a_j^{-1} \to e$ (resp. $a_j^{-1} n_j a_j \to e$),*
   then $n_j \to e$.

Proof. Let $\{X_i\}$ be a basis of $\mathfrak{n}^+$ such that each X_i belongs to $\mathfrak{g}_\alpha$ for some $\alpha \in \Delta_+$, which will be denoted by α_i. Note that it is possible to have $\alpha_i = \alpha_j$ for $i \neq j$. Any $n \in N^+$ can be written as $n = e^Y$ for some $Y \in \mathfrak{n}^+$ with $Y = \sum_i c_i X_i$. When n ranges over F, the coefficients c_i remain bounded. We have

$$a_j^{-1} n a_j = \exp\left[\mathrm{Ad}(a_j^{-1})Y\right] = \exp\left[\sum_i c_i \mathrm{Ad}(e^{-\log a_j})X_i\right]$$

$$= \exp\left[\sum_i c_i e^{-\alpha_i(\log a_j)} X_i\right].$$

Since $\lim_{j\to\infty} e^{-\alpha_i(\log a_j)} = 0$, this expression converges to e uniformly for $n \in F$ as $j \to \infty$. This proves part (a). Part (b) can be proved by either applying Θ to (a) or repeating essentially the same argument. To prove (c), let $n_j = \exp(Y_j)$ for $Y_j = \sum_i c_{ij} X_i$. It suffices to show that $\forall i$, $c_{ij} \to 0$ as $j \to \infty$. If not, by taking a subsequence, we may assume $\exists i$ and $\varepsilon > 0$ such that $\forall j$, $|c_{ij}| \geq \varepsilon$. Then

$$a_j n_j a_j^{-1} = \exp\left[\sum_h c_{hj} \mathrm{Ad}(a_j)X_h\right] = \exp\left[\sum_h c_{hj} e^{\mathrm{ad}(\log a_j)} X_h\right]$$

$$= \exp\left[\sum_h c_{hj} e^{\alpha_h(\log a_j)} X_h\right].$$

Since $\alpha_h(\log a_j) \to \infty$, $|c_{ij}| \geq \varepsilon$, and exp: $\mathfrak{n}^+ \to N^+$ is diffeomorphic, it is impossible to have $a_j n_j a_j^{-1} \to e$. This prove (c) for $n_j \in N^+$. The conclusion for $n_j \in N^-$ is proved by a similar argument. $\qquad\square$

Let $\pi: G \to G/(MAN^+)$ be the natural projection. By part (b) of Lemma 6.1, if $a_j \in A$ is positive contracting and if $x = n(MAN^+) \in \pi(N^-)$, where $n \in N^-$, then

$$a_j x = a_j n(MAN^+) = a_j n a_j^{-1}(MAN^+) \to \pi(e)$$

as $j \to \infty$. Therefore, via the left action of G on $G/(MAN^+)$, a positive contracting sequence in A shrinks $\pi(N^-)$ to the single point $\pi(e)$. Note that, by Proposition 5.19, $\pi(N^-) = \pi(N^-M)$ is an open subset of $G/(MAN^+)$ with a lower dimensional complement.

Lemma 6.2. *Let $\{p_j\}$ be a sequence contained in a compact subset of N^+ and let $\{a_j\}$ be a positive contracting sequence in A. Then $p_j a_j$ has a Cartan decomposition $p_j a_j = \xi_j a_j^+ \eta_j$ such that the three sequences $(a_j^{-1} a_j^+), \xi_j$, and η_j all converge to e. In particular, a_j^+ is positive contracting.*

Proof. Note that the two K-components in the Cartan decomposition are not unique. Let $p_j a_j = x_j a_j^+ y_j$ be a Cartan decomposition. First assume $x_j \to x$ in K. By the Bruhat decomposition (5.16), $x \in N^+ m' N^- AM$ for some $m' \in M'$. By Proposition 5.18, $m' N^- m'^{-1} \subset N^+ N^-$. Since $m' \in M'$ normalizes M and A, $N^+ m' N^- AM = N^+(m' N^- m'^{-1})AMm' \subset N^+ N^- AMm'$, which is an open subset of G. Hence, x and x_j for sufficiently large j all belong to $N^+ N^- AMm'$. Therefore, we may write $x = nn'bmm'$ and $x_j = n_j n'_j b_j m_j m'$ with $N^+ \ni n_j \to n$, $N^- \ni n'_j \to n'$, $A \ni b_j \to b$, and $M \ni m_j \to m$. We have

$$\begin{aligned}
a_j^{-1} p_j a_j &= a_j^{-1} x_j a_j^+ y_j \\
&= \left(a_j^{-1} n_j a_j\right)\left(a_j^{-1} n'_j a_j\right)\left[a_j^{-1} b_j (m_j m') a_j^+ (m_j m')^{-1}\right] m_j m' y_j.
\end{aligned}$$

The right-hand side of this equation is expressed as a decomposition $N^+ N^- AMm' y_j$. By Lemma 6.1(a), $a_j^{-1} p_j a_j \to e$ and $a_j^{-1} n_j a_j \to e$. For the moment, assume $y_j \to y$ in K. By Theorem 5.4, the map

$$N^+ \times N^- \times A \times M \ni (n, n', a, m) \mapsto nn'am \in N^+ N^- AM$$

is a diffeomorphism. It follows that $a_j^{-1} n'_j a_j \to u$, $[a_j^{-1} b_j (m_j m') a_j^+ (m_j m')^{-1}] \to c$ and $m_j \to v$ for some $u \in N^-$, $c \in A$, and $v \in M$. Then

$e = uc(vm'y)$. By the uniqueness of the Iwasawa decomposition $G = N^- AK$, this implies that $u = c = vm'y = e$. We have proved

(i) $a_j^{-1} n_j a_j \to e$,

(ii) $a_j^{-1} n'_j a_j \to e$,

(iii) $a_j^{-1} b_j (m_j m') a_j^+ (m_j m')^{-1} \to e$, and

(iv) $m_j m' y_j \to e$.

Although the assumption $y_j \to y$ is used to prove these claims, it now can be removed because of the compactness of K.

By (ii) and Lemma 6.1(c), $n'_j \to e$. Because a_j is positive contracting and $b_j \to b$ in A, it follows that for sufficiently large j, $a_j^{-1} b_j \in \exp(-\mathfrak{a}_+)$. Therefore, by (iii), $(m_j m') a_j^+ (m_j m')^{-1}$ is positive contracting and is contained in $A_+ = \exp(\mathfrak{a}_+)$ for sufficiently large j. Since the action of the Weyl group on $\mathfrak{a}$ permutes and is simply transitive on the set of the Weyl chambers, it follows that $(m_j m') a_j^+ (m_j m')^{-1} = a_j^+ \in \mathfrak{a}_+$ and $m_j m' \in M$. Using (iii) one more time, we see that $a_j^{-1} a_j^+ \to b^{-1}$.

Let $\xi_j = x_j (m_j m')^{-1} = n_j n'_j b_j$ and $\eta_j = m_j m' y_j$. Then $p_j a_j = x_j a_j^+ \times y_j = \xi_j a_j^+ \eta_j$, $\xi_j \to nb$ and, by (iv), $\eta_j \to e$. Since $\xi_j \in K$, $n \in N^+$, and $b \in A$, one must have $n = b = e$. This implies that $\xi_j \to e$ and $a_j^{-1} a_j^+ \to e$. The lemma is proved under the additional assumption that x_j converges.

In general, applying the result to any convergent subsequence of x_j, since $m_j m' \in M$ and $n = n' = b = e$ are shown in the preceding argument, we see that any limiting point of x_j must be contained in M. It follows that there exist $\xi_j \in K$ and $m_j \in M$ such that $\xi_j \to e$ and $x_j = \xi_j m_j$. Since $p_j a_j = x_j a_j^+ y_j = \xi_j a_j^+ (m_j y_j)$, the result under the convergence of x_j applies. The proof of the lemma is completed. $\qquad \square$

As in Guivarc'h and Raugi [24], a sequence $\{g_j\}$ in G will be called contracting if, in its Cartan decomposition $g_j = \xi_j a_j^+ \eta_j$, $\{a_j^+\}$ is positive contracting. Note that a contracting sequence contained in A is not necessarily positive contracting. A subset of G is called bounded if it is contained in a compact subset of G.

Proposition 6.1. *Let x_j and y_j be two bounded sequences in G. If g_j is a contracting sequence in G, then $x_j g_j y_j$ is also contracting.*

Proof. Let $g_j = \xi_j a_j^+ \eta_j$ be a Cartan decomposition and let $x_j \xi_j = k_j n_j a_j$ be the Iwasawa decomposition $G = K N^+ A$. Then n_j and a_j are sequences contained in compact subsets of N^+ and A, respectively. It follows that $a_j a_j^+$

is a positive contracting sequence in A and, by Lemma 6.2, $n_j(a_j a_j^+) = h_j b_j^+ h_j'$, where $h_j, h_j' \in K$ and $b_j^+ \in \overline{A}_+$ is positive contracting. Therefore, $x g_j = (k_j h_j) b_j^+ (h_j' \eta_j)$ is contracting. Because for any contracting sequence g_j with Cartan decomposition $g_j = \xi_j a_j^+ \eta_j$, $\Theta(g_j)^{-1} = \eta_j^{-1} a_j^+ \xi_j^{-1}$ is also contracting, then $g_j y_j = \Theta[\Theta(y_j)^{-1} \Theta(g_j)^{-1}]^{-1}$ is contracting. □

Proposition 6.2. *Let $\{g_j\}$ be a sequence in G. Then the following two statements and are equivalent:*

(a) *There is a Cartan decomposition $g_j = \xi_j a_j^+ \eta_j$ such that a_j^+ is positive contracting and $\xi_j M \to \xi_\infty M$ for some $\xi_\infty \in K \cap (N^- M A N^+)$.*

(b) *Let $g_j = n_j a_j k_j$ be the Iwasawa decomposition $G = N^- A K$. Then a_j is positive contracting and $n_j \to n_\infty$ for some $n_\infty \in N^-$.*

Moreover, these statements imply that $\xi_\infty M A N^+ = n_\infty M A N^+$, $(a_j^+)^{-1} a_j$ converges in A, and $k_j = \eta_j' \eta_j$ with $K \ni \eta_j' \to m$ for some $m \in M$.

Note that, although the ξ_j component in the Cartan decomposition $g_j = \xi_j a_j^+ \eta_j$ in (a) is not unique, $\xi_j M$ is uniquely determined by g_j because $a_j^+ \in \mathfrak{a}_+$ for large j. Note also that the convergence $\xi_j M \to \xi_\infty M$ is equivalent to $\xi_j M A N^+ \to \xi_\infty M A N^+$ by the Iwasawa decomposition $G = K A N^+$.

Proof. Assume (a). By taking a subsequence, we may assume $\xi_j \to \xi_\infty$. Note that, although ξ_∞ may depend on the subsequence, the limit n_∞ in (b) will be independent of the subsequence because it is uniquely determined by $\xi_\infty M A N^+ = n_\infty M A N^+$ because $N^- M A N^+$ is a diffeomorphic product.

For sufficiently large j, $\xi_j \in N^- M A N^+$. Let $\xi_j = p_j^- m_j b_j p_j$ be the decomposition $N^- M A N^+$. Then $\xi_j a_j^+ = p_j^- a_j^+ m_j b_j (a_j^+)^{-1} p_j a_j^+$ with $N^- \ni p_j^- \to p^-$, $A \ni b_j \to b$, $M \ni m_j \to m$, $N^+ \ni p_j \to p$, and $\xi_\infty = p^- m b p$. By Lemma 6.1(a), $(a_j^+)^{-1} p_j a_j^+ \to e$; hence, $m_j b_j (a_j^+)^{-1} p_j a_j^+ \to m b \in M A = A M$. Let $n_j' a_j' \eta_j'$ be the Iwasawa decomposition $G = N^- A K$ of $m_j b_j (a_j^+)^{-1} p_j a_j^+$. Then $N^- \ni n_j' \to e$, $A \ni a_j' \to b$, and $K \ni \eta_j' \to m$. Now $\xi_j a_j^+ = p_j^- a_j^+ n_j' a_j' \eta_j' = p_j^- a_j^+ n_j' (a_j^+)^{-1} a_j^+ a_j' \eta_j'$. Since $a_j^+ n_j' (a_j^+)^{-1} \to e$, part (b) and all the other claims are proved by setting $n_j = p_j^- a_j^+ n_j' (a_j^+)^{-1}$ and $a_j = a_j^+ a_j'$.

Now assume (b). Let $n_j = x_j n_j' b_j$ be the Iwasawa decomposition $G = K N^+ A$ with $K \ni x_j \to x$, $N^+ \ni n_j' \to n'$, and $A \ni b_j \to b$. Then $b_j a_j$ is positive contracting and $n_\infty(M A N^+) = x(M A N^+)$. By Lemma 6.2, $n_j' b_j a_j$ has a Cartan decomposition $n_j' b_j a_j = x_j' a_j^+ y_j'$ such that a_j^+ is positive contracting and $\lim_j x_j' = \lim_j y_j' = e$. Let $\xi_j = x_j x_j'$ and $\eta_j = y_j' k_j$. Then

$g_j = n_j a_j k_j = \xi_j a_j^+ \eta_j$ and the Cartan decomposition $g_j = \xi_j a_j^+ \eta_j$ satisfies the properties in (a).   □

6.2. Limiting Properties: A Special Case

Let g_t be a continuous left Lévy process in G satisfying the following stochastic differential equation:

$$dg_t = \sum_{i=1}^{p} X_i^l(g_t) \circ dW_t^i + c \sum_{i=p+1}^{d} X_i^l(g_t) \circ dW_t^i, \qquad (6.1)$$

where c is a constant and $\{X_1, \ldots, X_d\}$ is an orthonormal basis of $\mathfrak{g}$ with respect to the inner product $\langle \cdot, \cdot \rangle$ induced by the Killing form, defined by (5.3), such that $X_1, \ldots, X_p$ form a basis of $\mathfrak{p}$ and $X_{p+1}, \ldots, X_d$ form a basis of $\mathfrak{k}$. Note that, when $c = 1$, the orthonormal basis may be chosen arbitrarily because a suitable orthogonal transformation maps this basis into another one that can be divided into a basis of $\mathfrak{p}$ and a basis of $\mathfrak{k}$ as required here, and such a transformation will also transform the standard Brownian motion W_t into another standard Brownian motion. In this case, (6.1) becomes Equation (2.33) with $X_0 = 0$, and by Propositions 2.5 and 2.6, g_t is a Riemannian Brownian motion in G with respect to the left invariant metric induced by $\langle \cdot, \cdot \rangle$. When $c = 0$, the process g_t is called the horizontal diffusion in [43].

In this section, we obtain the limiting properties of the process g_t determined by the stochastic differential equation (6.1). Because of the symmetry possessed by this equation, the results may be established by a rather elementary method. These properties hold also for a general Lévy process, but more difficult proofs are required.

For $g, h \in G$ and $X \in \mathfrak{g}$, we will write gXh for $DL_g \circ DR_h(X)$. If G is a matrix group, then, by the discussion in Section 1.5, gXh may be understood as a matrix product. The computation that follows is carried out in this shorthand notation. One may consider working on a matrix group G; then the notation should be clearly understood. However, the computation is valid on a general Lie group G with proper interpretation.

Let $g_t = n_t a_t k_t$ be the Iwasawa decomposition $G = N^- AK$. We have

$$dg_t = (\circ dn_t) a_t k_t + n_t (\circ da_t) k_t + n_t a_t (\circ dk_t).$$

Equation (6.1) may be written as

$$dg_t = \sum_{i=1}^{d} c_i g_t X_i \circ dW_t^i,$$

where $c_i = 1$ for $1 \leq i \leq p$ and $c_i = c$ for $p + 1 \leq i \leq d$. Multiplying both sides of this equation by $a_t^{-1} n_t^{-1}$ on the left and k_t^{-1} on the right, we obtain

$$\mathrm{Ad}(a_t^{-1})(n_t^{-1} \circ dn_t) + a_t^{-1} \circ da_t + (\circ dk_t)k_t^{-1} = \sum_{i=1}^{d} c_i \mathrm{Ad}(k_t) X_i \circ dW_t^i.$$

For $X \in \mathfrak{g}$, let $X = X_\mathfrak{n} + X_\mathfrak{a} + X_\mathfrak{k}$ be the direct sum decomposition $\mathfrak{g} = \mathfrak{n}^- \oplus \mathfrak{a} \oplus \mathfrak{k}$. Since $\mathrm{Ad}(a)\mathfrak{n}^- \subset \mathfrak{n}^-$ for $a \in A$, we see that (6.1) is equivalent to the following three equations:

$$dk_t = \sum_{i=1}^{d} c_i [\mathrm{Ad}(k_t) X_i]_\mathfrak{k} k_t \circ dW_t^i, \tag{6.2}$$

$$da_t = \sum_{i=1}^{p} a_t [\mathrm{Ad}(k_t) X_i]_\mathfrak{a} \circ dW_t^i, \tag{6.3}$$

and

$$dn_t = \sum_{i=1}^{p} n_t \mathrm{Ad}(a_t) [\mathrm{Ad}(k_t) X_i]_\mathfrak{n} \circ dW_t^i. \tag{6.4}$$

From (6.2), we see that k_t is a diffusion process in K. Applying Ito's formula to $k_t k_t^{-1} = e$ and $\mathrm{Ad}(k_t) X = k_t X k_t^{-1}$ for $X \in \mathfrak{g}$, we obtain

$$dk_t^{-1} = -\sum_{i=1}^{d} c_i k_t^{-1} [\mathrm{Ad}(k_t) X_i]_\mathfrak{k} \circ dW_t^i$$

and

$$\begin{aligned} d\mathrm{Ad}(k_t) X &= d(k_t X k_t^{-1}) \\ &= \sum_{i=1}^{d} c_i [\mathrm{Ad}(k_t) X_i]_\mathfrak{k} k_t X k_t^{-1} \circ dW_t^i \\ &\quad - \sum_{i=1}^{d} c_i k_t X k_t^{-1} [\mathrm{Ad}(k_t) X_i]_\mathfrak{k} \circ dW_t^i \\ &= \sum_{i=1}^{d} c_i [[\mathrm{Ad}(k_t) X_i]_\mathfrak{k}, [\mathrm{Ad}(k_t) X]] \circ dW_t^i. \end{aligned} \tag{6.5}$$

Since A is abelian, $da_t = a_t \circ d \log a_t$. By (6.3),

$$d \log a_t = \sum_{i=1}^{p} [\mathrm{Ad}(k_t) X_i]_\mathfrak{a} \circ dW_t^i. \tag{6.6}$$

By (6.5) and (6.6), we have

$$\log a_t = \log a_0 + M_t + \int_0^t \left\{ \frac{1}{2} \sum_{i=1}^{p} [[\mathrm{Ad}(k_t)X_i]_{\mathfrak{k}}, \mathrm{Ad}(k_t)X_i]_{\mathfrak{a}} \right\} dt, \quad (6.7)$$

where $M_t = \sum_{i=1}^{p} \int_0^t [\mathrm{Ad}(k_{s-})X_i]_{\mathfrak{a}} \, dW_s^i$ is an $\mathfrak{a}$-valued L^2-martingale. Note that $(1/t)M_t \to 0$ almost surely as $t \to \infty$.

Let $\alpha_1, \ldots, \alpha_r$ be the list of all the positive roots, each repeated as many times as its multiplicity, and let H_j be the element of $\mathfrak{a}$ representing α_j under the inner product $\langle \cdot, \cdot \rangle$; that is, $\forall H \in \mathfrak{a}$, $\langle H, H_j \rangle = \alpha_j(H)$. For each j, let X'_j be a nonzero root vector of α_j; that is, $X'_j \in \mathfrak{g}$ and $\forall H \in \mathfrak{a}$, $[H, X'_j] = \alpha_j(H)X'_j$. Let $X_j = Y'_j + Z'_j$, where $Y'_j \in \mathfrak{p}$ and $Z'_j \in \mathfrak{k}$. Then $Y'_1, \ldots, Y'_r, Z'_1, \ldots, Z'_r$ are mutually orthogonal, and $\forall H \in \mathfrak{a}$, $[H, Y'_j] = \alpha_j(H)Z'_j$ and $[H, Z'_j] = \alpha_j(H)Y'_j$. Let $X'_{-j} = Y'_j - Z'_j = -\theta(X'_j)$. Then X'_{-j} is a root vector of the negative root $-\alpha_j$ and $(Y'_j)_{\mathfrak{k}} = Z'_j$. We will choose X'_j to satisfy $\langle X'_j, X'_j \rangle = 2$. Then by Proposition 5.6, $\langle Y'_j, Y'_j \rangle = \langle Z'_j, Z'_j \rangle = 1$, $[X'_j, X'_{-j}] = -2H_j$, $[Z'_i, Y'_i] = H_i$, and $[Z'_i, Y'_j]_{\mathfrak{a}} = 0$ for $i \neq j$. Define

$$H_\rho = \frac{1}{2} \sum_{i=1}^{r} H_i. \tag{6.8}$$

Let $\rho = (1/2) \sum_{i=1}^{r} \alpha_i$, the famous half sum of the positive roots (counting multiplicities). Then H_ρ is the vector in $\mathfrak{a}$ representing ρ under the inner product $\langle \cdot, \cdot \rangle$.

By the discussion in Section 5.4, for $G = SL(d, \mathbb{R})$, the positive roots are given by α_{ij} for $1 \leq i < j \leq d$, where $\alpha_{ij}(H) = \lambda_i - \lambda_j$ for $H = \mathrm{diag}\{\lambda_1, \ldots, \lambda_d\} \in \mathfrak{a}$. Since $\langle X, Y \rangle = 2d \, \mathrm{Trace}(XY')$ for $X, Y \in \mathfrak{g} = \mathfrak{sl}(d, R)$, the element of $\mathfrak{a}$ representing α_{ij} is $\frac{1}{2d} H_{ij}$, where H_{ij} is the diagonal matrix that has 1 at place (i, i), -1 at place (j, j), and 0 elsewhere. It follows that, for $G = SL(d, \mathbb{R})$,

$$H_\rho = \frac{1}{4d} \mathrm{diag}\{(d - 1), (d - 3), \ldots, -(d - 3), -(d - 1)\}. \tag{6.9}$$

For $G = SO(1, d)_+$, $\mathfrak{a}$ is one dimensional and can be taken to be the linear span of ξ_y given by (5.23) for any $y \in \mathbb{R}^d$ with unit Euclidean norm. There is only one positive root α of multiplicity $\dim(\mathfrak{n}^+) = d - 1$, given by $\alpha(\xi_y) = 1$. By (5.25), $B(\xi_x, \xi_y) = 2(d - 1)(x \cdot y)$ for $x, y \in \mathbb{R}^d$, where $(x \cdot y)$ is the Euclidean inner product. It follows that this root is represented by $\xi_y/[2(d - 1)]$. Therefore, for $G = SO(1, d)_+$,

$$H_\rho = \frac{1}{4} \xi_y. \tag{6.10}$$

In the preceding two examples, H_ρ is contained in the Weyl chamber $\mathfrak{a}_+$. This is in fact true in general.

Proposition 6.3. $H_\rho \in \mathfrak{a}_+$.

Proof. We want to show $\alpha(H_\rho) > 0$ for any simple root α. Let $H_\alpha \in \mathfrak{a}$ represent α. Without loss of generality, we may assume $\alpha_1 = \alpha$. Let s_α be the reflection in $\mathfrak{a}$ about the hyperplane $\alpha = 0$. By Proposition 5.13, s_α permutes the H_is for which α_i is not proportional to α. Therefore, $s_\alpha(H_\rho) = (1/2)\sum_{i=1}^r s_\alpha(H_i) = -cH_\alpha + H_\rho$ for some $c > 0$ and

$$\alpha(H_\rho) = \langle H_\rho, H_\alpha \rangle = \langle s_\alpha(H_\rho), s_\alpha(H_\alpha) \rangle = \langle -cH_\alpha + H_\rho, -H_\alpha \rangle$$
$$= c\langle H_\alpha, H_\alpha \rangle - \langle H_\rho, H_\alpha \rangle.$$

This implies $\alpha(H_\rho) > 0$. The argument is taken from [43]. $\qquad\square$

For $1 \le i \le p$ and $k \in K$, $\mathrm{Ad}(k)X_i = \sum_{j=1}^p a_{ij}(k)X_j$ for some $p \times p$ orthogonal matrix $\{a_{ij}(k)\}$. Then

$$\frac{1}{2}\sum_{i=1}^p [[\mathrm{Ad}(k)X_i]_{\mathfrak{k}}, \mathrm{Ad}(k)X_i]_{\mathfrak{a}} = \frac{1}{2}\sum_{i=1}^p [(X_i)_{\mathfrak{k}}, X_i]_{\mathfrak{a}}.$$

Let $Y'_{r+1}, \ldots, Y'_p$ be an orthonormal basis of $\mathfrak{a}$. Then $Y'_1, \ldots, Y'_p$ is an orthonormal basis of $\mathfrak{p}$. Without loss of generality, we may assume $X_i = Y'_i$ for $1 \le i \le p$. Since $(Y'_i)_{\mathfrak{k}} = Z'_i$ for $1 \le i \le r$ and $(Y'_i)_{\mathfrak{k}} = 0$ for $r < i \le p$, we have

$$\frac{1}{2}\sum_{i=1}^p [(X_i)_{\mathfrak{k}}, X_i]_{\mathfrak{a}} = \frac{1}{2}\sum_{i=1}^r [Z'_i, Y'_i]_{\mathfrak{a}} = \frac{1}{2}\sum_{i=1}^r H_i = H_\rho.$$

By (6.7),

$$\lim_{t\to\infty} \frac{1}{t}\log a_t = H_\rho. \tag{6.11}$$

We now prove the almost sure convergence of n_t as $t \to \infty$. For $X \in \mathfrak{g}_\alpha$,

$$\mathrm{Ad}(a_t)X = e^{\mathrm{ad}(\log a_t)}X = e^{\alpha(\log a_t)}X \approx e^{\alpha(H_\rho)t}X.$$

We see that the coefficients in (6.4), $\mathrm{Ad}(a_t)[\mathrm{Ad}(k_t)X_i]_{\mathfrak{n}}$, tend to zero exponentially fast as $t \to \infty$. Because N^- can be identified with a Euclidean space through the diffeomorphic exponential map $\exp: \mathfrak{n}^- \to N^-$ and it can be shown that the coefficients of the Itô form of Equation (6.4) also tend to zero exponentially fast, the almost sure convergence of n_t follows directly from the following lemma. This type of argument to prove the convergence of the nilpotent component was found in Taylor [56].

Lemma 6.3. *Let* $z_t = (z_t^1, \ldots, z_t^d)$ *be a process in* $\mathbb{R}^d$ *satisfying the Itô stochastic differential equation*

$$dz_t^i = \sum_{j=1}^{r} \sum_{k=1}^{d} a_{ijk}(t, \cdot) z_t^k \, dW_t^j + \sum_{k=1}^{d} b_{ik}(t, \cdot) z_t^k \, dt,$$

where the coefficients $a_{ijk}(t, \omega)$ *and* $b_{ik}(t, \omega)$ *are continuous processes adapted to the filtration generated by the standard Browian motion* $W_t = (W_t^1, \ldots, W_t^r)$. *Assume almost surely* $a_{ijk}(t, \omega)$ *and* $b_{ik}(t, \omega)$ *converge to 0 exponentially as* $t \to \infty$ *in the sense that* $\exists \delta > 0$ *such that, almost surely,* $|a_{ijk}(t, \cdot)| \leq e^{-\delta t}$ *and* $|b_{ik}(t, \cdot)| \leq e^{-\delta t}$ *for sufficiently large* $t > 0$. *Then, almost surely,* z_t *converges in* $\mathbb{R}^d$ *as* $t \to \infty$.

Proof. For any integer $n > 0$, let $A_n = \{(t, \omega); \forall s \leq t, |a_{ijk}(s, \omega)| \leq ne^{-\delta s}$ and $|b_{ik}(s, \omega)| \leq ne^{-\delta s}\}$ and let $\tau_n(\omega) = \inf\{t; (t, \omega) \notin A_n\}$. It is clear that the τ_n form an increasing sequence of stopping times and, by the assumption, for almost all ω, $\tau_n(\omega) = \infty$ for sufficiently large n. By stopping the process at τ_n, we may assume the coefficients $a_{ijk}(t, \cdot)$ and $b_{ik}(t, \cdot)$ are uniformly bounded by $Ce^{-\delta t}$ for some constant $C > 0$. In this proof, C will be a positive constant that may change from formula to formula.

The Itô stochastic differential equation is equivalent to the following Itô stochastic integral equation:

$$z_t^i = z_0^i + \int_0^t \sum_{j,k} a_{ijk}(s, \cdot) z_s^k \, dW_s^j + \int_0^t \sum_k b_{ik}(s, \cdot) z_s^k \, ds.$$

Let $|z_t|$ be the Euclidean norm of z_t and let $z_t^* = \sup_{0 \leq s \leq t} |z_t|$. Then

$$E[(z_t^*)^2] \leq C \left\{ |z_0|^2 + \int_0^t e^{-2\delta s} E[(z_s^*)^2] \, ds \right\}.$$

By Gronwall's inequality, $E[(z_t^*)^2] \leq C \exp\left(\int_0^t e^{-2\delta s} ds\right)$; hence, $E[(z_\infty^*)^2] \leq C$. Let $y_n = \sup_{n \leq s \leq n+1} |z_s - z_n|$. It suffices to show $\sum_n y_n < \infty$ almost surely. From

$$z_s^i - z_n^i = \int_n^s \sum_{j,k} a_{ijk}(u, \cdot) z_u^k dW_u^j + \int_n^s \sum_k b_{ik}(u, \cdot) z_u^k du,$$

we obtain

$$P\left(y_n \geq e^{-\delta n/2}\right) \leq e^{\delta n} E\left(y_n^2\right) \leq C e^{\delta n} \int_n^{n+1} e^{-2\delta u} E[(z_\infty^*)^2]du \leq C e^{-\delta n}.$$

It follows that $\sum_n P(y_n \geq e^{-\delta n/2}) < \infty$, and by the Borel–Cantelli lemma, $\sum_n y_n < \infty$ almost surely. □

Let $g_t = \xi_t a_t^+ \eta_t$ be a Cardan decomposition. By Proposition 6.2 and (6.11), the almost sure convergence of n_t implies that almost surely $\xi_t M$ converges and $(1/t) \log a_t^+ \to H_\rho$.

Let o denote the point eK in G/K. Because $[\mathfrak{p}, \mathfrak{p}] \subset \mathfrak{k}$, the restriction of $\langle \cdot, \cdot \rangle$ on $\mathfrak{p}$ satisfies the condition (2.35). By Proposition 2.7, we see that g_t is right K-invariant and if $g_0 = e$, then $x_t = g_t o$ is a Riemannian Brownian motion in G/K with respect to the G-invariant metric induced by the restriction of $\langle \cdot, \cdot \rangle$ on $\mathfrak{p}$. To summarize, we have the following result:

Theorem 6.1. *Let g_t be a left invariant diffusion process satisfying the stochastic differential equation (6.1) with the Cartan decomposition $g_t = \xi_t a_t^+ \eta_t$ and the Iwasawa decomposition $g_t = n_t a_t k_t$ of $G = N^- A K$. Then, almost surely, n_t and $\xi_t M$ converge as $t \to \infty$, and*

$$H_\rho = \lim_{t \to \infty} (1/t) \log a_t^+ = \lim_{t \to \infty} (1/t) \log a_t,$$

where H_ρ is defined by (6.8). Moreover, g_t is right K-invariant and if $g_0 = e$, then $g_t o$ is a Riemannian Brownian motion in G/K with respect to the G-invariant metric induced by the restriction of $\langle \cdot, \cdot \rangle$ to $\mathfrak{p}$, where $o = eK$.

The limiting properties stated in Theorem 6.1 hold also for a general left Lévy process under a certain nondegeneracy condition (see Theorems 6.4 and 7.1). These properties may be stated for the process $x_t = g_t o$ in the symmetric space G/K under the polar decomposition (see Proposition 5.16). Because $H_\rho \in \mathfrak{a}_+$, $x_t \in \pi(KA_+)$ at least for sufficiently large $t > 0$, where $\pi: G \to G/K$ is the natural projection, therefore, the process x_t possesses uniquely defined "radial component" $H_t \in \mathfrak{a}_+$ and "angular component" $\bar{\xi}_t \in K/M$; that is, $x_t = \xi_t \exp(H_t)o$ with $\bar{\xi}_t = \xi_t M$. As $t \to \infty$, $(1/t)H_t \to H_\rho$ and $\bar{\xi}_t \to \bar{\xi}_\infty$ for some random $\bar{\xi}_\infty \in K/M$. This means that, as $t \to \infty$, the process $x_t = g_t o$ in G/K converges to ∞ at nonrandom exponential rates (H_ρ) and in a random limiting direction ($\xi_\infty M$).

6.3. A Continuous Lévy Process in $GL(d, \mathbb{R})_+$

As an application of Theorem 6.1, consider the diffusion process $\bar{g}_t$ in the connected general linear group $\bar{G} = GL(d, \mathbb{R})_+$ satisfying the stochastic differential equation

$$d\bar{g}_t = \sum_{i,j=1}^{d} \bar{g}_t E_{ij} \circ dW_t^{ij}, \tag{6.12}$$

where E_{ij} is the $d \times d$ matrix that has 1 at place (i, j) and 0 elsewhere, and $W_t = \{W_t^{ij}\}$ is a d^2-dimensional standard Brownian motion. As mentioned in Section 5.4, $\bar{G}$ is the product group $\mathbb{R}_+ \times G$, where $G = SL(d, \mathbb{R})$, and possesses the same Cartan and Iwasawa decompositions as G except that A should be replaced by $\bar{A}$, the group of all the $d \times d$ diagonal matrices with positive diagonal elements (not necessarily of determinant 1), and A_+ by $\bar{A}_+ = \{a = \text{diag}\{a_1, \ldots, a_d\} \in \bar{A}; a_1 > a_2 > \cdots > a_d\}$.

The inner product $\langle X, Y \rangle = 2d\,\text{Trace}\,(XY')$ on the Lie algebra $\mathfrak{sl}(d, R)$ of $SL(d, \mathbb{R})$ determined by the Killing form extends directly to be an inner product on the Lie algebra $\bar{\mathfrak{g}} = \mathfrak{gl}(d, \mathbb{R})$ of $\bar{G} = GL(d, \mathbb{R})_+$. However, it seems to be more natural to use another inner product on $\bar{\mathfrak{g}}$ defined by $\langle X, Y \rangle_0 = \text{Trace}(XY')$. Under $\langle \cdot, \cdot \rangle_0$, $\{E_{ij}\}$ is an orthonormal basis of $\bar{\mathfrak{g}}$. By Propositions 2.5, 2.6, and 2.7, $\bar{g}_t$ is a Riemannian Brownian motion in $\bar{G}$ with respect to the left invariant metric on $\bar{G}$ induced by $\langle \cdot, \cdot \rangle_0$ and is right $SO(d)$-invariant.

There is an orthogonal transformation in $\bar{\mathfrak{g}}$ that maps the orthonormal basis $\{E_{ij}\}$ into another orthonormal basis $\{X_1, \ldots, X_{d^2}\}$ such that $X_1 = (1/\sqrt{d})I_d$ and $\{X_2, \ldots, X_{d^2}\}$ is a basis of $\mathfrak{g} = \mathfrak{sl}(d, R)$. By changing to another d^2-dimensional standard Brownian motion, denoted by $(W_t^1, \ldots, W_t^{d^2})$, we can write the stochasitc differential equation (6.12) as

$$d\bar{g}_t = \bar{g}_t X_1 \circ dW_t^1 + \sum_{i=2}^{d^2} \bar{g}_t X_i \circ dW_t^i.$$

Let $\bar{g}_t = u_t g_t$ with $u_t \in \mathbb{R}_+$ and $g_t \in SL(d, \mathbb{R})$. Since $d\bar{g}_t = (\circ du_t)g_t + u_t \circ dg_t$, we obtain $du_t = u_t \circ dW_t^1/\sqrt{d}$ and $dg_t = \sum_{i=2}^{d^2} g_t X_i \circ dW_t^i$. Then $u_t = u_0 \exp(W_t^1/\sqrt{d})$ and g_t is a Riemannian Brownian motion in $G = SL(d, \mathbb{R})$ with respect to the left invariant metric induced by $\langle \cdot, \cdot \rangle_0$ restricted to $\mathfrak{g} = \mathfrak{sl}(d, \mathbb{R})$. By the scaling property mentioned in Section 2.3, $g_{t/(2d)}$ is a Riemannian Brownian motion in G with respect to the left invariant metric induced by $\langle \cdot, \cdot \rangle = 2d\langle \cdot, \cdot \rangle_0$. It follows that, if $g_t = \xi_t a_t^+ \eta_t$ is a Cartan decomposition, then $\xi_t M \to \xi_\infty M$ and $(1/t) \log a_t^+ \to 2dH_\rho$ almost surely as $t \to \infty$, where H_ρ is given by (6.9). Because $(1/t) \log u_t \to 0$, it follows that $\lim_{t\to\infty}(1/t) \log \bar{a}_t^+ = \lim_{t\to\infty}(1/t) \log a_t^+ = 2dH_\rho$.

To summarize, we obtain the following result:

Theorem 6.2. *Let* $\bar{g}_t$ *be a left invariant diffusion process in* $\bar{G} = GL(d, \mathbb{R})_+$ *satisfying the stochastic differential equation (6.12) with the Cartan decomposition* $\bar{g}_t = \xi_t \bar{a}_t^+ \eta_t$ *and the Iwasawa decomposition* $\bar{g}_t = n_t \bar{a}_t k_t$ *of* $\bar{G} = N^- \bar{A} K$.

(a) *Almost surely,* $\xi_t M \to \xi_\infty M$ *for some* K*-valued random variable* ξ_∞, $n_t \to n_\infty$ *for some* N^-*-valued random variable* n_∞, *and*

$$\lim_{t \to \infty} \frac{1}{t} \log \bar{a}_t^+ = \lim_{t \to \infty} \frac{1}{t} \log \bar{a}_t$$
$$= \frac{1}{2} \text{diag} \{d - 1, d - 3, \dots, -(d - 3), -(d - 1)\}.$$
$$(6.13)$$

(b) $\bar{g}_t$ *is right* $SO(d)$*-invariant and is a Riemannian Brownian motion in* $\bar{G}$ *with respect to the left invariant metric induced by the inner product* $\langle X, Y \rangle_0 = \text{Trace} \,(XY')$ *on* $\bar{\mathfrak{g}} = \mathfrak{gl}(d, \mathbb{R})$.

(c) $\bar{g}_t = u_t g_t$, *where* $u_t = u_0 \exp(W_t^1 / \sqrt{d})$ *is a process in* $\mathbb{R}_+$ *for some one-dimensional standard Brownian motion* W_t^1 *and* g_t *is a Riemannian Brownian motion in* $G = SL(d, \mathbb{R})$ *with respect to the left invariant Riemannian metric on* G *induced by the inner product* $\langle \cdot, \cdot \rangle_0$ *restricted to* $\mathfrak{g} = \mathfrak{sl}(d, \mathbb{R})$. *Moreover, the two processes* u_t *and* g_t *are independent.*

Let $s_t = \bar{g}_t \bar{g}_t' = \xi_t (\bar{a}_t^+)^2 \xi_t'$, where $\bar{a}_t^+ = u_t a_t^+$ is a diagonal matrix. The process s_t takes values in the space S of positive definite symmetric matrices. For $x = (x_1, x_2, \dots, x_d) \in \mathbb{R}^d$ considered as a row vector, the equation $x s_t x' = 1$ describes an ellipsoid in $\mathbb{R}^d$ centered at the origin; hence, s_t may be called a random ellipsoid. Its orientation in the space $\mathbb{R}^d$ is determined by $\xi_t M$ and the half lengths of its axes are given by $l_i(t) = 1/\bar{a}_i^+(t)$ for $1 \le i \le d$, where $\bar{a}_i^+(t)$ is the ith diagonal element of $\bar{a}_t^+$. The "angular" convergence $\xi_t M \to \xi_\infty M$ implies that the orientation of the random ellipsoid s_t converges to a limit almost surely as $t \to \infty$, whereas the existence of the "radial" $\lim_{t \to \infty} (1/t) \log \bar{a}_t^+$ means that the lengths of the axes of s_t grows or decays exponentially at nonrandom rates:

$$\lim_{t \to \infty} \frac{1}{t} \log l_i(t) = -\lim_{t \to \infty} \frac{1}{t} \log \bar{a}_i^+(t) = -\frac{d - 2i + 1}{2} \quad \text{for } 1 \le i \le d.$$

Note that the space S can be identified with G/K via the map $G \ni g \mapsto gg' \in S$ with kernel K.

6.4. Invariant Measures and Irreducibility

In Section 1.1, the convolution of two probability measures μ_1 and μ_2 on G is defined to be a probability measure $\mu_1 * \mu_2$ on G given by $\mu_1 * \mu_2(f) = \int_G f(gh)\mu_1(dg)\mu_2(dh)$ for $f \in \mathcal{B}(G)_+$. We now define the convolution of a probability measure μ on G and a probability measure ν on a manifold X on which G acts on the left (resp. right) by

$$\mu * \nu(f) = \int_G \int_X f(gx)\mu(dg)\nu(dx)$$

$$\times \left(\text{resp.} \nu * \mu(f) = \int_G \int_X f(xg)\nu(dx)\mu(dg) \right)$$

for $f \in \mathcal{B}(X)_+$. Equivalently, this may also be defined as $\mu * \nu(B) = \int_G \nu(g^{-1}B)\mu(dg)$ (resp. $\nu * \mu(B) = \int_G \nu(Bg^{-1})\mu(dg)$) for $B \in \mathcal{B}(X)$. The measure ν is called μ-invariant if $\nu = \mu * \nu$ (resp. $\nu = \nu * \mu$). In the following, the manifold X is often taken to be the homogeneous space G/Q for the left action or $Q\backslash G$ for the right action, where Q is a closed subgroup of G.

The homogeneous space $G/(MAN^+)$ can be identified with K/M via the Iwasawa decomposition $G = KAN^+$. Let $\pi: G \to G/(MAN^+)$ be the natural projection. Recall that $\pi(N^-)$ is an open subset of $G/(MAN^+)$ with a lower dimensional complement. A probability measure ν on $G/(MAN^+)$ will be called irreducible if it is supported by $g\pi(N^-) = gN^-(MAN^+)$ for any $g \in G$, where (MAN^+) denotes the point $\pi(e)$ in $G/(MAN^+)$. In other words, $\nu[g\pi(N^-)] = 1$ for any $g \in G$.

Similarly, a probability measure ν on $(MAN^-)\backslash G$ will be called irreducible if it is supported by $(MAN^-)N^+g$ for any $g \in G$.

It is clear that if ν is a probability measure on $X = G/(MAN^+)$ (resp. $X = (MAN^-)\backslash G$) that does not charge any lower dimensional submanifold of X, then it is irreducible.

A subset H of G will be called totally left irreducible on G if there do not exist $g_1, \ldots, g_r, x \in G$ for some integer $r > 0$ such that

$$H \subset \bigcup_{i=1}^{r} g_i(N^-MAN^+)^c x, \qquad (6.14)$$

where the superscript c denotes the complement in G. It will be called totally right irreducible on G if there do not exist $x, g_1, \ldots, g_r \in G$ for some integer $r > 0$ such that

$$H \subset \bigcup_{i=1}^{r} x(N^-MAN^+)^c g_i. \qquad (6.15)$$

Because N^-MAN^+ is an open subset of G with a lower dimensional complement, therefore, if H is not lower dimensional, then it is both totally left and totally right irreducible.

It is well known that a Lie group is an analytic manifold. Later we will show that a connected analytic submanifold H of G is totally left (resp. right) irreducible on G if and only if there do not exist $x, y \in G$ such that $H \subset x(N^- M A N^+)^c y$. In this case, the total left and total right irreducibilities of H mean the same thing and we may simply call H totally irreducible. Because a Lie subgroup of G is an analytic submanifold, this in particular applies to a connected Lie subgroup H of G.

For the moment, the equivalence of the left and right total irreducibilities will be established for $G = SL(d, \mathbb{R})$ in the following proposition. Recall that, for any square matrix g, $g[k]$ denotes the submatrix formed by the first k rows and the first k columns of g.

Proposition 6.4. *Let $G = SL(d, \mathbb{R})$ and let H be a connected analytic submanifold of G. Then H is totally left (resp. right) irreducible if and only if there do not exist an integer k with $1 \leq k < d$ and two matrices $x, y \in G$ such that $\det((xhy)[k]) = 0$ for all $h \in H$.*

Proof. If H is not totally left irreducible, then $\exists g_1, \ldots, g_r, x \in G$ and subsets $H_1, \ldots, H_r$ of H such that $H = \cup_{i=1}^r H_i$ and $g_i^{-1} H_i x^{-1} \subset (N^- M A N^+)^c$. By Proposition 5.20, there exist subsets H_{ik} of H_i for $k = 1, \ldots, (d-1)$ such that $H_i = \cup_{k=1}^{d-1} H_{ik}$ and the function $F_{ik}(h) = \det\{(g_i^{-1} h x^{-1})[k]\}$ vanishes for $h \in H_{ik}$. As an analytic function on H, if F_{ik} vanishes on a set of positive measure, it must vanish identically on H. This proves the claim for the total left irreducibility. The total right irreducibility is proved similarly. $\square$

For any probability measure μ on G, let G_μ be the smallest closed subgroup of G containing $\text{supp}(\mu)$, the support of μ. If G_μ is totally left (resp. right) irreducible, then μ will be called totally left (resp. right) irreducible. Recall that $\Theta: G \to G$ is the Cartan involution. Let $\Theta(\cdot)^{-1}$ denote the map $G \ni g \mapsto [\Theta(g)]^{-1} \in G$. Note that it is not equal to $\Theta^{-1} = \Theta$.

Proposition 6.5.

   (a) *Let μ be a probability measure on G. Then μ is totally left (resp. right) irreducible if and only if $[\Theta(\cdot)^{-1}\mu]$ is totally right (resp. left) irreducible.*
   (b) *The map $\phi: (MAN^-)\backslash G \to G/(MAN^+)$ defined by $(MAN^-)g \mapsto \Theta(g)^{-1}(MAN^+)$ is a diffeomorphism.*
   (c) *Let μ and ν be probability measures on G and on $(MAN^-)\backslash G$ respectively. Then ν is μ-invariant if and only if $\phi\nu$ is $[\Theta(\cdot)^{-1}\mu]$-invariant. Moreover, ν is irreducible on $(MAN^-)\backslash G$ if and only if $\phi\nu$ is irreducible on $G/(MAN^+)$.*

Proof. Since $\Theta^{-1}(\cdot)$ maps $\mathrm{supp}\,(\mu)$ onto $\mathrm{supp}\,[\Theta(\cdot)^{-1}\mu]$, it also maps G_μ onto $G_{\Theta(\cdot)^{-1}\mu}$. Now (a) follows from

$$\Theta[g(N^- MAN^+)h]^{-1} = \Theta(h)^{-1}(N^- MAN^+)\Theta(g)^{-1}.$$

Because $\Theta(MAN^-)^{-1} = N^+ AM = MAN^+$, it is easy to check that the map ϕ in (b) is well defined and is one-to-one and onto. It is also easy to check that it is regular at the point $o = (MAN^-)$. It remains to show that ϕ is regular at every point of $(MAN^-)\backslash G$. For $g \in G$, let r_g: $(MAN^-)\backslash G \to (MAN^-)\backslash G$ be the map $(MAN^-)g' \mapsto (MAN^-)g'g$ and let $l_g \colon G/(MAN^+) \to G/(MAN^+)$ be the map $g'(MAN^+) \mapsto gg'(MAN^+)$. Then r_g and l_g are diffeomorphisms. Since $\phi \circ r_g = l_h \circ \phi$ for $h = \Theta(g)^{-1}$, it follows that ϕ is regular at every point on $(MAN^-)\backslash G$. This proves (b). Let μ and ν be probability measures on G and on $(MAN^-)\backslash G$ respectively. Then for any $f \in \mathcal{B}(G/(MAN^+))_+$,

$$\nu * \mu(f \circ \phi) = \int f(\phi(xg))\nu(dx)\mu(dg) = \int f(\Theta(g)^{-1}\phi(x))\mu(dg)\nu(dx)$$

$$= \int f(hy)[\Theta(\cdot)^{-1}\mu](dh)(\phi\nu)(dy) = [\Theta(\cdot)^{-1}\mu] * (\phi\nu)(f).$$

It follows that ν is μ-invariant if and only if $\phi\nu$ is $[\Theta(\cdot)^{-1}\mu]$-invariant. Moreover, for any $g \in G$, ϕ maps $(MAN^-)N^+g$ onto $\Theta(g)^{-1}N^-(MAN^+)$; therefore, ν is irreducible on $(MAN^-)\backslash G$ if and only if $\phi\nu$ is irreducible on $G/(MAN^+)$. This proves (c). $\qquad\square$

The notion of an irreducible measure on $G/(MAN^+)$ or on $(MAN^-)\backslash G$ was introduced in Guivarc'h and Raugi [24]. A totally left irreducible subset of G defined here was called totally irreducible in [24] without the word "left." The following theorem provides a connection between these two different notions of irreducibility.

Theorem 6.3. *Let μ be a probability measure on G. Then*

(a) *G_μ is totally left irreducible on G if and only if any μ-invariant probability measure ν on $G/(MAN^+)$ is irreducible;*
(b) *G_μ is totally right irreducible on G if and only if any μ-invariant probability measure on $(MAN^-)\backslash G$ is irreducible.*

In [24], this theorem was first proved for $G = SL(d, \mathbb{R})$ by a rather elementary argument using exterior algebra and projective spaces, then the general case was proved using a result on the dimensions of certain algebraic subvarieties of $G/(MAN^+)$. In the following we will adapt the proof for $G = SL(d, \mathbb{R})$ to the general case.

Let $\{X_1, \ldots, X_d\}$ be a basis of $\mathfrak{g}$ and, for $g \in G$, let $\hat{g} = \{c_{ij}(g)\}$ denote the matrix defined by

$$\text{Ad}(g)X_j = \sum_{i=1}^{d} c_{ij}(g)X_i. \tag{6.16}$$

It is well known that any connected semi-simple Lie group G is unimodular in the sense that $\det(\hat{g}) = 1$ for any $g \in G$. To show this, note that the invariance of the Killing form B under $\text{Ad}(g)$ implies $\hat{g}'b\hat{g} = b$, where b is the matrix representing B under the basis of $\mathfrak{g}$. Then $[\det(\hat{g})]^2 = 1$. The connectness of G now implies that $\det(\hat{g}) = 1$.

Let $\hat{G} = SL(d, \mathbb{R})$ and let $J: G \to \hat{G}$ be the map defined by $J(g) = \hat{g}$. It is easy to see that J is a Lie group homomorphism from G into $\hat{G}$ and its kernel is the center Z of G.

As before, let $\langle \cdot, \cdot \rangle$ be the inner product on $\mathfrak{g}$ induced by the Killing form. Choose an orthonormal basis $\{H_1, \ldots, H_m\}$ of $\mathfrak{g}_0$ such that $\{H_1, \ldots, H_a\}$ is a basis of $\mathfrak{a}$. We may assume that $H_1 \in \mathfrak{a}_+$ and $\alpha(H_1) \neq \beta(H_1)$ for any two distinct roots α and β. Let Δ_0 be the set of all the roots and the zero functional on $\mathfrak{a}$. We can introduce an order on Δ_0 by setting $\alpha < \beta$ if $\alpha(H_1) < \beta(H_1)$. Then, for any positive root α, $\alpha > 0 > -\alpha$. Let $\alpha_1 > \alpha_2 > \cdots > \alpha_r$ be the complete set of positive roots. If we set $\alpha_{r+1} = 0, \alpha_{r+2} = -\alpha_r, \alpha_{r+3} = -\alpha_{r-1}, \ldots, \alpha_{2r+1} = -\alpha_1$, then $\Delta_0 = \{\alpha_1 > \alpha_2 > \cdots > \alpha_{2r+1}\}$. For $1 \leq i \leq r$, let $\{X_{1i}, X_{2i}, \ldots, X_{d_i i}\}$ be an orthonormal basis of $\mathfrak{g}_{\alpha_i}$. Let $X_{j(r+1)} = H_j$ for $1 \leq j \leq m$ and let $X_{j(2r+2-i)} = \theta(X_{ji})$ for $1 \leq j \leq d_i$ and $1 \leq i \leq r$. Then, for $1 \leq i \leq (2r + 1)$, $\{X_{1i}, X_{2i}, \ldots, X_{d_i i}\}$ is an orthonormal basis of $\mathfrak{g}_{\alpha_i}$, where $d_{r+1} = m$ and $d_{2r+2-i} = d_i$ for $1 \leq i \leq r$. Let $\{X_1, X_2, \ldots, X_d\}$ be the ordered set given by

$$\{X_{11}, X_{21}, \ldots, X_{d_1 1}, X_{12}, X_{22}, \ldots, X_{d_2 2}, \ldots, X_{1(2r+1)}, \\ X_{2(2r+1)}, \ldots, X_{d_{2r+1}(2r+1)}\} \tag{6.17}$$

and let $k = 2r + 1$. Then $d = \sum_{i=1}^{k} d_i$. Let

$$h_q = \sum_{p=1}^{q} d_p \text{ for } 1 \leq q \leq k \text{ with } h_0 = 0.$$

Any matrix $b = \{c_{ij}\}_{i,j=1,\ldots,d}$ in $\hat{G}$ may be regarded as a linear map $\mathfrak{g} \to \mathfrak{g}$ given by $b(X_j) = \sum_{i=1}^{d} c_{ij}X_i$. It can be expressed in the following block form:

$$b = \begin{bmatrix} b_{11} & b_{21} & \ldots & b_{k1} \\ b_{12} & b_{22} & \ldots & b_{2k} \\ \vdots & \vdots & \ldots & \vdots \\ b_{1k} & b_{2k} & \ldots & b_{kk} \end{bmatrix}, \tag{6.18}$$

where b_{pq} is the $d_p \times d_q$ matrix $\{c_{ij}\}_{h_{p-1}+1 \le i \le h_p, \, h_{q-1}+1 \le j \le h_q}$ for $p, q = 1, 2, \ldots, k$. The submatrix b_{pq} may be regarded as a linear map $\mathfrak{g}_{\alpha_q} \to \mathfrak{g}_{\alpha_p}$ given by $b_{pq}(X_j) = \sum_{i=h_{p-1}+1}^{h_p} c_{ij} X_i$ for $h_{q-1} + 1 \le j \le h_q$. In particular, $b_{r+1\,r+1}$ is a linear endomorphism on $\mathfrak{g}_{\alpha_{r+1}} = \mathfrak{g}_0$.

Let $\hat{K} = SO(d) \subset \hat{G}$ and let

$$\hat{A} = \{b \in \hat{G}; \; b_{pq} = c_p \delta_{pq} I_{d_p} \text{ for some } c_p > 0\},$$
$$\hat{N}^+ = \{b \in \hat{G}; \; b_{pq} = 0 \text{ for } p > q \text{ and } b_{pp} = I_{d_p}\},$$
$$\hat{N}^- = \{b \in \hat{G}; \; b_{pq} = 0 \text{ for } p < q \text{ and } b_{pp} = I_{d_p}\},$$
$$\hat{M} = \{b \in SO(d); \; b_{pq} = 0 \text{ for } p \ne q \text{ and } b_{r+1\,r+1} \text{ fixes elements of } \mathfrak{a}\},$$

and

$$\hat{M}' = \{b \in SO(d); \; \exists \text{ a permutation } \sigma \text{ on } \{1, 2, \ldots, k\} \text{ with}$$
$$\sigma(r + 1) = r + 1 \text{ and } d_p = d_{\sigma(p)} \text{ for } 1 \le p \le k \text{ such that } b_{pq} = 0 \text{ for}$$
$$q \ne \sigma(p) \text{ and } b_{r+1\,r+1} \text{ leaves } \mathfrak{a} \text{ invariant}\}.$$

It is clear that $\hat{K}$, $\hat{A}$, $\hat{N}^+$, $\hat{N}^-$, $\hat{M}$, and $\hat{M}'$ are closed subgroups of $\hat{G}$ with $\hat{M} \subset \hat{M}' \subset \hat{K}$.

For $g \in G$, let $\{b_{pq}(g)\}$ be the matrix in the block form (6.18) given by $J(g)$.

Lemma 6.4. $J^{-1}(\hat{K}) = K$, $J^{-1}(\hat{A}) = ZA$, $J^{-1}(\hat{N}^+) = ZN^+$, $J^{-1}(\hat{N}^-) = ZN^-$, $J^{-1}(\hat{M}) = M$, and $J^{-1}(\hat{M}') = M'$. Consequently, $J(K) \subset \hat{K}$, $J(A) \subset \hat{A}$, $J(N^+) \subset \hat{N}^+$, $J(N^-) \subset \hat{N}^-$, $J(M) \subset \hat{M}$, and $J(M') \subset \hat{M}'$.

Proof. Let $k \in K$. Since $\{X_1, X_2, \ldots, X_d\}$ is an orthonormal basis of $\mathfrak{g}$ under the $\mathrm{Ad}(K)$-invariant inner product $\langle \cdot, \cdot \rangle$, it follows that $J(k) \in \hat{K}$. Each X_i belongs to $\mathfrak{g}_\alpha$ for some $\alpha \in \Delta_0$. Denote this α as γ_i. If $a \in A$, then

$$\mathrm{Ad}(a)X_j = \mathrm{Ad}(e^{\log a})X_j = e^{\mathrm{ad}(\log a)}X_j = e^{\gamma_j(\log a)}X_j.$$

From this it follows that $b_{pq}(a) = c_p(a)\delta_{pq} I_{d_p}$ with $c_p(a) = e^{\alpha_p(\log a)}$; hence, $J(a) \in \hat{A}$. Note that, for two distinct positive roots α and β, the equation $\alpha = \beta$ determines a proper subspace of $\mathfrak{a}$; therefore, there exists $a \in A$ such that $\alpha(\log a) \ne 0$ and $\alpha(\log a) \ne \beta(\log a)$ for any root α and any root β

different from α. For such $a \in A$, the $c_p(a)$ are distinct for $1 \le p \le k$. Because $[\mathfrak{g}_\alpha, \mathfrak{g}_\beta] \subset \mathfrak{g}_{\alpha+\beta}$, for $n \in N^+$, $\mathrm{Ad}(n)X_j = e^{\mathrm{ad}(\log n)}X_j = X_j + R_j$, where $R_j \in \sum_{\beta > \beta_j} \mathfrak{g}_\beta$. By (6.16), this implies that $c_{ij}(n) = 0$ if $\gamma_i < \gamma_j$ and $c_{ij}(n) = \delta_{ij}$ if $\gamma_i = \gamma_j$; hence, $J(n) \in \hat{N}^+$. Similarly, if $n^- \in N^-$, then $J(n^-) \in \hat{N}^-$.

We have proved that $J(K) \subset \hat{K}$, $J(A) \subset \hat{A}$, $J(N^+) \subset \hat{N}^+$, and $J(N^-) \subset \hat{N}^-$. Let $g \in G$ and let $g = kan$ be the Iwasawa decomposition $G = KAN^+$. If $J(g) \in \hat{K}$, then $J(an) \in \hat{K} \cap (\hat{A}\hat{N}^+)$. By the uniqueness of the Iwasawa decomposition on $\hat{G} = SL(d, \mathbb{R})$, $J(an) = I_d$ and $an \in Z \subset K$. This implies $an = e$ and $g \in K$; hence, $J^{-1}(\hat{K}) \subset K$. By $J(K) \subset \hat{K}$, we have $J^{-1}(\hat{K}) = K$. Now suppose $J(g) \in \hat{A}$. Then $J(k) = J(n) = I_d$; hence, both k and n must belong to Z, which forces $n = e$ and $g \in ZA$. This proves $J^{-1}(\hat{A}) = ZA$. Similarly, one can prove $J^{-1}(\hat{N}^+) = ZN^+$ and $J^{-1}(\hat{N}^-) = ZN^-$.

For $m \in M$ and $a \in A$, $ma = am$; hence, $b_{pq}(m)c_q(a) = c_p(a)b_{pq}(m)$. We have shown earlier that $a \in A$ may be chosen such that the $c_p(a)$ are distinct. Therefore, $b_{pq}(m) = 0$ for $p \ne q$. Because $\mathrm{Ad}(m)H = H$ for $H \in \mathfrak{a}$, $b_{r+1\,r+1}(m)$ fixes the elements of $\mathfrak{a}$. It follows that $J(M) \in \hat{M}$. By Proposition 5.11(d), any $m' \in M'$ permutes the root spaces $\mathfrak{g}_\alpha$ and leaves $\mathfrak{g}_0$ invariant. Let $a \in A$ with distinct $c_p(a) = e^{\alpha_p(\log a)}$. Then $J(a) = \mathrm{diag}\{c_1(a)I_{d_1}, \dots, c_k(a)I_{d_k}\}$ and

$$J(m')J(a)J(m')^{-1} = J(m'am'^{-1}) = \mathrm{diag}\{c_{\sigma(1)}I_{d_1}, \dots, c_{\sigma(k)}I_{d_k}\},$$

where σ is a permutation of $\{1, 2, \dots, k\}$ with $\sigma(r+1) = r+1$. It follows that $b_{pq}(m')c_q(a) = c_{\sigma(p)}(a)b_{pq}(m')$; hence, $b_{pq}(m') = 0$ for $q \ne \sigma(p)$. Since $\mathrm{Ad}(m')$ leaves $\mathfrak{a}$ invariant, this shows $J(M') \subset \hat{M}'$. Finally, by the definition of $\hat{M}$ (resp. $\hat{M}'$) and because $J^{-1}(\hat{K}) = K$, if $g \in G$ and $J(g) \in \hat{M}$ (resp. $J(g) \in \hat{M}'$), then $g \in K$ and $\mathrm{Ad}(g)$ fixes elements in $\mathfrak{a}$ (resp. leaves $\mathfrak{a}$ invariant). This shows that $g \in M$ (resp. $g \in M'$); hence, $J^{-1}(\hat{M}) = M$ (resp. $J^{-1}(\hat{M}') = M'$). $\qquad\square$

For $1 \le i \le d$, let $\bigwedge_i \mathbb{R}^d$ be the vector space spanned by the exterior products

$$u_1 \wedge u_2 \wedge \cdots \wedge u_i, \quad \text{where } u_1, u_2, \dots, u_i \in \mathbb{R}^d.$$

Let $\{e_1, e_2, \dots, e_d\}$ be the standard basis of $\mathbb{R}^d$. Then

$$e_{j_1} \wedge e_{j_2} \wedge \cdots e_{j_i}, \quad \text{where } 1 \le j_1 < j_2 < \cdots < j_i \le d,$$

form a basis of $\bigwedge_i \mathbb{R}^d$. It is well known that $u_1, \dots, u_i \in \mathbb{R}^d$ are linearly independent if and only if $u_1 \wedge \cdots \wedge u_i \ne 0$. Moreover, if $w = \{w_{jk}\}_{j,k=1,2,\dots,i}$

is an $i \times i$ matrix and $v_k = \sum_{j=1}^{i} w_{jk} u_j$ for $1 \leq k \leq i$, then $v_1 \wedge \cdots \wedge v_i = \det(w) u_1 \wedge \cdots \wedge u_i$. Consequently, any two linearly independent subsets $\{u_1, \ldots, u_i\}$ and $\{v_1, \ldots, v_i\}$ of $\mathbb{R}^d$ span the same subspace of $\mathbb{R}^d$ if and only if $u_1 \wedge \cdots \wedge u_i = c(v_1 \wedge \cdots \wedge v_i)$ for some nonzero $c \in \mathbb{R}$. Let $P(\bigwedge_i \mathbb{R}^d)$ be the projective space on $\bigwedge_i \mathbb{R}^d$, which by definition is the set of the one-dimensional subspaces of $\bigwedge_i \mathbb{R}^d$ and can be identified with $O(n)/[O(n-1) \times O(1)]$, where $n = \dim(\bigwedge_i \mathbb{R}^d) = d!/[i!(d-i)!]$. For a nonzero $u \in \bigwedge_i \mathbb{R}^d$, let $\bar{u} \in P(\bigwedge_i \mathbb{R}^d)$ be the one-dimensional subspace of $\bigwedge_i \mathbb{R}^d$ containing u.

Any $b \in \hat{G} = SL(d, \mathbb{R})$ acts on $\bigwedge_i \mathbb{R}^d$ by $u_1 \wedge \cdots \wedge u_i \mapsto b(u_1 \wedge \cdots \wedge u_i) = (bu_1) \wedge \cdots \wedge (bu_i)$ and on $P(\bigwedge_i \mathbb{R}^d)$ by $\bar{u} \mapsto b\bar{u} = \overline{bu}$.

Recall $h_q = \sum_{p=1}^{q} d_p$ for $1 \leq q \leq k$. Let $\Lambda_q = \bigwedge_{h_q} \mathbb{R}^d$ and $f_q = e_1 \wedge e_2 \cdots \wedge e_{h_q}$. If $b \in \hat{M}\hat{A}\hat{N}^+$, then $b_{pq} = 0$ for $p > q$ and b_{pp} is invertible; hence, it fixes the point $\overline{f_q}$ in $P(\Lambda_q)$. It follows that the map

$$F_q : \ G/(MAN^+) \to P(\Lambda_q) \ \text{ given by } \ g(MAN^+) \mapsto J(g)\overline{f_q}$$

is well defined. For $y \in G$ and $1 \leq q < k$, let

$$H_q^y = \{\bar{u} \in P(\Lambda_q); \ u \in \Lambda_q, u \neq 0, \text{ and } u \wedge [J(y)f_q'] = 0\},$$

where $f_q' = e_{h_q+1} \wedge e_{h_q+2} \wedge \cdots \wedge e_d$. Let $\pi: G \to G/(MAN^+)$ be the natural projection.

Lemma 6.5. *For $y \in G$,*

$$y[\pi(N^-)]^c = \bigcup_{q=1}^{k-1} F_q^{-1}(H_q^y),$$

where $[\pi(N^-)]^c$ is the complement of $\pi(N^-)$ in $G/(MAN^+)$.

Proof. For any matrix $b \in \hat{G}$ and $1 \leq i \leq d$, let $b[i]$ be the submatrix of b formed by the first i rows and the first i columns of b. Because, for $b \in \hat{M}\hat{A}\hat{N}^+$, $b_{pq} = 0$ for $p > q$ and b_{pp} is invertible, it follows that, for $1 \leq q \leq k$,

$$\forall b' \in \hat{G} \text{ and } b \in \hat{M}\hat{A}\hat{N}^+, \quad (b'b)[h_q] = b'[h_q]\, b[h_q]. \tag{6.19}$$

From this we see that if $b' \in \hat{N}^-$ and $b \in \hat{M}\hat{A}\hat{N}^+$, then $\det((b'b)[h_q]) \neq 0$ for $1 \leq q \leq k-1$. Therefore,

$$\hat{N}^-\hat{M}\hat{A}\hat{N}^+ \subset \{b \in \hat{G}; \ \det(b[h_q]) \neq 0 \text{ for } 1 \leq q \leq k-1\}. \tag{6.20}$$

By the definition of $\hat{M}'$, any $b \in \hat{M}'$ determines a permutation σ on the set $\{1, 2, \ldots, k\}$. However, if $s = m_s M$ is not the identity element e_W of W, then $\mathrm{Ad}(m_s)$ permutes the set of the spaces

$$\mathfrak{g}_{\alpha_1}, \ldots, \quad \mathfrak{g}_{\alpha_r}, \quad \mathfrak{g}_{\alpha_{r+1}} = \mathfrak{g}_0, \quad \mathfrak{g}_{\alpha_{r+2}} = \mathfrak{g}_{-\alpha_r}, \ldots, \quad \mathfrak{g}_{\alpha_{2r+1}} = \mathfrak{g}_{-\alpha_1}$$

nontrivially. Therefore, $J(m_s)$ determines a nontrivial permutation σ. This implies that, for any $b \in \hat{G}$ given by the block form (6.18), $bJ(m_s)$ is obtained from b by a nontrivial permutation of columns in (6.18). From this it is easy to show that, if $b \in \hat{N}^-$, then the determinant of $[bJ(m_s)][h_q]$ vanishes for some $1 \leq q \leq k - 1$. It follows from (6.19) that, if $s \neq e_W$, then

$$\hat{N}^- J(m_s) \hat{M} \hat{A} \hat{N}^+ \subset \{b \in \hat{G}; \ \det(b[h_q]) = 0 \text{ for some } q \text{ with}$$
$$1 \leq q \leq k - 1\}. \tag{6.21}$$

By the Bruhat decomposition, $G = \bigcup_{s \in W} N^- m_s MAN^+$ is a disjoint union; hence,

$$y[\pi(N^-)]^c = \bigcup_{s \in W, s \neq e_W} y\pi(N^- m_s) = \pi \left[\bigcup_{s \in W, s \neq e_W} y(N^- m_s MAN^+) \right].$$

Any $z \in G/(MAN^+)$ can be expressed as $z = \pi(g) = g(MAN^+)$ for some $g \in G$. Then $z \in y[\pi(N^-)]^c$ if and only if $g \in y(N^- m_s MAN^+)$ for some $s \neq e_W$. By Lemma 6.4, (6.20), and (6.21), and the fact that for any $b \in \hat{G}$, $(bf_q) \wedge f_q' = \det(b[h_q])(f_q \wedge f_q')$, we obtain the following equivalent relations:

$$z \in y[\pi(N^-)]^c \iff J(y^{-1}g) \in \hat{N}^- J(m_s) \hat{M} \hat{A} \hat{N}^+ \text{ for some } s \neq e_W$$
$$\iff \det(J(y^{-1}g)[h_q]) = 0 \text{ for some } 1 \leq q \leq k - 1$$
$$\iff [J(y^{-1}g)f_q] \wedge f_q' = 0 \text{ for some } 1 \leq q \leq k - 1$$
$$\iff [J(g)f_q] \wedge [J(y)f_q'] = 0 \text{ for some } 1 \leq q \leq k - 1$$
$$\iff z \in F_q^{-1}(H_q^y) \text{ for some } 1 \leq q \leq k - 1. \qquad \square$$

Proof of Theorem 6.3. Assume the μ-invariant probability measure ν on $G/(MAN^+)$ is not irreducible; that is, it charges $y[\pi(N^-)]^c$ for some $y \in G$. We want to show that (6.14) holds with $H = G_\mu$ for some $g_1, \ldots, g_r, x \in G$. By Lemma 6.5, ν must charge $F_q^{-1}(H_q^y)$ for some $q \in \{1, 2, \ldots, k - 1\}$, or equivalently, $F_q\nu(H_q^y) > 0$.

Let $\mathcal{L}$ be the set of the subspaces U of Λ_q such that $\exists \xi \in G, \forall u \in U, u \wedge [J(\xi)f_q'] = 0$. If $g \in G$ and $u \in \Lambda_q$, then $[J(g)u] \wedge [J(\xi)f_q'] = J(g)\{u \wedge [J(g^{-1}\xi)f_q']\}$. It follows that, for any $g \in G$ and $U \in \mathcal{L}$, $J(g)U \in \mathcal{L}$. For any subspace U of Λ_q, let $\overline{U} = \{\bar{u} \in P(\Lambda_q); 0 \neq u \in U\}$ and set $\overline{\{0\}} = \emptyset$. Define

$$U_\xi = \{u \in \Lambda_q; \ u \wedge [J(\xi)f_q'] = 0\}$$

for $\xi \in G$. Then U_ξ is a subspace contained in $\mathcal{L}$ and $\overline{U_\xi} = H_q^\xi$. Moreover, if U is a subspace contained in $\mathcal{L}$, then $U \subset U_\xi$ for some $\xi \in G$. Because $F_q \nu(\overline{U_y}) = F_q \nu(H_q^y) > 0$, $\mathcal{L}$ contains subspaces U with $F_q \nu(\overline{U}) > 0$. Let V be such a subspace of the minimal dimension. Note that $\dim(V) \geq 1$. Then, for $g, h \in G$, either $J(g)V = J(h)V$ or $\dim\{[J(g)V] \cap [J(h)V]\} < \dim(V)$, and in the latter case, $F_q \nu\{[J(g)\overline{V}] \cap [J(h)\overline{V}]\} = F_q \nu\{\overline{[J(g)V] \cap [J(h)V]}\} = 0$. Let $c_q = \sup_{g \in G} F_q \nu[J(g)\overline{V}]$. It is now easy to show that $\exists g_1, g_2, \ldots, g_r \in G$ for some integer $r > 0$ such that $F_q \nu[J(g_i)\overline{V}] = c_q$ for $1 \leq i \leq r$, $F_q \nu\{[J(g_i)\overline{V}] \cap [J(g_j)\overline{V}]\} = 0$ for $i \neq j$, and $\forall g \in G$, either $J(g)\overline{V} = J(g_i)\overline{V}$ for some $i \in \{1, 2, \ldots, r\}$ or $F_q \nu[J(g)\overline{V}] < c_q$.

Let $\Gamma = F_q^{-1}[\bigcup_{i=1}^r J(g_i)\overline{V}]$. Then $\forall g \in G$, either $g\Gamma = \Gamma$ or $\nu(g\Gamma) < \nu(\Gamma)$. Because ν is μ-invariant,

$$\nu(\Gamma) = \int_G \nu(g^{-1}\Gamma)\mu(dg).$$

It follows that $\forall g \in \operatorname{supp}(\mu)$, $g^{-1}\Gamma = \Gamma$; therefore, $G_\mu \Gamma = \Gamma$. In particular, for $z \in \Gamma$,

$$G_\mu z \subset \Gamma.$$

Note that $V \subset U_\xi$ for some $\xi \in G$; hence,

$$F_q^{-1}(\overline{V}) \subset F_q^{-1}(\overline{U_\xi}) = F_q^{-1}(H_q^\xi) \subset \xi[\pi(N^-)]^c$$

and

$$\Gamma = \bigcup_{i=1}^r g_i F_q^{-1}(\overline{V}) \subset \bigcup_{i=1}^r g_i \xi[\pi(N^-)]^c = \bigcup_{i=1}^r \pi[g_i \xi(N^- MAN^+)^c].$$

For $z \in \Gamma$, let $z = \pi(x^{-1})$ for some $x \in G$. Then

$$G_\mu x^{-1} \subset \bigcup_{i=1}^r g_i \xi(N^- MAN^+)^c,$$

which implies that $G_\mu \subset \bigcup_{i=1}^r g_i \xi(N^- MAN^+)^c x$. This is (6.14) if we replace H by G_μ and $g_i \xi$ by g_i.

We have proved that if G_μ is totally left irreducible on G, then any μ-invariant probability measure ν on $G/(MAN^+)$ is irreducible. Now we prove the converse. Let μ be a probability measure on G that is not totally left irreducible; that is, (6.14) holds for $H = G_\mu$ and some $g_1, g_2, \ldots, g_r, x \in G$. Let C be the closure of $\pi(G_\mu x^{-1})$ in $G/(MAN^+)$. Then C is contained in $\bigcup_{i=1}^r g_i [\pi(N^-)]^c$ and is G_μ-invariant; that is, $gC \subset C$ for any $g \in G_\mu$.

Let ν_0 be any probability measure supported by C and let $\nu_j = (1/j)[\nu_0 + \sum_{i=1}^{j-1} \mu^{*i} * \nu_0]$, where $\mu^{*i} = \mu * \cdots * \mu$ is an i-fold convolution. The G_μ-invariance of C implies the ν_j is supported by C. Because C is compact, a subsequence of ν_j converges weakly to a μ-invariant probability measure ν supported by C. This measure is evidently not irreducible. The proof of (a) of Theorem 6.3 is now completed.

By Proposition 6.5, it is easy to derive (b) of Theorem 6.3 from (a).   □

For a probability measure μ on G, let T_μ be the smallest closed semigroup in G containing supp(μ). Here, a subset C of G is called a semigroup if $e \in C$ and for any $g, h \in C$, $gh \in C$.

Note that, at the end of the proof of Theorem 6.3 when constructing a μ-invariant probability measure ν that is not irreducible, one may let μ be a probability measure on G such that T_μ is not totally left irreducible—that is, (6.14) holds for $H = T_\mu$. Let C be the closure of $\pi(T_\mu x^{-1})$ in $G/(MAN^+)$. Then C is T_μ-invariant and in the exactly same way one can show that there is a μ-invariant probability measure on $G/(MAN^+)$ that is not irreducible. Therefore, the irreducibility of μ-invariant probability measures on $G/(MAN^+)$ also implies the total left irreducibility of T_μ on G. Since $T_\mu \subset G_\mu$, the total left irreducibility of T_μ is apparently stronger than that of G_μ, and it follows that the two are in fact equivalent. The same holds for the total right irreducibility and we have following result:

Corollary 6.1. *For any probability measure μ on G, G_μ is totally left (resp. right) irreducible if and only if T_μ is totally left (resp. right) irreducible.*

We now establish the equivalence of the total left and total right irreducibilities for a connected analytic submanifold of G such as a connected Lie subgroup of G.

Proposition 6.6. *Let H be a connected analytic submanifold of G. Then H is totally left (resp. right) irreducible on G if and only if there do not exist $x, y \in G$ such that $H \subset x(N^- MAN^+)^c y$. Consequently, H is totally left irreducible if and only if it is totally right irreducible.*

Proof. By the Bruhat decomposition, (6.20), and (6.21), if H is not totally left irreducible, then there exist $g_1, \ldots, g_r, x \in G$ and $H_{1q}, \ldots, H_{rq} \subset H$ for $1 \leq q \leq (k-1)$ such that $H = \bigcup_{i,q} H_{iq}$ and the analytic function $\phi_{iq}: H \to \mathbb{R}$ defined by $g \mapsto \det(J(g_i^{-1} g x^{-1})[h_q])$ vanishes on H_{iq}. If H_{iq}

has a positive measure on H, then ϕ_{iq} vanishes identically on H. This implies that $g_i^{-1} H x^{-1} \subset (N^- M A N^+)^c$.   □

6.5. Limiting Properties of Lévy Processes

Let g_t be a left (resp. right) Lévy process in G with $g_t^e = g_0^{-1} g_t$ (resp. $g_t^e = g_t g_0^{-1}$) and let μ_t be the distribution of g_t^e, that is, $\mu_t = P_t(e, \cdot)$, where P_t is the transition semigroup of g_t. In Section 1.1, we have seen that $\{\mu_t\}_{t \in \mathbb{R}_+}$ is a continuous convolution semigroup of probability measures on G, that is, $\mu_t * \mu_s = \mu_{s+t}$ and $\mu_t \to \delta_e$ weakly as $t \to 0$.

Let Q be a closed subgroup of G. A probability measure ν on G/Q will be called a stationary measure of the process g_t on G/Q if it is μ_t-invariant, that is, if $\nu = \mu_t * \nu$, for any $t \in \mathbb{R}_+$. This is equivalent to $\nu(f) = E[g_t^e \nu(f)]$ for any $f \in \mathcal{B}(G/Q)_+$ and $t \in \mathbb{R}_+$. If G/Q is compact, then there is always a stationary measure of g_t on G/Q. In fact, if ν is a probability measure on G/Q, then a subsequence of $(1/n) \int_0^n \mu_t * \nu \, dt$ converges weakly to a stationary measure of g_t on G/Q as $n \to \infty$.

If g_t is a right Lévy process, then by Proposition 2.1, its one-point motion $x_t = g_t x$, $x \in G/Q$, is a Markov process. In this case, it is easy to see that a stationary measure of g_t on G/Q is also a stationary measure of the Markov process x_t as defined in Appendix B.1. Note the difference between the stationary measure of g_t on G/Q and the stationary measure of g_t as a Markov process in G.

Similarly, a probability measure ν on $Q \backslash G$ is called a stationary measure of the process g_t on $Q \backslash G$ if it is μ_t-invariant, that is, if $\nu = \nu * \mu_t$, for any $t \in \mathbb{R}_+$. Such a measure always exists on a compact $Q \backslash G$.

Define

$$\mu = \int_0^\infty e^{-t} dt \, \mu_t. \tag{6.22}$$

Then μ is a probability measure on G. As before, let G_μ and T_μ denote respectively the smallest closed subgroup and the smallest closed semigroup of G containing supp(μ).

Proposition 6.7. *supp(μ) is equal to the closure of $\bigcup_{t \in \mathbb{R}_+}$ supp(μ_t) in G and is a semigroup. Consequently, $T_\mu = $ supp(μ).*

Proof. By the right continuity of the process g_t^e, it is easy to show that $\bigcup_{t \in \mathbb{R}_+}$ supp$(\mu_t) \subset$ supp(μ). Now suppose $g \in$ supp(μ). Then for any $f \in C(G)_+$ with $f(g) > 0$, $\mu(f) > 0$ and $\mu_t(f) > 0$ for some $t \in \mathbb{R}_+$. It follows that any neighborhood of g intersects the closure of $\bigcup_{t \in \mathbb{R}_+} \mathrm{supp}(\mu_t)$, which implies $\mathrm{supp}(\mu)$ is equal to this closure. Because for any two probability measures ν_1 and ν_2 on G, $[\mathrm{supp}(\nu_1)][\mathrm{supp}(\nu_2)] \subset \mathrm{supp}(\nu_1 * \nu_2)$, it follows that $\bigcup_{t \in \mathbb{R}_+} \mathrm{supp}(\mu_t)$ is a semigroup and so is its closure. $\qquad \square$

By Proposition 6.7, we see that T_μ and G_μ are respectively the smallest closed semigroup and the smallest closed group containing $\mathrm{supp}(\mu_t)$ for any $t \in \mathbb{R}_+$; therefore, they will be called respectively the semigroup and the group generated by the process g_t^e.

In Section 6.1, we have introduced the notions of contracting sequences in G and positive contracting sequences in A. We now extend these definitions to functions defined on $\mathbb{R}_+$. A function $a \colon \mathbb{R}_+ \to A$ is called positive contracting if $\alpha(\log a(t)) \to \infty$ as $t \to \infty$ for any positive root α, and a function $g \colon \mathbb{R}_+ \to G$ is called contracting if, in its Cartan decomposition $g(t) = \xi(t)a^+(t)\eta(t)$, $a^+(t)$ is positive contracting.

Let Q be a closed subgroup of G and let ν be a measure on G/Q. By the left action of G on G/Q, any $g \in G$ may be regarded as a map $G/Q \to G/Q$; therefore, $g\nu$ is a measure on G/Q given by $g\nu(f) = \int f(gx)\nu(dx)$ for $f \in \mathcal{B}(G/Q)_+$. Similarly, if ν is a measure on $Q \backslash G$ and $g \in G$, then using the right action of G on $Q \backslash G$, νg is a measure on $Q \backslash G$ given by $\nu g(f) = \int f(xg)\nu(dx)$ for $f \in \mathcal{B}(Q \backslash G)_+$.

The basic limiting properties of a left Lévy process in G are given in the following theorem:

Theorem 6.4. *Let g_t be a left Lévy process in G, and let G_μ and T_μ be respectively the group and the semigroup generated by the process $g_t^e = g_0^{-1}g_t$. Assume G_μ is totally left irreducible and T_μ contains a contracting sequence in G.*

(a) *There is a random variable z taking values in $G/(MAN^+)$, independent of g_0, such that for any irreducible probability measure ν on $G/(MAN^+)$, P-almost surely, $g_t^e \nu$ converges to δ_z weakly as $t \to \infty$ (and, consequently, $g_t\nu \to \delta_{g_0 z}$ weakly). Moreover, the distribution of z is the unique stationary measure of g_t on $G/(MAN^+)$.*

(b) *Let $g_t = \xi_t a_t^+ \eta_t$ be a Cartan decomposition. Then, P-almost surely, a_t^+ is positive contracting and $\xi_t(MAN^+) \to g_0 z$ as $t \to \infty$.*

(c) *Let $g_t = n_t a_t k_t$ be the Iwasawa decomposition $G = N^- AK$. Then, P-almost surely, a_t is positive contracting and $n_t \to n_\infty$ as $t \to \infty$ for some random variable n_∞ taking values in N^-. Moreover, $n_\infty(MAN^+) = g_0 z$ P-almost surely.*

Note that the convergence $\xi_t(MAN^+) \to g_0 z$ in Theorem 6.4 is equivalent to $\xi_t M \to \xi_\infty M$ for some K-valued random variable ξ_∞ satisfying $\xi_\infty(MAN^+) = g_0 z$. Recall that (ξ_t, η_t) in the Cartan decomposition of $g_t = \xi_t a_t^+ \eta_t$ is not unique, but as a_t^+ is positive contracting, $a_t^+ \in \mathfrak{a}_+$ for large $t > 0$; hence, all the possible choices for (ξ_t, η_t) are given by $(\xi_t m, m^{-1}\eta_t)$ for $m \in M$. The convergence $\xi_t M \to \xi_\infty M$ is not affected by this nonuniqueness. Moreover, $m \in M$ may be chosen properly so that $\xi_t \to \xi_\infty$.

Before proving Theorem 6.4, we note that the corresponding results for a right Lévy process can be derived from this theorem using the transformation $g \mapsto \Theta(g)^{-1}$. Let g_t be a right Lévy process on G, and let $g_t = \xi_t a_t^+ \eta_t$ and $g_t = k_t a_t n_t$ be, respectively, a Cartan decomposition and the Iwasawa decomposition $G = KAN^+$. Then $g_t' = \Theta(g_t)^{-1}$ is a left Lévy process, $g_t' = \eta_t^{-1} a_t^+ \xi_t^{-1}$ is a Cartan decomposition, and $g_t' = n_t' a_t k_t^{-1}$ with $n_t' = \Theta(n_t)^{-1}$ is the Iwasawa decomposition $G = N^- AK$. By Proposition 6.5, we can easily derive the following results from Theorem 6.4.

Theorem 6.5. *Let g_t be a right Lévy process in G, and let G_μ and T_μ be respectively the group and the semigroup generated by the process $g_t^e = g_t g_0^{-1}$. Assume G_μ is totally right irreducible and T_μ contains a contracting sequence in G.*

(a) There is a random variable $\tilde{z}$ taking values in $(MAN^-)\backslash G$, independent of g_0, such that, for any irreducible probability measure $\tilde{\nu}$ on $(MAN^-)\backslash G$, P-almost surely, $\tilde{\nu} g_t^e \to \delta_{\tilde{z}}$ weakly as $t \to \infty$ (and, consequently, $\tilde{\nu} g_t \to \delta_{\tilde{z} g_0}$ weakly). Moreover, the distribution of $\tilde{z}$ is the unique stationary measure of g_t on $(MAN^-)\backslash G$.

(b) Let $g_t = \xi_t a_t^+ \eta_t$ be a Cartan decomposition. Then, P-almost surely, a_t^+ is positive contracting and $(MAN^-)\eta_t \to \tilde{z} g_0$ as $t \to \infty$.

(c) Let $g_t = k_t a_t n_t$ be the Iwasawa decomposition $G = KAN^+$. Then, P-almost surely, a_t is positive contracting and $n_t \to n_\infty$ as $t \to \infty$ for some random variable n_∞ taking values in N^+. Moreover, $(MAN^-)n_\infty = \tilde{z} g_0$ P-almost surely.

We note that the convergence $(MAN^-)\eta_t \to \tilde{z} g_0$ in Theorem 6.5(b) is equivalent to $M\eta_t \to M\eta_\infty$ for some K-valued random variable η_∞ satisfying $(MAN^-)\eta_\infty = \tilde{z} g_0$, and, by choosing η_t suitably in the coset $M\eta_t$, we may assume that $\eta_t \to \eta_\infty$.

The following lemma will be needed in the proof of Theorem 6.4.

Lemma 6.6. *Let g_j be a sequence in G and ν be an irreducible probability measure on $G/(MAN^+)$. If $g_j\nu \to \delta_z$ weakly for some $z \in G/(MAN^+)$, then, for any irreducible probability measure ν' on $G/(MAN^+)$, $g_j\nu' \to \delta_z$ weakly. Moreover, if $g_j = \xi_j a_j^+ \eta_j$ is a Cartan decomposition, then a_j^+ is positive contracting and $\xi_j(MAN^+) \to z$ in $G/(MAN^+)$.*

Proof. Let $\{X_1, X_2, \ldots, X_p\}$ be a basis of $\mathfrak{n}^-$ such that each X_i belongs to $\mathfrak{g}_{-\alpha}$ for some $\alpha \in \Delta_+$. This α will be denoted by α_i. Because K is compact, by taking a sub sequence, we may assume $\xi_j \to \xi$ and $\eta_j \to \eta$ in K, and for $1 \leq i \leq p, \alpha_i(\log a_j^+) \to \lambda_i$ as $j \to \infty$, where $\lambda_i \in [0, \infty]$. Let $\pi\colon G \to G/(MAN^+)$ be the natural projection and write (MAN^+) for the point $\pi(e)$ in $G/(MAN^+)$. Any $x \in \pi(\eta^{-1}N^-) = \eta^{-1}N^-(MAN^+)$ can be written as $x = \eta^{-1}\exp(\sum_{i=1}^p c_i(x)X_i)(MAN^+)$, where $c_i(x) \in \mathbb{R}$ may be used as local coordinates of x on the open subset $\eta^{-1}N^-(MAN^+)$ of $G/(MAN^+)$. We have, for any $x \in \eta^{-1}N^-(MAN^+)$,

$$\eta_j x = \exp\left[\sum_{i=1}^p c_{ij}(x)X_i\right](MAN^+),$$

where $c_{ij}(x) \to c_i(x)$ as $j \to \infty$, and

$$g_j x = \xi_j a_j^+ \eta_j x = \xi_j a_j^+ \exp\left[\sum_{i=1}^p c_{ij}(x)X_i\right](MAN^+)$$

$$= \xi_j \exp\left[\sum_{i=1}^p c_{ij}(x)\mathrm{Ad}(a_j^+)X_i\right]a_j^+(MAN^+)$$

$$= \xi_j \exp\left[\sum_{i=1}^p c_{ij}(x)e^{-\alpha_i(\log a_j^+)}X_i\right](MAN^+)$$

$$\to \xi \exp\left[\sum_{i=1}^p c_i(x)e^{-\lambda_i}X_i\right](MAN^+) \qquad (6.23)$$

as $j \to \infty$. Because the probability measure ν is irreducible on $G/(MAN^+)$, it does not charge the complement of $\eta^{-1}N^-(MAN^+)$; therefore, for any $f \in C(G/(MAN^+))$,

$$g_j\nu(f)$$
$$= \int f(g_j x)\nu(dx) \to \int f\left(\xi \exp\left[\sum_{i=1}^p c_i(x)e^{-\lambda_i}X_i\right](MAN^+)\right)\nu(dx).$$

However, since $g_j \nu \to \delta_z$ weakly, $g_j \nu(f) \to f(z)$. Therefore, for $f \in C(G/(MAN^+))$,

$$f(z) = \int f \left(\xi \exp \left[\sum_{i=1}^{p} c_i(x) e^{-\lambda_i} X_i \right] (MAN^+) \right) \nu(dx).$$

It follows that $z = \xi \exp[\sum_{i=1}^{p} c_i(x) e^{-\lambda_i} X_i](MAN^+)$ for ν-almost all $x \in \eta^{-1} N^- (MAN^+)$. Let $\mathfrak{n}'$ be the subspace of $\mathfrak{n}^-$ spanned by the X_is with $\lambda_i = \infty$ and let Π be the set of indices i with $\lambda_i < \infty$. Then $z = \xi \exp[\sum_{i \in \Pi} c_i^0 e^{-\lambda_i} X_i](MAN^+)$ for some constants c_i^0 and ν must be supported by the set of $x \in \eta^{-1} N^- (MAN^+)$ such that $c_i(x) = c_i^0$ for $i \in \Pi$. Replacing ν by $h\nu$ for a properly chosen $h \in K$ and g_j by $g_j h^{-1}$, we may assume that the support of ν contains the point $\xi(MAN^+)$ in $G/(MAN^+)$. Note that these replacements will not change ξ and z. Then $c_i^0 = 0$ for $i \in \Pi$; hence, ν is supported by the set $\xi \exp(\mathfrak{n}')(MAN^+)$ and $z = \xi(MAN^+)$. If we can show $\mathfrak{n}' = \mathfrak{n}^-$, then a_j^+ is positive contracting and $\xi_j(MAN^+) \to z$. Moreover, by (6.23), it is easy to show that, for any irreducible probability measure ν' on $G/(MAN^+)$, $g_j \nu' \to \delta_z$ weakly.

It remains to show $\mathfrak{n}' = \mathfrak{n}^-$. If not, there is a finite λ_i, say λ_1. By decomposing the associated root α_1 into a sum of simple roots, we may assume that α_1 is simple. Let s_1 be the reflection in $\mathfrak{a}$ about the hyperplane $\alpha_1 = 0$. By Proposition 5.13(b), s_1 permutes the positive roots that are not proportional to α_1. If $\lambda_i = \infty$, then α_i cannot be proportional to α_1; hence, $\alpha_i \circ s_1 \in \Delta_+$. Regard s_1 as an element of the Weyl group $W = M'/M$ and let $m_1 \in M'$ represent s_1; that is, $s_1 = m_1 M$. By Proposition 5.11(d) and noting $s_1 = s_1^{-1}$, we have $\mathrm{Ad}(m_1)\mathfrak{g}_\alpha = \mathfrak{g}_{\alpha \circ s_1}$. This implies that $\mathrm{Ad}(m_1)\mathfrak{n}' \subset \mathfrak{n}^-$ and $m_1 \exp(\mathfrak{n}') \subset N^- m_1$. By the Bruhat decomposition $G = \bigcup_{s \in W} N^- m_s MAN^+$, we see that $m_1 \exp(\mathfrak{n}')$ is contained in the complement of $N^- MAN^+$ in G. Since ν is supported by

$$\xi \exp(\mathfrak{n}')(MAN^+) = \xi m_1^{-1}[m_1 \exp(\mathfrak{n}')](MAN^+) = \xi m_1^{-1} \pi[m_1 \exp(\mathfrak{n}')],$$

it follows that ν is supported by $\xi m_1^{-1}[\pi(N^-)]^c$, which contradicts the assumption that ν is irreducible on $G/(MAN^+)$. □

Proof of Theorem 6.4. Because $G/(MAN^+)$ is compact, there is a stationary measure ν of g_t on $G/(MAN^+)$. It is clear that ν is μ-invariant. By Theorem 6.3, ν is irreducible on $G/(MAN^+)$. If we can show that $g_t \nu \to \delta_u$ weakly as $t \to \infty$ for some random variable u taking values in $G/(MAN^+)$, then, by Lemma 6.6, $g_t \nu' \to \delta_u$ weakly for any irreducible probability measure ν'

on $G/(MAN^+)$. Because ν is a stationary measure of g_t on $G/(MAN^+)$, for any $f \in C(G/(MAN^+))$,

$$\nu(f) = E\big[g_t^e \nu(f)\big] = E\big[g_0^{-1} g_t \nu(f)\big] \to E\big[f(g_0^{-1} u)\big].$$

This shows that ν is the distribution of $z = g_0^{-1} u$; therefore, the stationary measure of g_t on $G/(MAN^+)$ is unique. Part (a) of Theorem 6.4 is proved. Part (b) follows from (a) and Lemma 6.6. Since ν is irreducible on $G/(MAN^+)$, $g_0 z \in N^-(MAN^+)$ almost surely. By Proposition 6.2, (c) follows from (b).

It remains to prove that $g_t \nu \to \delta_u$ weakly as $t \to \infty$, where ν is a stationary measure of g_t on $G/(MAN^+)$. This will be accomplished in the following three steps. Without loss of generality, we may and will assume that $g_0 = e$ in the rest of the proof.

Step 1. We first show that $g_t \nu$ converges weakly to some random measure ζ on $G/(MAN^+)$ as $t \to \infty$.

Let $\{\mathcal{F}_t\}$ be the completed natural filtration of the Lévy process g_t. We note that $g_{t+s} = g_t g_s'$, where $g_s' = g_t^{-1} g_{t+s}$ is independent of $\mathcal{F}_t$ and has the same distribution as g_s. For any $f \in C(G/(MAN^+))$, let $M_t = g_t \nu(f) = \int f(g_t x)\nu(dx)$. Then

$$E[M_{t+s} \mid \mathcal{F}_t] = \int E[f(g_t g_s' x) \mid \mathcal{F}_t]\nu(dx) = \int E[f(g g_s x)]_{g=g_t} \nu(dx)$$
$$= E[g_s \nu(f \circ g)]_{g=g_t} = \nu(f \circ g_t) = M_t.$$

This shows that M_t is a bounded martingale; hence, M_t converges to a limit, denoted by M_∞, as $t \to \infty$ almost surely. Note that M_∞ depends on $f \in C(G/(MAN^+))$. It is easy to see that almost surely M_∞ is a continuous linear functional on the Banach space $C(G/(MAN^+))$ equipped with the norm $\|f\| = \sup_x |f(x)|$. Hence, $M_\infty = \zeta(f)$ for some random measure ζ on $G/(MAN^+)$ and $g_t \nu \to \zeta$ weakly.

Step 2. Recall that μ_t is the distribution of $g_t = g_t^e$ and $\mu = \int_0^\infty dt e^{-t}\mu_t$. We now show that there is a sequence $t_i \to \infty$ such that, for $\mu \times P$-almost all $(h, \omega) \in G \times \Omega$, the sequence of measures $g(t_i, \omega)h\nu$ converges weakly to $\zeta(\omega)$. Here, we have written $g(t, \omega)$ for $g_t(\omega)$ for typographical convenience.

If we can show that, for any $f \in C(G/(MAN^+))$,

$$\int_G E\left\{\left[\int f(g_t hx)\nu(dx) - \int f(g_t x)\nu(dx)\right]^2\right\}\mu(dh) \to 0 \quad (6.24)$$

as $t \to \infty$, then there is a sequence $t_i \to \infty$ such that, for $\mu \times P$-almost all (h, ω), $\int f(g(t_i, \omega)hx)v(dx) - \int f(g(t_i, \omega)x)v(dx) \to 0$. By Step 1, $\int f(g_t x)v(dx) \to \zeta(f)$ almost surely. It follows that, for $\mu \times P$-almost all (h, ω),

$$g(t_i, \omega)hv(f) = \int f(g(t_i, \omega)hx)v(dx) \to \zeta(\omega)(f).$$

The integral in (6.24) is equal to

$$\int_0^\infty e^{-s}ds\, E\left\{\int_G \left[\int f(g_t hx)v(dx)\right]^2 \mu_s(dh)\right\} + E\left\{\left[\int f(g_t x)v(dx)\right]^2\right\}$$

$$-2\int_0^\infty e^{-s}ds\, E\left\{\int_G \left[\int f(g_t hx)v(dx)\right]\left[\int f(g_t x)v(dx)\right]\mu_s(dh)\right\}$$

$$=\int_0^\infty e^{-s}ds\, E\left\{\left[\int f(g_{t+s}x)v(dx)\right]^2\right\} + E\left\{\left[\int f(g_t x)v(dx)\right]^2\right\}$$

$$-2\int_0^\infty e^{-s}ds\, E\left\{\int f(g_{t+s}x)v(dx)\int f(g_t x)v(dx)\right\}$$

$$=\int_0^\infty e^{-s}ds\{E\left(M_{t+s}^2\right) + E\left(M_t^2\right) - 2E(M_{t+s}M_t)\}$$

$$=\int_0^\infty e^{-s}ds\{E\left(M_{t+s}^2\right) - E\left(M_t^2\right)\} \to 0$$

as $t \to \infty$ because $E(M_t^2) \to E(M_\infty^2)$.

Step 3. We now show that $\zeta = \delta_z$ for some random variable z taking values in $G/(MAN^+)$. Because $T_\mu = \mathrm{supp}(\mu)$, by Step 2, there is a sequence h_i dense in T_μ such that, for P-almost all ω,

$$\forall i, \quad \lim_{j\to\infty} g(t_j, \omega)h_i v = \lim_{j\to\infty} g(t_j, \omega)v = \zeta(\omega), \qquad (6.25)$$

where the limits are taken with respect to the weak convergence topology. Fix such an ω. As in the proof of Lemma 6.6, let $\{X_1, X_2, \ldots, X_p\}$ be a basis of $\mathfrak{n}^-$ such that each X_i belongs to $\mathfrak{g}_{-\alpha}$ for some positive root α, which is denoted by α_i. We may assume that $\alpha_1, \alpha_2, \ldots, \alpha_l$ are simple roots and $\alpha_{l+1}, \ldots, \alpha_p$ are other positive roots that are linear combinations of simple roots with non negative integer coefficients. Let $g_t = \xi_t a_t^+ \eta_t$ be a Cartan decomposition. By taking a subsequence of $\{t_j\}$ if necessary, we may assume that

$$\xi(t_j, \omega) \to \xi(\omega), \quad \eta(t_j, \omega) \to \eta(\omega), \quad \text{and} \quad \alpha_i(\log a^+(t_j, \omega)) \to \lambda_i(\omega)$$

as $j \to \infty$, where $\lambda_i(\omega) \in [0, \infty]$. Since simple roots form a basis of the dual space of $\mathfrak{a}$, any $a \in A$ is uniquely determined by $\{\alpha_i(\log a); 1 \le i \le l\}$ and conversely, given $c_i \in \mathbb{R}$ for $1 \le i \le l$, there is a unique $a \in A$ satisfying $c_i = \alpha_i(\log a)$ for $1 \le i \le l$. Moreover, $a \in A_+ = \exp(\mathfrak{a}_+)$ if and only if $\alpha_i(\log a) > 0$ for $1 \le i \le l$. Let $a^+(\omega) \in A_+$ be defined by

$$\alpha_i(\log a^+(\omega)) = \begin{cases} \lambda_i(\omega), & \text{if } \lambda_i(\omega) < \infty \\ 0, & \text{if } \lambda_i(\omega) = \infty \end{cases}$$

for $1 \le i \le l$. Any $x \in \pi(N^-)$ may be expressed as $x = \exp(\sum_{i=1}^{p} x_i X_i) \times (MAN^+)$ for some $(x_1, \ldots, x_p) \in \mathbb{R}^p$, where $\pi \colon G \to G/(MAN^+)$ is the natural projection. As $j \to \infty$,

$$a(t_j, \omega)x = \exp\left[\sum_{i=1}^{p} x_i \mathrm{Ad}(a(t_j, \omega))X_i\right] (MAN^+)$$

$$= \exp\left[\sum_{i=1}^{p} x_i e^{-\alpha_i(\log a(t_j, \omega))} X_i\right] (MAN^+)$$

$$\to \exp\left[\sum_{i=1}^{p} x_i e^{-\lambda_i(\omega)} X_i\right] (MAN^+)$$

$$= a^+(\omega) \exp\left[\sum_{i=1}^{p} \epsilon_i(\omega) x_i X_i\right] (MAN^+),$$

where $\epsilon_i(\omega) = 1$ if $\lambda_i(\omega) < \infty$ and $\epsilon_i(\omega) = 0$ otherwise. Therefore,

$$\forall x \in \pi(N^-), \quad a^+(t_j, \omega)x \to a^+(\omega)\tau(\omega)x$$

as $j \to \infty$, where $\tau(\omega)$ is the map $\pi(N^-) \to \pi(N^-)$ defined by

$$\exp\left(\sum_i x_i X_i\right) (MAN^+) \mapsto \exp\left[\sum_i \epsilon_i(\omega) x_i X_i\right] (MAN^+).$$

It is easy to see that $\tau(\omega)$ is a continuous map on $\pi(N^-)$ and, for any $x \in \eta(\omega)^{-1}\pi(N^-)$,

$$g(t_j, \omega)x \to \xi(\omega)a^+(\omega)\tau(\omega)\eta(\omega)x.$$

By the irreducibility of ν and (6.25),

$$\xi(\omega)a^+(\omega)\tau(\omega)\eta(\omega)h_i\nu = \xi(\omega)a^+(\omega)\tau(\omega)\eta(\omega)\nu = \zeta(\omega); \qquad (6.26)$$

hence, $\tau(\omega)\eta(\omega)h_i\nu = \tau(\omega)\eta(\omega)\nu$ for any i. Because $\{h_i\}$ is dense in T_μ, it follows that

$$\forall h \in T_\mu, \quad \tau(\omega)\eta(\omega)h\nu = \tau(\omega)\eta(\omega)\nu. \qquad (6.27)$$

Now let h_i be a contracting sequence in T_μ and let $h_i = u_i b_i v_i$ be a Cartan decomposition. We may assume $u_i \to u$ and $v_i \to v$ in K. We claim that $\exists h \in T_\mu$ such that $hu \in \eta(\omega)^{-1}(N^- M A N^+)$. If not, then $T_\mu u \subset \eta(\omega)^{-1}(N^- M A N^+)^c$ and $T_\mu \pi(u) \subset \eta(\omega)^{-1}[\pi(N^-)]^c$. Because $\eta(\omega)^{-1}[\pi(N^-)]^c$ is closed, it also contains $\overline{T_\mu \pi(u)}$, the closure of $T_\mu \pi(u)$. Since $\forall h \in T_\mu = \operatorname{supp}(\mu)$, $h T_\mu \pi(u) \subset T_\mu \pi(u)$, any weak limit of

$$(1/j) \left(\delta_{\pi(u)} + \sum_{i=1}^{j-1} \mu^{*i} * \delta_{\pi(u)} \right),$$

where $\mu^{*i} = \mu * \mu * \cdots * \mu$ is a i-fold convolution, is a μ-invariant probability measure on $G/(M A N^+)$ supported by $\overline{T_\mu \pi(u)} \subset \eta(\omega)^{-1}[\pi(N^-)]^c$. This is impossible by Theorem 6.3.

Choose $h \in T_\mu$ such that $hu \in \eta(\omega)^{-1}(N^- M A N^+)$. Since $h h_i \in T_\mu$, $\tau(\omega)\eta(\omega) h h_i v = \tau(\omega)\eta(\omega) v$. Because b_i is positive contracting,

$$\forall x \in \pi(v^{-1} N^-), \quad \eta(\omega) h h_i x \to \eta(\omega)\pi(hu) \text{ as } i \to \infty.$$

It follows that $\eta(\omega) h h_i v \to \delta_{\eta(\omega)\pi(hu)}$ weakly. Since $\tau(\omega)$ is continuous on $\pi(N^-)$ and $\eta(\omega)\pi(hu) \in \pi(N^-)$, it follows that $\tau(\omega)\eta(\omega) h h_i v \to \delta_{\tau(\omega)[\eta(\omega)\pi(hu)]}$ weakly and, hence, $\tau(\omega)\eta(\omega) v = \delta_{\tau(\omega)[\eta(\omega)\pi(hu)]}$. By the second equality in (6.26), $\zeta(\omega) = \delta_{\xi(\omega) a^+(\omega)\tau(\omega)[\eta(\omega)\pi(hu)]}$. Theorem 6.4 is proved. $\square$

Let g_t be a left Lévy process in G satisfying the hypotheses of Theorem 6.4 and let ρ be the unique stationary measure of g_t on $G/(M A N^+)$ given in Theorem 6.4(a). For any $x, y \in G$, because $y\rho$ is an irreducible probability measure on $G/(M A N^+)$, $x g_t y \rho \to x \delta_u$ weakly as $t \to \infty$, where $u = g_0 z$. Now assume that $t \mapsto x_t$ and $t \mapsto y_t$ are G-valued functions on $\mathbb{R}_+$ such that $x_t \to x$ and $y_t \to y$ as $t \to \infty$. Since $G/(M A N^+)$ is compact, for any $f \in C(G/(M A N^+))$, $f(x_t g y_t z) \to f(xgyz)$ as $t \to \infty$ uniformly for $(g, z) \in G \times (G/(M A N^+))$. It follows that $x_t g_t y_t \rho \to x \delta_u = \delta_{xu}$ weakly. As in the proof of Theorem 6.4, using Lemma 6.6 and Proposition 6.2, one may obtain the convergence results for the process $x_t g_t y_t$ under the Cartan decomposition and the Iwasawa decomposition $G = N^- A K$. Applying the transformation $g \mapsto \Theta(g)^{-1}$, we can derive similar results for a right Lévy process g_t in G satisfying the hypotheses of Theorem 6.5. We obtain the following two corollaries of Theorems 6.4 and 6.5.

Corollary 6.2. *Let g_t be a left Lévy process in G satisfying the hypotheses of Theorem 6.4, and let $t \mapsto x_t$ and $t \mapsto y_t$ be G-valued functions such that $x_t \to x$ and $y_t \to y$ as $t \to \infty$.*

(a) *For any irreducible probability measure v on $G/(MAN^+)$, $x_t g_t y_t v \to$*
   δ_{xg_0z} weakly as $t \to \infty$, where z is given in Theorem 6.4.

(b) *Let $\xi_t a_t^+ \eta_t$ be a Cartan decomposition of $x_t g_t y_t$. Then, almost surely,*
   a_t^+ is positive contracting and $\xi_t(MAN^+) \to xg_0z$ as $t \to \infty$.

(c) *Let $n_t a_t k_t$ be the Iwasawa decomposition $G = N^- AK$ of $x_t g_t y_t$. Then,*
   almost surely, a_t is positive contracting and $n_t \to n_\infty$ as $t \to \infty$
   for some random variable n_∞ taking values in N^-. Moreover,
   $n_\infty(MAN^+) = xg_0z$ almost surely.

Corollary 6.3. *Let g_t be a right Lévy process in G satisfying the hypotheses*
of Theorem 6.5, and let $t \mapsto x_t$ and $t \mapsto y_t$ be G-valued functions such that
$x_t \to x$ and $y_t \to y$ as $t \to \infty$.

(a) *For any irreducible probability measure $\tilde{v}$ on $(MAN^-)\backslash G$, $\tilde{v} x_t g_t y_t \to$*
   $\delta_{\tilde{z}g_0y}$ weakly as $t \to \infty$, where $\tilde{z}$ is given in Theorem 6.5.

(b) *Let $\xi_t a_t^+ \eta_t$ be a Cartan decomposition of $x_t g_t y_t$. Then, almost surely,*
   a_t^+ is positive contracting and $(MAN^-)\eta_t \to \tilde{z}g_0y$ as $t \to \infty$.

(c) *Let $k_t a_t n_t$ be the Iwasawa decomposition $G = KAN^+$ of $x_t g_t y_t$. Then,*
   almost surely, a_t is positive contracting and $n_t \to n_\infty$ as $t \to \infty$
   for some random variable n_∞ taking values in N^+. Moreover,
   $(MAN^-)n_\infty = \tilde{z}g_0y$ almost surely.

In the preceding results, nothing is said about the limiting properties of the
components η_t and k_t (resp. ξ_t and k_t) of a left (resp. right) Lévy process g_t.
The following proposition provides this information.

Proposition 6.8. *Let g_t be a left (resp. right) Lévy process in G and let*
the notations in Corollary 6.2 (resp. Corollary 6.3) be used here. Assume
that G_μ is totally right (resp. left) irreducible and T_μ contains a contracting
sequence. Then g_t has a unique station measure $\tilde{\rho}$ (resp. ρ) on $(MAN^-)\backslash G$
(resp. $G/(MAN^+)$) such that $(MAN^-)\eta_t$ (resp. $\xi_t(MAN^+)$) converges in
distribution to $\tilde{\rho}y$ (resp. $x\rho$). Moreover, if G_μ is also totally left (resp. right)
irreducible, then $(MAN^-)k_t$ (resp. $k_t(MAN^+)$) also converges in distribution
to $\tilde{\rho}y$ (resp. $x\rho$).

Proof. We will only prove the case when g_t is a left Lévy process. The
proof for a right Lévy process g_t is similar. Let $\bar{g}_t$ be a right Lévy process
in G that has the same marginal distributions as g_t^e and is independent of
g_t. The two processes $\bar{g}_t^e = \bar{g}_t$ and g_t^e have the same G_μ and T_μ. Recall that

$\xi_t a_t^+ \eta_t$ is a Cartan decomposition and $n_t a_t k_t$ is the Iwasawa decomposition $G = N^- AK$ of $x_t g_t y_t$. Let $\bar{\xi}_t \bar{a}_t^+ \bar{\eta}_t$ be a Cartan decomposition and let $\bar{n}'_t \bar{a}_t \bar{k}_t$ be the Iwasawa decomposition $G = N^- AK$ of $x_t(g_0 \bar{g}_t) y_t = (x_t g_0) \bar{g}_t y_t$. By Corollary 6.3 (noting that g_0 is independent of the process $\bar{g}_t$), we have $(MAN^-)\bar{\eta}_t \to \bar{z}y$, where $\bar{z}$ is a random variable taking values in $(MAN^-)\backslash G$ whose distribution is the unique stationary measure of $\bar{g}_t$ on $(MAN^-)\backslash G$. Note that a stationary measure of g_t on $(MAN^-)\backslash G$ is also a stationary measure of $\bar{g}_t$ on $(MAN^-)\backslash G$, and vice versa. This implies that the stationary measure $\tilde{\rho}$ of g_t on $(MAN^-)\backslash G$ is unique and is the distribution of $\bar{z}$. Since, for each $t \in \mathbb{R}_+$, $g_0 \bar{g}_t$ has the same distribution as g_t and hence $\bar{\eta}_t$ is identical in distribution to η_t, it follows that $(MAN^-)\eta_t \overset{d}{\to} \tilde{\rho}y$ (convergence in distribution). If G_μ is also totally left irreducible, then by Corollary 6.2, $\xi_t(MAN^+) \to xg_0 z$, where z is a random variable taking values in $G/(MAN^+)$ whose distribution ρ is irreducible on $G/(MAN^+)$. The irreducibility of ρ implies $xg_0 z \in N^-(MAN^+)$. By Proposition 6.2, $k_t = \eta'_t \eta_t$, where η'_t is a process in K such that all its limiting points as $t \to \infty$ belong to M. It follows that $(MAN^-)k_t \overset{d}{\to} \tilde{\rho}y$. $\qquad\square$

By Theorems 6.4 and 6.5, a left or a right Lévy process g_t is contracting in the sense that $\alpha(\log a_t^+) \to \infty$ almost surely for any positive root α as $t \to \infty$, where a_t^+ is the $\overline{A_+}$-component in the Cardan decomposition of g_t. Under weaker hypotheses, g_t may be partially contracting in the sense that there is a subset F of the set Σ of simple roots such that, for any $\alpha \in F$, $\alpha(\log a_t^+) \to \infty$ almost surely as $t \to \infty$. In this case, we will say that g_t is F-contracting. Some discussion of this more general contracting behavior can be found in [24, section 2 D].

6.6. Some Sufficient Conditions

The hypotheses of Theorems 6.4 and 6.5 seem to be quite general, but they are not stated in a form that can be easily verified. We now derive some sufficient conditions, in terms of the vector fields and the Lévy measure contained in the generator of the Lévy process, that are easier to verify in practice.

Let P_t and L be respectively the transition semigroup and the generator of a left Lévy process g_t in G. We have

$$Lf = \frac{d}{dt} P_t f \mid_{t=0} \tag{6.28}$$

for $f \in D(L)$, the domain of L. By Theorem 1.1 and (1.24), $C_0^{2,l}(G) \subset D(L)$ and, for $f \in C_0^{2,l}(G)$,

$$
\begin{aligned}
Lf(g) &= \frac{1}{2} \sum_{j,k=1}^{d} a_{jk} X_j^l X_k^l f(g) + \sum_{i=1}^{d} c_i X_i^l f(g) \\
&\quad + \int_G \left[f(gh) - f(g) - \sum_{i=1}^{d} x_i(h) X_i^l f(g) \right] \Pi(dh), \\
&= \frac{1}{2} \sum_{j=1}^{m} Y_j^l Y_j^l f(g) + Z^l f(g) \qquad\qquad (6.29) \\
&\quad + \int_G \left[f(gh) - f(g) - \sum_{i=1}^{d} x_i(h) X_i^l f(g) \right] \Pi(dh), \quad (6.30)
\end{aligned}
$$

where a_{jk} and c_i are constants with $\{a_{jk}\}$ being a nonnegative definite symmetric matrix, $\{X_1, \ldots, X_d\}$ is a basis of $\mathfrak{g}$, $x_1, \ldots, x_d \in C_c^\infty(G)$ are the coordinate functions associated to the basis, Y_j and Z are defined by (1.22), and Π is the Lévy measure of the process g_t. For the generator of a right Lévy process, X_i^l, Y_j^l, Z^l, and gh should be replaced by X_i^r, Y_j^r, Z^r, and hg, respectively.

In this section, we assume that g_t is a left or right Lévy process in G with Lévy measure Π. In the actual computation, we concentrate on left Lévy processes; the results, however, hold also for right Lévy processes.

Proposition 6.9. supp(Π) *is contained in* T_μ. *Consequently, if* supp$(\Pi) = G$, *then the hypotheses of Theorem 6.4 and Theorem 6.5 are satisfied.*

Proof. Let $g \in$ supp(Π) with $g \neq e$. We want to show $g \in T_\mu$. Let $f \in C_c(G)_+$ be such that $f(g) > 0$ and $f = 0$ in a neighborhood of e. If $\mu_t(f) = P_t f(e) > 0$ for some $t > 0$, then by the right continuity of $t \mapsto g_t^e$, $\mu(f) > 0$. This being true for any function f satisfying the condition specified here implies $g \in$ supp$(\mu) = T_\mu$. Therefore, it suffices to show $P_t f(e) > 0$ for some $t > 0$. By (6.29), for such a function f, $Lf(e) = \Pi(f) > 0$. Now by (6.28), $P_t f(e) > 0$ for some $t > 0$. $\qquad\square$

Let

$$
\mathcal{L} = \left\{ \sum_{j,k=1}^{d} a_{jk} X_j \xi_k; \quad (\xi_1, \ldots, \xi_d) \in \mathbb{R}^d \right\}. \qquad (6.31)
$$

Proposition 6.10. *The definition of $\mathcal{L}$ is independent of the choice of the basis $\{X_1, \ldots, X_d\}$ of $\mathfrak{g}$ and the associated coefficients a_{ij}. Moreover,*

$$\mathcal{L} = \mathrm{span}(Y_1, Y_2, \ldots, Y_m).$$

Consequently, $\mathrm{Lie}(\mathcal{L}) = \mathrm{Lie}(Y_1, Y_2, \ldots, Y_m)$; that is, the Lie algebras generated by $\mathcal{L}$ and by $\{Y_1, Y_2, \ldots, Y_m\}$ coincide.

Proof. By Proposition 1.3, the differential operator $\sum_{j,k=1}^{d} a_{jk} X_j X_k$ is independent of the basis $\{X_1, \ldots, X_d\}$. If $\{X_1', \ldots, X_d'\}$ is another basis of $\mathfrak{g}$ and $\{a_{jk}'\}$ is the associated matrix, then $X_j = \sum_{p=1}^{d} b_{jp} X_p'$ and $a_{pq}' = \sum_{j,k=1}^{d} a_{jk} b_{jp} b_{kq}$ for some invertible matrix $\{b_{pq}\}$, and, for any $(\eta_1, \ldots, \eta_d) \in \mathbb{R}^d$,

$$\sum_{p,q=1}^{d} a_{pq}' X_p' \eta_q = \sum_{j,k,p,q=1}^{d} a_{jk} b_{jp} b_{kq} X_p' \eta_q = \sum_{j,k=1}^{d} a_{jk} X_j \left(\sum_{q=1}^{d} b_{kq} \eta_q \right).$$

This proves that the space $\mathcal{L}$ defined by (6.31) is independent of the choice for the basis of $\mathfrak{g}$ and the associated coefficients a_{ij}. Now let $V = \mathrm{span}\{Y_1, Y_2, \ldots, Y_m\}$ and let $\sigma = \{\sigma_{ij}\}$ be the $m \times d$ matrix in (1.22). Then $a = \sigma'\sigma$, where $a = \{a_{jk}\}$. It follows that $\dim(V) = \mathrm{Rank}(\sigma) = \mathrm{Rank}(a) = \dim(\mathcal{L})$. However, $\sum_{j,k=1}^{d} a_{jk} X_j \xi_k = \sum_{p=1}^{m} (\sum_{k=1}^{d} \sigma_{pk} \xi_k) Y_p$, which implies $\mathcal{L} \subset V$. Therefore, $\mathcal{L} = V$. $\qquad\square$

The following proposition provides a sufficient condition for $G_\mu = G$, which implies the first hypothesis in Theorems 6.4 and 6.5, namely, the total left and right irreducibility of G_μ on G. Let $\mathfrak{g}_\mu$ be the Lie algebra of the closed subgroup G_μ of G.

Proposition 6.11. *$\mathrm{Lie}(\mathcal{L}) \subset \mathfrak{g}_\mu$. Consequently, if $\mathrm{Lie}(\mathcal{L}) = \mathfrak{g}$, then $G_\mu = G$.*

Proof. Since $\{a_{jk}\}$ is a nonnegative definite symmetric matrix, there is $\{b_{jk}\} \in O(d)$ such that $\sum_{j,k=1}^{d} a_{jk} b_{jp} b_{kq} = \lambda_p \delta_{pq}$, where $\lambda_1 \geq \lambda_2 \geq \cdots \lambda_d$ are the eigenvalues of $\{a_{jk}\}$ that are all nonnegative. Let $\{c_{jk}\} = \{b_{jk}\}^{-1}$ and let $V_i = \sum_{j=1}^{d} c_{ij} X_j$. Then $\sum_{j,k=1}^{d} a_{jk} X_j^l X_k^l = \sum_{i=1}^{d} \lambda_i V_i^l V_i^l$. Suppose $\lambda_1 \geq \cdots \lambda_r > 0$ and $\lambda_{r+1} = \cdots = \lambda_d = 0$. Then $\mathcal{L}$ is spanned by $V_1, \ldots, V_r$.

Let $\langle X, Y \rangle$ be the inner product on $\mathfrak{g}$ determined by the Killing form B, given by (5.3). Fix an arbitrary vector Y in $\mathfrak{g}$ orthogonal to $\mathfrak{g}_\mu$ with respect to this inner product. It suffices to show $\langle Y, V_i \rangle = 0$ for $i \leq r$.

We may assume that $X_1, \ldots, X_d$ form an orthonormal basis of $\mathfrak{g}$ and $X_1, \ldots, X_p$ span $\mathfrak{g}_\mu$ for $p \leq d$. Since Y is orthogonal to $\mathfrak{g}_\mu$, $Y = \sum_{j>p} c_j X_j$. Let $f \in C_c^\infty(G)_+$ be equal to $(\sum_{j>p} c_j x_j)^2$ near e. By the properties of the coordinate functions x_i, we have $f(e) = 0$, $V_i f(e) = 0$, and

$$V_i^l V_i^l f(e) = 2 \left(\sum_{j>r} c_j c_{ij} \right)^2 = 2 \langle Y, V_i \rangle^2.$$

Therefore,

$$\frac{d}{dt} P_t f(e) \mid_{t=0} = L f(e) = \sum_i \lambda_i \langle Y, V_i \rangle^2 + \Pi(f).$$

We may assume that the coordinate functions x_i are defined by $g = \exp[\sum_i x_i(g) X_i]$ for g contained in a neighborhood of e. Then $x_j = 0$ on $\exp(\mathfrak{g}_\mu) \bigcap U$ for $j > p$, and hence $f = 0$ on $\exp(\mathfrak{g}_\mu) \bigcap U$, where U is a sufficiently small neighborhood of e. Since G_μ is closed, $G_\mu \bigcap U = \exp(\mathfrak{g}_\mu) \bigcap U$ when U is small enough. We may modify the value of f so that it vanishes outside U, and hence vanishes on G_μ. Then $P_t f(e) = 0$ for all t. This implies $\langle Y, V_i \rangle = 0$ for $i \le r$. $\qquad\square$

Let g_t be a left Lévy process in G. By the discussion in Section 1.4, if g_t has a finite Lévy measure, then its generator L takes the simpler form (1.11); that is, for $f \in C_0^{2,l}(G)$,

$$Lf(g) = \frac{1}{2} \sum_{j,k=1}^d a_{jk} X_j^l X_k^l f(g) + \sum_{i=1}^d b_i X_i^l f(g) + \int_G [f(gh) - f(g)] \Pi(dh)$$

$$= \frac{1}{2} \sum_{i=1}^m Y_i^l Y_i^l f(g) + Y_0^l f(g) + \int_G [f(gh) - f(g)] \Pi(dh), \qquad (6.32)$$

where $b_i = c_i - \int_G x_i(h) \Pi(dh)$ and $Y_0 = \sum_{i=1}^d b_i X_i$. In this case, g_t may be obtained as the solution of the stochastic differential equation

$$dg_t = \sum_{i=1}^m Y_i^l(g_t) \circ dW_t^i + Y_0^l(g_t) dt \qquad (6.33)$$

with random jumps at exponentially spaced random times that are independent of the driving Brownian motion W_t. The following proposition offers a slight improvement over Proposition 6.11, in the case of a finite Π, by adding Y_0 to $\mathcal{L}$.

Proposition 6.12. *Assume Π is finite and let $Y_0, Y_1, \ldots, Y_m \in \mathfrak{g}$ be given as before. Then*

$$\mathrm{Lie}(Y_0, Y_1, \ldots, Y_m) \subset \mathfrak{g}_\mu.$$

Consequently, if $\mathrm{Lie}(Y_0, Y_1, \ldots, Y_m) = \mathfrak{g}$, then $G = G_\mu$.

Proof. By Proposition 6.11, $Y_i \in \mathfrak{g}_\mu$ for $1 \leq i \leq m$. It suffices to show $Y_0 \in \mathfrak{g}_\mu$. We will continue to use the notations introduced in the proof of Proposition 6.11. Let Z be an arbitrary element of $\mathfrak{g}$ that is orthogonal to $\mathfrak{g}_\mu$. We need only to show $\langle Y_0, Z \rangle = 0$.

By Lemma 2.1, $X_j^l X_k^l x_i(e) = (1/2) C_{jk}^i$ and

$$\sum_{j,k} a_{jk} X_j^l X_k^l x_i(e) = \frac{1}{2} \sum_{j,k} a_{jk} C_{jk}^i = 0 \qquad (6.34)$$

because $a_{jk} = a_{kj}$ and $C_{jk}^i = -C_{kj}^i$. Since Z is orthogonal to $\mathfrak{g}_\mu$, $Z = \sum_{i>p} c_i X_i$. Let $f \in C_c^\infty(G)_+$ be equal to $\sum_{i>p} c_i x_i$ near e. By (6.32) and (6.34), we get

$$\frac{d}{dt} P_t f(e)\, |_{t=0} = Lf(e) = \sum_{i>p} b_i c_i + \Pi(f) = \langle Y_0, Z \rangle + \Pi(f).$$

As in the proof of Proposition 6.11, we may choose f to be supported by a small neighborhood of e so that f vanishes on G_μ. Then $P_t f(e) = 0$, which implies $\langle Y_0, Z \rangle \leq 0$. Replacing Z by $-Z$ yields $\langle Y_0, -Z \rangle \leq 0$; hence, $\langle Y_0, Z \rangle = 0$. $\qquad \square$

We now consider the second hypothesis in Theorems 6.4 and 6.5, namely, the assumption that T_μ contains a contracting sequence.

By Proposition 5.15, $\mathrm{Ad}(K)\mathfrak{a}_+$ is an open subset of $\mathfrak{p} = \mathrm{Ad}(K)\mathfrak{a}$ with a lower dimensional complement. The elements of $\mathfrak{p}$ contained in $\mathrm{Ad}(K)\mathfrak{a}_+$ are called regular. For $G = SL(d, \mathbb{R})$, $Y \in \mathfrak{p}$ is regular if and only if the symmetric matrix Y has distinct eigenvalues.

Let $Y \in \mathfrak{p}$ be regular with $Y = \mathrm{Ad}(k)H$ for some $H \in \mathfrak{a}_+$, and let $g = e^Y$. Then, for $j = 1, 2, \ldots$, the j-fold products $g^j = g \cdots g$ are equal to $ke^{jH}k^{-1}$ and form a contracting sequence. Therefore, if T_μ contains some e^Y, where $Y \in \mathfrak{p}$ is regular, then T_μ contains a contracting sequence. By Proposition 6.9, $\mathrm{supp}(\Pi) \subset T_\mu$; hence, if $\mathrm{supp}(\Pi)$ contains e^Y for some regular $Y \in \mathfrak{p}$, then T_μ contains a contracting sequence.

If the Lévy measure Π is finite, then g_t^e is the solution of the stochastic differential equation (6.33) for $t < T$, where the first jumping time T is an exponential random variable independent of the driving Brownian motion W_t. In this case, we will show that, for arbitrary constants $c_0, c_1, \ldots, c_m \in \mathbb{R}$ with $c_0 \geq 0$, T_μ contains $\exp(\sum_{i=1}^m c_i Y_i + c_0 Y_0)$.

We now recall the support theorem for the solution of a stochastic differential equation on $\mathbb{R}^n$ of the following form:

$$dx_t = \sum_{i=1}^{m} F_i(x_t) \circ dW_t^i + F_0(x_t)\, dt, \tag{6.35}$$

where $F_i = \sum_{j=1}^{d} c_{ij}(x)(\partial/\partial x_j)$ for $i = 0, 1, \ldots, m$ are vector fields on $\mathbb{R}^d$ whose coefficients c_{ij} are bounded smooth functions with bounded derivatives. Let x_t be the solution of this equation satisfying the initial condition $x_0 = 0$. Fix a constant $\tau > 0$, and let H_τ be the Banach space of the continuous maps $x \colon [0, \tau] \to \mathbb{R}^d$ with $x(0) = 0$ equipped with the norm $\|x\| = \sup_{0 \le t \le \tau} |x(t)|$. The process x_t for $t \in [0, \tau]$ may be regarded as an H_τ-valued random variable. Let Q_z be its distribution. Let $C_{p,\tau}$ be the space of all the piecewise smooth $\mathbb{R}^m$-valued functions on $[0, \tau]$. For $\phi = (\phi_1, \ldots, \phi_m) \in C_{p,\tau}$, let x^ϕ be the solution of the ordinary differential equation

$$dx^\phi(t) = \left[\sum_{i=1}^{m} F_i(x^\phi(t))\phi_i(t) + F_0(x^\phi(t)) \right] dt$$

for $t \in [0, \tau]$ satisfying the initial condition $x^\phi(0) = 0$. By Ikeda and Watanabe [33, chapter VI, theorem 8.1], the support of the process x_t over the time interval $[0, \tau]$, namely, the support of Q_z, is equal to the closure in H_τ of the set

$$\{x^\phi; \quad \phi \in C_{p,\tau}\}.$$

Let U be a neighborhood of e that is a diffeomorphic image of an open subset of $\mathfrak{g}$ under the exponential map. We may identify U with an open subset of $\mathbb{R}^d$ and e with 0 in $\mathbb{R}^d$; hence, we may regard (6.33) as a stochastic differential equation on $\mathbb{R}^d$ by suitably extending the left invariant vector fields Y_i^l from U to $\mathbb{R}^d$. Let $\tilde{Y}_i$ denote the extension of Y_i^l, let g_t^l denote the solution of (6.33) regarded as an equation on G, and let $\tilde{g}_t$ denote the solution of (6.33) regarded as an equation on $\mathbb{R}^d$ with Y_i^l replaced by $\tilde{Y}_i$, satisfying the initial conditions $g_0^l = \tilde{g}_0 = e$. Then $g_t^l = g_t^e$ for $t < T$ and $g_t^l = \tilde{g}_t$ if either g_s^l or $\tilde{g}_s$, for $s \in [0, t]$, is entirely contained in U. We can apply the support theorem to (6.33) regarded as an equation on $\mathbb{R}^d$ to conclude that, for any $\tau > 0$ and $\phi \in C_{p,\tau}$, g^ϕ is contained in the support of the process $\tilde{g}_t$ over the time interval $[0, \tau]$, where $g^\phi(t)$ for $t \in [0, \tau]$ is the solution of the ordinary differential equation

$$dg^\phi(t) = \left[\sum_{i=1}^{m} \tilde{Y}_i(g^\phi(t))\phi_i(t) + \tilde{Y}_0(g^\phi(t)) \right] dt \tag{6.36}$$

satisfying the initial condition $g^\phi(0) = e$. If $\tau > 0$ is sufficiently small, then $g^\phi(t)$ for $t \in [0, \tau]$ is entirely contained in U; therefore, g^ϕ is also contained in the support of the process g'_t on the time interval $[0, \tau]$. If ϕ is equal to a constant element b of $C^\infty([0, t], \mathbb{R}^m)$, then the solution $g^b(t) = g^\phi(t)$ of (6.36) can be explicitly calculated as

$$g^b(t) = \exp\left[\left(\sum_{i=1}^m b_i Y_i + Y_0\right) t\right]. \tag{6.37}$$

Since the first jumping time T of g_t is independent of the driving Brownian motion W_t in (6.33) and $P(T > \tau) > 0$ for any $\tau > 0$, it follows that, for any $b \in \mathbb{R}^m$, g^b is contained in the support of the Lévy process g_t on the time interval $[0, \tau]$ for small $\tau > 0$. This implies $g^b(t) \in \text{supp}(\mu_t)$; hence, $g^b(t) \in T_\mu$, for small $t > 0$, where μ_t as before is the distribution of g_t. Therefore, T_μ contains $g(t) = \exp[(\sum_{i=1}^m b_i Y_i + Y_0)t]$ for small $t > 0$. Since T_μ is a semigroup, it in fact contains $g(t)$ for any $t > 0$. This proves that T_μ contains

$$\exp\left(\sum_{i=1}^m c_i Y_i + c_0 Y_0\right)$$

for arbitrary real numbers $c_0, c_1, \ldots, c_m$ with $c_0 \geq 0$.

Lemma 6.7. *For any* $X \in \text{Lie}\,(Y_1, \ldots, Y_m)$ *and* $c \geq 0$, $\exp(X + cY_0) \in T_\mu$.

Proof. Let $X, Y \in \mathfrak{g}$. Using the Taylor expansions of $f(e^{tX}e^{tY})$ and $f(e^{-tX}e^{-tY}e^{tX}e^{tY})$ for $f \in C_c^\infty(G)$, we can show (see [26, chapter II, lemma 1.8]) that

$$e^{tX}e^{tY} = \exp\left\{t(X + Y) + \frac{t^2}{2}[X, Y] + O(t^3)\right\} \tag{6.38}$$

and

$$e^{-tX}e^{-tY}e^{tX}e^{tY} = \exp\{t^2[X, Y] + O(t^3)\}. \tag{6.39}$$

Let $\mathcal{L} = \text{span}\,(Y_1, \ldots, Y_m)$ as before. We have proved before the statement of the lemma that if $X, Y \in \mathcal{L}$, then $e^{\pm tX}$ and $e^{\pm tY}$ belong to T_μ for any $t > 0$. By (6.39), there is a continuous function $Z: \mathbb{R} \to \mathfrak{g}$ such that $Z(0) = [X, Y]$ and $e^{tZ(t)} \in T_\mu$. For fixed $t > 0$, the set $\{e^{ntZ(t)}; n = 1, 2, \ldots\} \subset T_\mu$. As $t \to 0$, this set converges to the ray $\{e^{t[X,Y]}; t > 0\}$. It follows that $e^{t[X,Y]} \in T_\mu$ for any $t > 0$. Now we may add $[X, Y]$ to $\mathcal{L}$ and repeat the same argument to show that $e^{tX} \in T_\mu$ for any $X \in \text{Lie}\,(Y_1, \ldots, Y_m)$ and $t > 0$. By (6.38), for

$X \in \mathrm{Lie}\,(Y_1, \ldots, Y_m)$ and $c \geq 0$, $e^{tX} e^{tcY_0} = e^{t(X+cY_0)+O(t^2)}$. This implies that $e^{t(X+cY_0)} \in T_\mu$ for $t > 0$. $\qquad\qquad\qquad\qquad\qquad\qquad\qquad\qquad\qquad\qquad\qquad$ □

It follows that T_μ contains a contracting sequence if there exist $X \in \mathrm{Lie}\,(Y_1, \ldots, Y_m)$ and $c \geq 0$ such that $X + c_0 Y_0$ is a regular element of $\mathfrak{p}$.

To summarize, we have the following conclusions:

Proposition 6.13. T_μ *contains a contracting sequence if either of the following two conditions holds.*

(i) $\mathrm{supp}(\Pi)$ *contains* e^Y *for some regular* $Y \in \mathfrak{p}$.
(ii) Π *is finite and there exist* $X \in \mathrm{Lie}\,(Y_1, \ldots, Y_m)$ *and* $c \geq 0$ *such that* $X + cY_0$ *is a regular element of* $\mathfrak{p}$.

By Proposition 6.6, a connected Lie subgroup of G is totally left irreducible if and only if it is totally right irreducible, so it may be simply called totally irreducible. Let $\mathfrak{g}_\mathfrak{p}$ be the Lie sub algebra of $\mathfrak{g}$ generated by $\mathfrak{p}$ and let $G_\mathfrak{p}$ be the Lie subgroup of G generated by $\mathfrak{g}_\mathfrak{p}$. The following proposition is useful in verifying the hypotheses of Theorems 6.4 and 6.5. For example, together with Propositions 6.11 and 6.13 it implies that if $\mathrm{Lie}\,(\mathcal{L}) \supset \mathfrak{p}$ and if Π is finite, then these hypotheses are satisfied. In particular, it implies that the continuous Lévy process g_t determined by the stochastic differential equation (6.1) satisfies these hypotheses even if $c = 0$.

Proposition 6.14. $G_\mathfrak{p}$ *is totally irreducible and contains a contracting sequence in* G.

Proof. Because $\exp(\mathfrak{p}) \subset G_p$, it is clear that G_p contains a contracting sequence in G. Since, for any positive root α and $H \in \mathfrak{a}_+$, the linear map $\mathrm{ad}(H)$: $\mathfrak{p}_\alpha \to \mathfrak{k}_\alpha$ is bijective, where $\mathfrak{p}_\alpha$ and $\mathfrak{k}_\alpha$ are defined by (5.9) and (5.10) respectively, it follows that $\mathfrak{g}_\mathfrak{p}$ contains both $\mathfrak{g}_\alpha$ and $\mathfrak{g}_{-\alpha}$. It clearly also contains $\mathfrak{a}$; hence, $\mathfrak{n}^- \oplus \mathfrak{a} \oplus \mathfrak{n}^+ \subset \mathfrak{g}_\mathfrak{p}$ and $\mathfrak{g} = \mathfrak{g}_\mathfrak{p} + \mathfrak{m}$ (which is not necessarily a direct sum). Since $[\mathfrak{m}, \mathfrak{p}] \subset \mathfrak{p} \subset \mathfrak{g}_\mathfrak{p}$, by the Jacobi identity, $[\mathfrak{m}, \mathfrak{g}_\mathfrak{p}] \subset \mathfrak{g}_\mathfrak{p}$. It follows that $\mathfrak{g}_\mathfrak{p}$ is an ideal in $\mathfrak{g}$. Let $\tilde{\mathfrak{m}}$ be the orthogonal complement of $\mathfrak{g}_\mathfrak{p}$ in $\mathfrak{g}$ with respect to the Killing form B; that is, $\tilde{\mathfrak{m}} = \{Z \in \mathfrak{g}; B(Z, X) = 0 \text{ for } X \in \mathfrak{g}_\mathfrak{p}\}$. Then $\tilde{\mathfrak{m}} \subset \mathfrak{m}$. By proposition 6.1 in [26, chapter II], $\tilde{\mathfrak{m}}$ is an ideal of $\mathfrak{g}$ and $\mathfrak{g} = \mathfrak{g}_\mathfrak{p} \oplus \tilde{\mathfrak{m}}$ is a direct sum. It follows that $[\mathfrak{g}_\mathfrak{p}, \tilde{\mathfrak{m}}] = 0$. Let $\tilde{M}$ be the Lie subgroup of G generated by $\tilde{\mathfrak{m}}$. Then $\tilde{M} \subset M$ and $gm = mg$ for $g \in G_\mathfrak{p}$ and $m \in \tilde{M}$. Moreover, $G = G_\mathfrak{p} \tilde{M} = \tilde{M} G_\mathfrak{p}$. If $G_\mathfrak{p}$ is not totally irreducible, then by Proposition 6.6, there is $m \in \tilde{M}$ such that $G_\mathfrak{p} m \subset (N^- MAN^+)^c$.

By the Bruhat decomposition (5.15), $M(N^- M A N^+)^c \subset (N^- M A N^+)^c$. Then $G_p \subset (N^- M A N^+)^c$, but this is impossible because $N^- A N^+ \subset G_p \cap (N^- M A N^+)$. $\square$

The following proposition is useful for checking whether a connected subgroup of $G = SL(d, \mathbb{R})$ is totally irreducible. As before, for any square matrix g, let $g[i]$ denote the submatrix of g formed by the first i rows and i columns.

Proposition 6.15. *Let $G = SL(d, \mathbb{R})$ and let H be a connected Lie subgroup of G. Assume H contains a positive contracting sequence in A. Then H is not totally irreducible if and only if there exist $m_1, m_2 \in M'$ and $1 \le i \le (d/2)$ such that $\det ((m_1 h m_2)[i]) = 0$ for all $h \in H$.*

Proof. By Proposition 6.4, it is easy to show that H is not totally irreducible if and only if $\exists x_1, x_2 \in G$ and $1 \le i \le (d-1)$ such that

$$\forall h \in H, \quad \det((x_1 h x_2)[i]) = 0.$$

By the Bruhat decomposition, we may write $x_1 = p_1 m_1 n_1$ and $x_2 = n_2 m_2 p_2$, where $p_1 \in (M A N^-), p_2 \in (M A N^+), n_1 \in N^+, n_2 \in N^-$, and $m_1, m_2 \in M'$. Let $a_j \in A$ be a positive contracting sequence contained in H. Then

$$a_j^{-1} n_1 a_j \to e \quad \text{and} \quad a_j n_2 a_j^{-1} \to e.$$

Since

$$x_1 a_j h a_j x_2 = p_1 (m_1 a_j m_1^{-1}) m_1 (a_j^{-1} n_1 a_j) h (a_j n_2 a_j^{-1}) m_2 (m_2^{-1} a_j m_2) p_2,$$

$\det ((x_1 a_j h a_j x_2)[i]) = 0$, $p_1(m_1 a_j m_1^{-1}) \in (M A N^-)$, and $(m_2^{-1} a_j m_2) p_2 \in (M A N^+)$, by (5.20), we see that $\det ((m_1 h m_2)[i]) = 0$ for any $h \in H$.

The proposition is proved if $i \le d/2$. Assume $i > d/2$. Note that $\det ((m_1 h m_2)[i]) = 0$ means that the vectors

$$(m_1 h m_2) e_1, \ (m_1 h m_2) e_2, \ \ldots, \ (m_1 h m_2) e_i, \ e_{i+1}, \ e_{i+2}, \ \ldots, \ e_d$$

are not linearly independent, where $\{e_1, \ldots, e_d\}$ is the standard basis of $\mathbb{R}^d$. This is equivalent to

$$[(m_1 h m_2)(e_1 \wedge \cdots \wedge e_i)] \wedge (e_{i+1} \wedge \cdots \wedge e_d) = 0.$$

Let γ be the matrix that has 1 along the second diagonal (that is, the diagonal from the upper right corner to the lower left corner) and 0 elsewhere. Then

$\gamma e_i = e_{d-i+1}$ for $1 \leq i \leq d$. The preceding relation can be written as

$$[(m_1 h m_2)\gamma(e_{d-i+1} \wedge \cdots \wedge e_d)] \wedge [\gamma(e_1 \wedge \cdots \wedge e_{d-i})] = 0,$$

which is equivalent to

$$[\gamma^{-1} m_2^{-1} h^{-1} m_1^{-1} \gamma(e_1 \wedge \cdots \wedge e_{d-i})] \wedge (e_{d-i+1} \wedge \cdots \wedge e_d) = 0.$$

This is the same as

$$\det\left((\gamma^{-1} m_2^{-1} h^{-1} m_1^{-1} \gamma)[d - i]\right) = 0.$$

Note that $\gamma^{-1} = \gamma$. For any matrix $g \in G$, $\gamma g \gamma$ is the matrix obtained by rearranging both the rows and the columns of g by the permutation $(1, 2, \ldots, d) \mapsto (d, d - 1, \ldots, 1)$. The proposition is thus proved. □

To apply Proposition 6.15 to verify the total irreducibility of H, one just needs to check that all the submatrices of size $\leq d/2$ cannot be identically equal to zero on H. We now present an example of a "small" closed subgroup H of $G = SL(3, \mathbb{R})$ that is totally irreducible and contains a contracting sequence. Let

$$X = \begin{bmatrix} 1 & 0 & 0 \\ 0 & 0 & 0 \\ 0 & 0 & -1 \end{bmatrix}, \quad Y = \begin{bmatrix} 0 & 1 & 0 \\ 0 & 0 & 1 \\ 0 & 0 & 0 \end{bmatrix}, \quad \text{and} \quad Z = \begin{bmatrix} 0 & 0 & 0 \\ 1 & 0 & 0 \\ 0 & 1 & 0 \end{bmatrix}.$$

Then $[X, Y] = Y$, $[X, Z] = -Z$, and $[Y, Z] = X$. Therefore, X, Y, and Z span a three-dimensional Lie subalgebra $\mathfrak{h}$ of $\mathfrak{g} = \mathfrak{sl}(3, R)$. Let H be the Lie subgroup of $G = SL(3, \mathbb{R})$ generated by $\mathfrak{h}$. Then H contains the positive contracting sequence $\{e^{jX}; j = 1, 2, \ldots\}$ in A. It is easy to see that, for small $t > 0$, all the entries of the matrix e^{tY} on and above the diagonal, as well as those of the matrix e^{tZ} on and below the diagonal, are nonzero. By Proposition 6.15, H is totally irreducible on $G = SL(3, \mathbb{R})$.

To prove that H is closed, note that its Lie algebra $\mathfrak{h}$ is invariant under the Cartan involution $\theta : V \mapsto -V'$ on $\mathfrak{g}$; hence, H is invariant under the Cartan involution Θ on G. A direct computation shows that the Killing form of $\mathfrak{h}$ is given by

$$B_{\mathfrak{h}}(xX + yY + zZ, x'X + y'Y + z'Z) = 2(xx' + yz' + zy'),$$

which is equal to $\text{Trace}[(xX + yY + zZ)(x'X + y'Y + z'Z)]$ and hence is proportional to the restriction of the Killing form of $\mathfrak{g}$ on $\mathfrak{h}$. Therefore, H is semi-simple of noncompact type, the Cartan decomposition of $\mathfrak{h}$ is given by $\mathfrak{h} = \mathfrak{k}' \oplus \mathfrak{p}'$, where $\mathfrak{k}' = \text{span}(Y - Z)$ and $\mathfrak{p}' = \text{span}(X, Y + Z)$, and $\mathfrak{a}' = \text{span}(X)$ is a maximal abelian subspace of $\mathfrak{p}'$. There are only two roots

$\pm\alpha$, given by $\alpha(X) = 1$, and with $\mathfrak{a}'_+ = \{cX; c > 0\}$ as the Weyl chamber, the positive root space is $\mathfrak{n}'^+ = \text{span}(Y)$. Let A', N'^+, and K' be respectively the one-dimensional Lie subgroups of H generated by X, Y, and $Y - Z$. Then A', N'^+, and K' are closed subgroups of $G = SL(3, \mathbb{R})$, and $H = K'A'N'^+$ is the Iwasawa decomposition of H. It follows that H is a closed subgroup of G.

Remark. Gol'dsheid and Margulis [23] and Guivarc'h and Raugi [25] have shown that, when G is the special linear group $SL(m, R)$, T_μ contains a contracting sequence if and only if G_μ contains a contracting sequence. In this case, if $G_\mu = G$, then the hypotheses of Theorems 6.4 and 6.5 are satisfied.

7

Rate of Convergence

In this chapter, we consider the rate of convergence for the abelian and nilpotent components of a Lévy process g_t as $t \to \infty$. At first the discussion is concentrated on a left Lévy process g_t. In Section 7.1, we mention some useful facts on Iwasawa decomposition and obtain stochastic integral equations for both the (abelian) A- and K-components of g_t in its Iwasawa decomposition. In Section 7.2, using ergodic theory, we show that, if the Lévy measure Π satisfies a certain integrability condition, then the A-component of g_t converges to ∞ exponentially at rates determined by a nonrandom element in the Weyl chamber $\mathfrak{a}_+$. More precisely, if $g_t = \xi_t a_t^+ \eta_t$ is the Cartan decomposition and $g_t = n_t a_t k_t$ is the Iwasawa decomposition $G = N^- A K$, then, almost surely, $\lim_{t\to\infty}(1/t)\log a_t = \lim_{t\to\infty}(1/t)\log a_t^+ = H^+$ for some $H^+ \in \mathfrak{a}_+$. When the normalized Haar measure on K is a stationary measure, a more explicit formula for the rate vector H^+ of g_t is obtained in Section 7.3, from which one sees immediately that if g_t is continuous, then H^+ is proportional to the vector H_ρ in $\mathfrak{a}_+$ representing the half sum of positive roots. The convergence rate of the (nilpotent) N^--component n_t is considered in Section 7.4. In the last section, the corresponding results for a right Lévy process are derived. The results of this chapter will play a pivotal role when discussing the dynamical behaviors of the induced stochastic flows in the next chapter.

7.1. Components under the Iwasawa Decomposition

Let G be a connected semi-simple Lie group of noncompact type and of a finite center. We will continue to use the notations introduced in Chapters 5 and 6.

For $X, Y \in \mathfrak{g}$, let $\langle X, Y \rangle$ be the inner product on $\mathfrak{g}$ induced by the Killing form B of G, defined by (5.3), and let $\|X\| = \sqrt{\langle X, Y \rangle}$ be the associated norm.

Let $\pi \colon G \to G/K$ be the natural projection and let o be the point eK in G/K. By the discussion in Section 5.1, the inner product $\langle \cdot, \cdot \rangle$ restricted to $\mathfrak{p}$

induces a G-invariant Riemannian metric $\{\langle \cdot, \cdot \rangle_x; x \in G/K\}$ on the symmetric space G/K satisfying

$$\forall g \in G \text{ and } X, Y \in \mathfrak{p}, \quad \langle gD\pi(X), gD\pi(Y) \rangle_{gK} = \langle X, Y \rangle,$$

where we have written g for Dg for the sake of simplicity, and any geodesic ray starting at o and parametrized by arclength is given by $\pi(e^{tY}) = e^{tY}o$ for $Y \in \mathfrak{p}$ with $\|Y\| = 1$. For $x, y \in G/K$, let $\mathrm{dist}(x, y)$ be the distance between x and y determined by the Riemannian metric. Since each geodesic ray from o can be extended indefinitely, it follows that the Riemannian metric on G/K is complete; hence, any two points x and y in G/K can be joined by a geodesic whose length is equal to $\mathrm{dist}(x, y)$. By Theorem 5.1(c), the map $K \times \mathfrak{p} \ni (k, Y) \mapsto ke^Y \in G$ is a diffeomorphism; therefore, the map $\mathfrak{p} \ni Y \mapsto e^Y o \in G/K$ is a diffeomorphism. This implies that

$$\forall Y \in \mathfrak{p}, \quad \mathrm{dist}(e^Y o, o) = \|Y\|. \tag{7.1}$$

Proposition 7.1. *For $a \in A$ and $n \in N^+$ (resp. $n \in N^-$) with $n \neq e$,*

$$\mathrm{dist}(ao, o) < \mathrm{dist}(ano, o) \quad \text{and} \quad \mathrm{dist}(ao, o) < \mathrm{dist}(nao, o).$$

Proof. We only prove the first inequality $\mathrm{dist}(ao, o) < \mathrm{dist}(ano, o)$ with $n \in N^+$. Because $ana^{-1} \in N^+$, the second inequality follows from the first one. The assertion about $n \in N^-$ can be proved by a similar argument.

We now show that the orbits N^+o and Ao are orthogonal at o; that is, if $v_1 \in T_o(N^+o)$ and $v_2 \in T_o(Ao)$, then $\langle v_1, v_2 \rangle_o = 0$. There are $X \in \mathfrak{n}^+$ and $Y \in \mathfrak{a}$ such that $v_1 = D\pi(X)$ and $v_2 = D\pi(Y)$. Let Z be the projection of X into $\mathfrak{p}$ via the decomposition $\mathfrak{g} = \mathfrak{k} \oplus \mathfrak{p}$. Then $v_1 = D\pi(Z)$ and $X - Z \in \mathfrak{k}$. It follows that $\langle X - Z, Y \rangle = 0$ and $\langle v_1, v_2 \rangle_o = \langle Z, Y \rangle = \langle X, Y \rangle = 0$. The claim is proved. Since $T_o(G/K) = T_o(Ao) \oplus T_o(N^+o)$, the submanifold Ao contains all the geodesics intersecting N^+o orthogonally at o.

Let $n \in N^+$ such that the point ano minimizes the distance from o to the submanifold aN^+o. It suffices to show $n = e$. The minimizing property implies that the geodesic segment γ in G/K joining o to ano is orthogonal to aN^+o. Since the submanifold Ao contains all the geodesics intersecting N^+o orthogonally at o, $anAo$ contains all the geodesics intersecting $aN^+o = anN^+o$ orthogonally at ano. It follows that $\gamma \subset anAo$; therefore, there exists $a' \in A$ such that $ana'o = o$. By the uniqueness of the Iwasawa decomposition $G = N^+AK$, this implies $n = e$. $\square$

For any $g \in G$, let

$$g = (g_N)(g_A)(g_K) \quad \text{be the Iwasawa decomposition } G = N^- AK \quad (7.2)$$

and, for any $X \in \mathfrak{g}$, let

$$X = X_{\mathfrak{n}} + X_{\mathfrak{a}} + X_{\mathfrak{k}} \quad \text{be the direct sum decomposition } \mathfrak{g} = \mathfrak{n}^- \oplus \mathfrak{a} \oplus \mathfrak{k}. \quad (7.3)$$

Note that $X_{\mathfrak{a}}$ is also equal to the orthogonal projection of X into the subspace $\mathfrak{a}$.

We have, for $g, u \in G$,

$$(gu)_N = g_N g_A (g_K u)_N g_A^{-1}, \quad (gu)_A = g_A (g_K u)_A, \quad \text{and} \quad (gu)_K = (g_K u)_K. \quad (7.4)$$

To prove this, let $g = nak$ be the Iwasawa decomposition $G = N^- AK$; then $gu = na(ku) = [na(ku)_N a^{-1}][a(ku)_A][(ku)_K]$.

For $X \in \mathfrak{g}$ and $s \in \mathbb{R}$ with small $|s|$, the Iwasawa decomposition $G = N^- AK$ implies that $e^{sX} = e^{sY + O(s^2)} e^{sH + O(s^2)} e^{sZ + O(s^2)}$ for $Y \in \mathfrak{n}^-$, $H \in \mathfrak{a}$, and $Z \in \mathfrak{k}$. Comparing this with $e^{tX} = \exp(tX_{\mathfrak{n}} + tX_{\mathfrak{a}} + tX_{\mathfrak{k}})$ yields $Y = X_{\mathfrak{n}}$, $H = X_{\mathfrak{a}}$, and $Z = X_{\mathfrak{k}}$. Therefore,

$$(e^{sX})_N = e^{sX_{\mathfrak{n}} + O(s^2)}, \quad (e^{sX})_A = e^{sX_{\mathfrak{a}} + O(s^2)}, \quad \text{and} \quad (e^{sX})_K = e^{sX_{\mathfrak{k}} + O(s^2)}. \quad (7.5)$$

For any $g \in G$, we define $g_{\mathfrak{p}} \in \mathfrak{p}$ by

$$g = k \exp(g_{\mathfrak{p}}), \quad \text{where } k \in K. \quad (7.6)$$

Proposition 7.2. $\sup_{k \in K} \| \log[(kg)_A] \| = \| g_{\mathfrak{p}} \|$.

Proof. Let $g = he^Y$ for $h \in K$ and $Y \in \mathfrak{p}$. Then $kg = khe^Y = \exp[\mathrm{Ad}(kh)Y]kh$. Choose $k \in K$ such that $\mathrm{Ad}(kh)Y \in \mathfrak{a}$; then $\log[(kg)_A] = \mathrm{Ad}(kh)Y$. Since $\|\mathrm{Ad}(kh)Y\| = \|Y\| = \|g_{\mathfrak{p}}\|$, it suffices to prove that $\| \log[(kg)_A] \| \leq \|g_{\mathfrak{p}}\|$ for any $k \in K$.

Let $kg = nah$ be the Iwasawa decomposition $G = N^- AK$. Then $a = (kg)_A$, $(kg)_{\mathfrak{p}} = g_{\mathfrak{p}}$, and $kgo = nao$. By (7.1) and Proposition 7.1,

$$\| \log[(kg)_A] \| = \| \log a \| = \mathrm{dist}(ao, o) \leq \mathrm{dist}(nao, o) = \mathrm{dist}(kgo, o)$$
$$= \| g_{\mathfrak{p}} \|. \qquad \square$$

Via the Iwasawa decomposition $G = KAN^+$, K may be naturally identified with $G/(AN^+)$. The left action of G on $G/(AN^+)$ becomes a left action of G on K given by $K \ni k \mapsto g * k \in K$ for $g \in G$, where $g * k$ is the K-component of gk in the Iwasawa decomposition $G = KAN^+$. Similarly, via the Iwasawa decomposition $G = N^- AK$, K can be naturally identified with $(N^- A) \backslash G$, and the right action of G on $(N^- A) \backslash G$ becomes a right action of G on K given by $K \ni k \mapsto k * g \in K$, where $k * g$ is the K component of kg in the Iwasawa decomposition $G = N^- AK$. Note that $k * g = (kg)_K$, and if $k, h \in K$, then $h * k = hk$ regardless of whether $*$ is the left action on $G/(AN^+)$ or the right action on $(N^- A) \backslash G$.

Let g_t be a left Lévy process in G and let $g_t = n_t a_t k_t$ be the Iwasawa decomposition $G = N^- AK$. By Proposition 2.1, the process $x_t = (N^- A) g_t$ is a Feller process in $(N^- A) \backslash G$ with transition semigroup $P_t' f(x) = E[f(xg_t^e)]$. Since x_t is identified with k_t, it follows that k_t is a Feller process in K with transition semigroup $P_t^K f(k) = E[f(k * g_t^e)]$. Suppose g_t is also right K-invariant. Then, for any $k \in K$, kg_t^e has the same distribution as $g_t^e k$. For $f \in \mathcal{B}(K)_+$, $P_t^K f(k) = E[f((kg_t^e)_K)] = E[f((g_t^e k)_K)] = E[f(k_t^e k)]$, where $k_t^e = (g_t^e)_K$. This shows that k_t is a right Lévy process in K. To summarize, we have the following result:

Proposition 7.3. *Let g_t be a left Lévy process in G and let $g_t = n_t a_t k_t$ be the Iwasawa decomposition $G = N^- AK$. Then k_t is a Feller process in K with transition semigroup $P_t^K f(k) = E[f(k * g_t^e)]$. Moreover, if g_t is also right K-invariant, then k_t is a right Lévy process in K.*

Let g_t be a left Lévy process in G. Suppose a basis $\{X_1, \ldots, X_d\}$ of $\mathfrak{g}$ together with coordinate functions x_i is chosen. Then the generator L of g_t, restricted to $C_0^{2,l}(G)$, is given by (1.7) with coefficients a_{ij} and c_i and the Lévy measure Π. Let N be the counting measure of the right jumps of g_t defined by (1.12), which is a Poisson random measure on $\mathbb{R}_+ \times G$ with characteristic measure Π, and let $Y_1, \ldots, Y_m, Z \in \mathfrak{g}$ be given by (1.22). Then g_t satisfies the stochastic integral equation (1.23) for $f \in C_b(G) \cap C^2(G)$, where $W_t = (W_t^1, \ldots, W_t^m)$ is an m-dimensional standard Brownian motion independent of N and $\tilde{N}$ is the compensated random measure of N. Moreover, if Π satisfies the finite first moment condition (1.10), then g_t satisfies the simpler stochastic integral equation (1.25) for $f \in C^2(G)$, where $Y_0 = Z - \sum_{i=1}^d [\int_G x_i(u) \Pi(du)] X_i$.

Proposition 7.4. *Let g_t be a left Lévy process in G satisfying the stochastic integral equation (1.23) and let $g_t = n_t a_t k_t$ be the Iwasawa decomposition*

$G = N^- A K$. *Assume that the Lévy measure* Π *satisfies the following integrability condition:*

$$\int_G \|g_{\mathfrak{p}}\| \, \Pi(dg) < \infty. \tag{7.7}$$

Assume also that the basis $\{X_1, \ldots, X_d\}$ *of* $\mathfrak{g}$ *is chosen so that* $X_1, \ldots, X_p$ *form a basis of* $\mathfrak{p}$ *and* $X_{p+1}, \ldots, X_d$ *form a basis of* $\mathfrak{k}$. *Then* $\int_G |x_i(u)| \Pi(du) < \infty$ *for* $1 \le i \le p$ *and*

$$\log a_t = \log a_0 + \int_0^t \sum_{i=1}^m [\mathrm{Ad}(k_{s-})Y_i]_{\mathfrak{a}} \, dW_s^i$$
$$+ \int_0^t \left\{ \frac{1}{2} \sum_{i=1}^m [[\mathrm{Ad}(k_s)Y_i]_{\mathfrak{k}}, \mathrm{Ad}(k_s)Y_i]_{\mathfrak{a}} + [\mathrm{Ad}(k_s)Z_0]_{\mathfrak{a}} \right\} ds$$
$$+ \int_0^t \int_G \log[(k_{s-}u)_A] N(ds \, du), \tag{7.8}$$

where $Z_0 = Z - \sum_{i=1}^p [\int_G x_i(u) \Pi(du)] X_i$. *Moreover, if* Π *satisfies the finite first moment condition (1.10), then* Z_0 *in (7.8) may be replaced by* $Y_0 = Z - \sum_{i=1}^d [\int_G x_i(u) \Pi(du)] X_i$.

Note that the first integral in (7.8), namely,

$$M_t = \int_0^t \sum_{i=1}^m [\mathrm{Ad}(k_{s-})Y_i]_{\mathfrak{a}} \, dW_s^i, \tag{7.9}$$

is a stochastic Itô integral. Because k_t is a process in the compact subgroup K, the integrand $\sum_{i=1}^m [\mathrm{Ad}(k_{s-})Y_i]_{\mathfrak{a}}$ is bounded, so the integral exists and M_t is a continuous L^2-martingale. The second integral in (7.8) exists because it also has a bounded integrand. By (7.7) and Proposition 7.2, $\int_G \sup_{k \in K} \|\log[(ku)_A]\| \Pi(du) < \infty$. This implies that the third integral in (7.8) exists and has a finite expectation. Consequently, $E[\|\log a_t\|] < \infty$.

Proof of Proposition 7.4. The coordinate functions $x_1, \ldots, x_d$ may be determined by $g = \exp[\sum_{i=p+1}^d x_i(g) X_i] \cdot \exp[\sum_{i=1}^p x_i(g) X_i]$ for g contained in some neighborhood U of e. Then $g_{\mathfrak{p}} = \sum_{i=1}^p x_i(g) X_i$ for $g \in U$. By (7.7), $\int |x_i(u)| \Pi(du) < \infty$ for $1 \le i \le p$. This is true for any choice of coordinate functions associated to the same basis of $\mathfrak{g}$ because of (1.6).

Let $g = nak$ be the Iwasawa decomposition $G = N^- A K$ of $g \in G$. By (7.4) and (7.5), for $Y \in \mathfrak{g}$ and small $s > 0$,

$$(ge^{sY})_A = a \exp\{s[\mathrm{Ad}(k)Y]_{\mathfrak{a}} + O(s^2)\} \quad \text{and} \quad (ge^{sY})_K$$
$$= \exp\{s[\mathrm{Ad}(k)Y]_{\mathfrak{k}} + O(s^2)\} k. \tag{7.10}$$

Hence, if $f \in C_c^\infty(G)$ depends only on the A-component of the Iwasawa decomposition $G = N^- AK$, that is, if $f(g) = f(a)$, then, for $Y \in \mathfrak{g}$,

$$Y^l f(g) = \frac{d}{ds} f\left(ge^{sY}\right) \mid_{s=0} = [\mathrm{Ad}(k)Y]_\mathfrak{a}^l f(a) \qquad (7.11)$$

and

$$Y^l Y^l f(g) = \frac{d}{ds} Y^l f\left(ge^{sY}\right) \mid_{s=0} = \frac{d}{ds} [\mathrm{Ad}(e^{s[\mathrm{Ad}(k)Y]_\mathfrak{k}}k)Y]_\mathfrak{a}^l f(ae^{s[\mathrm{Ad}(k)Y]_\mathfrak{a}}) \mid_{s=0}$$

$$= [[\mathrm{Ad}(k)Y]_\mathfrak{k}, \mathrm{Ad}(k)Y]_\mathfrak{a}^l f(a) + \frac{d^2}{ds^2} f(ae^{sH}) \mid_{s=0}, \qquad (7.12)$$

where $H = [\mathrm{Ad}(k)Y]_\mathfrak{a}$.

We may regard $\mathfrak{a}$ as a Euclidean space and we may apply (1.23) to an $\mathfrak{a}$-valued function f on G. Fix a constant $C > 0$. Let $\phi: \mathfrak{a} \to \mathfrak{a}$ be a smooth map with a compact support such that $\phi(X) = X$ for $X \in \mathfrak{a}$ with $\|X\| \leq C$ and $\|\phi(X) - \phi(Y)\| \leq \|X - Y\|$ for any $X, Y \in \mathfrak{a}$. Let $f(g) = \phi(\log g_A)$ for $g \in G$, and let $V = \{g \in G; \|\log g_A\| \leq C\}$. Choose an open neighborhood U of e such that $UU \subset V$ and let

$$\tau = \inf\{t > 0; \ g_t \in (G - U)\}.$$

Then, for $t < \tau$ and $u \in U$, $f(g_t) = \log a_t$ and $f(g_t u) - f(g_t) = \log[(k_t u)_A]$. By (1.23), (7.11), and (7.12), if $t < \tau$, then

$$\log a_t = \log a_0 + M_t + \int_0^t \left\{ \frac{1}{2} \sum_{i=1}^m [[\mathrm{Ad}(k_s)Y_i]_\mathfrak{k}, \mathrm{Ad}(k_s)Y_i]_\mathfrak{a} + [\mathrm{Ad}(k_s)Z]_\mathfrak{a} \right\} ds$$

$$+ \int_0^t \int_U \{\log[(k_{s-}u)_A]\} \tilde{N}(dsdu)$$

$$+ \int_0^t \int_U \left\{ \log[(k_{s-}u)_A] - \sum_{i=1}^d x_i(u)[\mathrm{Ad}(k_{s-})X_i]_\mathfrak{a} \right\} ds \Pi(du) + R,$$

where M_t is given by (7.9) and

$$R = \int_0^t \int_{G-U} [f(g_{s-}u) - f(g_{s-})] N(dsdu)$$

$$- \int_0^t \int_{G-U} \left[\sum_{i=1}^d x_i(u)[\mathrm{Ad}(k_{s-})X_i] \right]_\mathfrak{a} ds \Pi(du).$$

Since $\|f(g_{s-}u) - f(g_{s-})\| = \|\phi(\log(g_{s-}u)_A) - \phi(\log(g_{s-})_A)\|$ is bounded by

$$\|\log(g_{s-}u)_A - \log(g_{s-})_A\| = \|\log[(k_{s-}u)_A]\|,$$

it follows that

$$
E[\|R\|] \leq E\left\{ \int_0^t \int_{G-U} \|\log[(k_{s-}u)_A]\| ds\, \Pi(du) \right.
$$
$$
\left. + \int_0^t \int_{G-U} \left\| \sum_{i=1}^d x_i(u)[\mathrm{Ad}(k_{s-})X_i]_{\mathfrak{a}} \right\| ds\, \Pi(du) \right\}
$$
$$
\leq t \int_{G-U} \|g_{\mathfrak{p}}\| \Pi(dg) + C't \Pi(G-U)
$$

for some constant $C' > 0$. We may let $C \uparrow \infty$ and $U \uparrow G$. Then $\tau \uparrow \infty$ and, by (7.7), $E[\|R\|] \to 0$. This proves (7.8). If Π satisfies the finite first moment condition, Y_0 can be defined. Although $Y_0 \neq Z_0$, $[\mathrm{Ad}(k_s)Y_0]_{\mathfrak{a}} = [\mathrm{Ad}(k_s)Z_0]_{\mathfrak{a}}$ because $[\mathrm{Ad}(k)X_i]_{\mathfrak{a}} = 0$ for $i > p$. □

To end this section, we derive a stochastic integral equation for the K-component of g_t in the Iwasawa decomposition $G = N^- AK$. By Proposition 7.3, the K-component k_t is a Feller process in K. Let $f \in C^2(K)$. We may regard f as a function on G defined by $f(g) = f(g_K)$. Then $f \in C_b(G) \cap C^2(G)$. Let $g \in G$ and let k be its K-component in the Iwasawa decomposition $G = N^- AK$. By (7.4), $(gu)_K = (ku)_K$ for $u \in G$ and then, by (7.10), for $Y \in \mathfrak{g}$,

$$Y^l f(g) = \frac{d}{ds} f\left(ge^{sY}\right)\big|_{s=0} = [\mathrm{Ad}(k)Y]_{\mathfrak{k}}^r f(k).$$

It follows from (1.23) that, for $f \in C^2(K)$,

$$
f(k_t) = f(k_0) + \sum_{i=1}^m \int_0^t [\mathrm{Ad}(k_{s-})Y_i]_{\mathfrak{k}}^r f(k_{s-}) \circ dW_s^i + \int_0^t [\mathrm{Ad}(k_s)Z]_{\mathfrak{k}}^r f(k_s) ds
$$
$$
+ \int_0^t \int_G [f((k_{s-}u)_K) - f(k_{s-})] \tilde{N}(dsdu)
$$
$$
+ \int_0^t \int_G \left\{ f((k_s u)_K) - f(k_s) \right.
$$
$$
\left. - \sum_{j=1}^d x_j(u)[\mathrm{Ad}(k_s)X_j]_{\mathfrak{k}}^r f(k_s) \right\} ds\, \Pi(du). \tag{7.13}
$$

Here, the basis $\{X_1, \ldots, X_d\}$ of $\mathfrak{g}$ does not have to satisfy the condition specified in Proposition 7.4. Moreover, if Π satisfies the finite first moment condition (1.10), using (1.25) instead of (1.23), we obtain the following simpler stochastic integral equation for k_t:

$$f(k_t) = f(k_0) + \sum_{i=1}^{m} \int_0^t [\mathrm{Ad}(k_{s-})Y_i]_{\mathfrak{k}}^r f(k_{s-}) \circ dW_s^i + \int_0^t [\mathrm{Ad}(k_s)Y_0]_{\mathfrak{k}}^r f(k_s) ds$$

$$+ \int_0^t \int_G [f((k_{s-}u)_K) - f(k_{s-})] N(ds\,du) \tag{7.14}$$

for any $f \in C^2(K)$, where $Y_0 = Z - \sum_{i=1}^{d} [\int_G x_i(u) \Pi(du)] X_i$.

7.2. Rate of Convergence of the Abelian Component

We now recall some basic elements of ergodic theory. Let $(S, \mathcal{S}, \mu)$ be a probability measure space. A measurable map $\tau \colon S \to S$ is called μ-preserving, or a measure-preserving transformation on $(S, \mathcal{S}, \mu)$, if $\tau\mu = \mu$. A set $B \in \mathcal{S}$ is called τ-invariant if $\tau^{-1}(B) = B$. All the τ-invariant sets form a σ-algebra $\mathcal{I}$, called the invariant σ-algebra of τ. A real-valued $\mathcal{S}$-measurable function f is called τ-invariant if $f \circ \tau = f$. This is equivalent to the $\mathcal{I}$-measurability of f. If $\mathcal{I}$ is μ-trivial, that is, $\mu(B) = 0$ or 1 for any $B \in \mathcal{I}$, then either τ will be called ergodic (with respect to μ) or μ will be called ergodic (with respect to τ). By the Birkhoff ergodic theorem (see, for example, [34, theorem 9.6]), if τ is μ-preserving and if $f \in L^1(\mu)$, then $(1/n) \sum_{i=0}^{n-1} f \circ \tau^i$ converges to an τ-invariant function $\bar{f}$ μ-almost surely and in $L^1(\mu)$ as $n \to \infty$ with $\mu(f) = \mu(\bar{f})$. Moreover, if μ is ergodic, then $\bar{f} = \mu(f)$ μ-almost surely. By the ergodic decomposition theorem ([34, theorem 9.12]), any probability measure μ preserved by τ can be expressed as $\int \nu(dm) m$, where ν is a probability measure on the set of ergodic probability measures (defined with respect to τ).

Let $\{\tau_t; t \in \mathbb{R}_+\}$ be a semigroup of measure-preserving transformations on $(S, \mathcal{S}, \mu)$. The invariant sets, invariant functions, and ergodicity can still be defined by replacing τ by τ_t for all $t \in \mathbb{R}_+$. The continuous-time ergodic theorem states that if $f \in L^1(\mu)$, then $(1/t) \int_0^t f \circ \tau_s ds$ converges to some invariant function $\bar{f}$ μ-almost surely and in $L^1(\mu)$ as $t \to \infty$ with $\mu(f) = \mu(\bar{f})$, and if $\{\tau_t\}$ is also ergodic, then $\bar{f} = \mu(f)$ μ-almost surely.

Let T be one of the index sets $\mathbb{R}_+ = [0, \infty)$, $\mathbb{R} = (-\infty, \infty)$, $\mathbb{Z}_+ = \{0, 1, 2, \ldots, \}$, or $\mathbb{Z} = \{\ldots, -2, -1, 0, 1, 2, \ldots\}$. A process $\xi = \{\xi_t\}_{t \in T}$ is called stationary if, for any $t_1, \ldots, t_n, t \in T$, the two families $\{\xi_{t_1}, \ldots, \xi_{t_n}\}$ and

$\{\xi_{t_1+t}, \ldots, \xi_{t_n+t}\}$ have the same distribution. A stationary process ξ_n indexed by $T = \mathbb{Z}_+$ will be called a stationary sequence and one indexed by $T = \mathbb{Z}$ will be called a two-sided stationary sequence. Any stationary sequence can be extended to be a two-sided stationary sequence ([34, lemma 9.2]). If ξ is an S-valued random variable whose distribution μ is preserved by τ and $f:$ $S \to S'$ is a measurable map from S into another measurable space S', then $\eta_n = f(\tau^n \xi)$ is a stationary sequence in S'. In fact, any stationary sequence η_n can be obtained in this way ([34, lemma 9.1]). The ergodic theorem can be stated for an S-valued stationary sequence ξ_n (resp. stationary process ξ_t for $t \in \mathbb{R}_+$) as follows: Let $\mu = \xi_0 P$. If $f \in L^1(\mu)$, then $(1/n) \sum_{i=0}^{n-1} f(\xi_i) \to \eta$ (resp. $(1/t) \int_0^t f(\xi_s) ds \to \eta$) P-almost surely and in $L^1(P)$ as $n \to \infty$ (resp. $t \to \infty$) for some real-valued random variable η with $E(\eta) = \mu(f)$. In particular, if ξ_n is a real-valued stationary sequence and $E(|\xi_0|) < \infty$, then $(1/n)(\xi_0 + \xi_1 + \cdots + \xi_{n-1}) \to \bar{\xi}$ almost surely and in L^1 as $n \to \infty$ for some $\bar{\xi}$ with $E(\bar{\xi}) = E(\xi_0)$.

The following lemma is taken from Guivarc'h and Raugi [24, lemme 3.6].

Lemma 7.1. *Let τ be a measure-preserving transformation on a probability measure space $(S, \mathcal{S}, \mu)$. If $f \in L^1(\mu)$ satisfies $\sum_{i=0}^n f \circ \tau^i \to \infty$ μ-almost surely as $n \to \infty$, then $\mu(f) > 0$.*

Proof. By the ergodic decomposition theorem, we may assume that τ is ergodic with respect to μ. We can show that there exist a measure-preserving transformation θ on another probability measure space $(X, \mathcal{X}, Q)$ with a measurable inverse θ^{-1} and a measurable map $\pi: X \to S$ such that $\pi Q = \mu$ and $\pi \circ \theta(x) = \tau \circ \pi(x)$ for Q-almost all $x \in X$. Then replacing f by $f \circ \pi$ and τ by θ, we may assume that τ has a measurable inverse τ^{-1}. To show the existence of $(X, \mathcal{X}, Q, \theta, \pi)$, let ξ be an S-valued random variable with distribution μ and let $\{\eta_n; n \in \mathbb{Z}\}$ be a two-sided stationary sequence that extends $\{\xi_n = \tau^n \xi; n \in \mathbb{Z}_+\}$. Let $X = S^{\mathbb{Z}}$ be equipped with the product σ-algebra $\mathcal{X}$, let Q be the distribution of $\{\eta_n\}$ considered as a random variable taking values in X, let θ be the shift operator on X defined by $(\theta\{\eta_n\})_k = \eta_{k+1}$ for any $k \in \mathbb{Z}$, and let $\pi: X \to S$ be defined by $\pi(\{\eta_n\}) = \eta_0$. Then θ is a measure-preserving map on $(X, \mathcal{X}, \nu)$ with a measurable inverse that has the desired properties.

Let $s_n(x) = \sum_{i=0}^{n-1} f \circ \tau^i(x)$. Then $s_n(x) \to \infty$ as $n \to \infty$ for μ-almost all x. However, by the ergodicity assumption, $(1/n)s_n \to \mu(f)$ μ-almost surely. Consequently, $\mu(f) \geq 0$. Let $Y = S \times \mathbb{R}$ and let $T: Y \to Y$ be the map defined by $T(x, r) \mapsto (\tau x, r + f(x))$. Then T preserves the product measure

$\nu = \mu \times \lambda$, where λ is the Lebesque measure on $\mathbb{R}$, and has a measurable inverse given by $T^{-1}(x, r) = (\tau^{-1}x, r - f(\tau^{-1}x))$. For $\varepsilon > 0$, let $Y_\varepsilon = S \times [-\varepsilon, \varepsilon]$. Since

$$T^n(x, r) = (\tau^n x, r + s_n(x)),$$

it suffices to show that if $\mu(f) = 0$, then ν-almost all $(x, r) \in Y_\varepsilon$ will return to Y_ε under T infinitely many times in the sense that $T^n(x, r) \in Y_\varepsilon$ for infinitely many $n > 0$.

A measurable subset A of Y is called errant if $\nu(A) > 0$ and if $T^k(A)$ for $k > 0$ are ν-disjoint, that is, $\nu[T^k(A) \cap T^j(A)] = 0$ for $j \neq k$. We can show that if $\mu(f) = 0$, then there is no errant subset. Otherwise, if A is errant, because $(1/n)(r + s_n(x)) \to 0$ for ν-almost all (r, x), there is a subset B of A such that $\nu(B) > 0$ and $(1/n)(r + s_n(x)) \to 0$ uniformly for $(r, x) \in B$. Therefore, for any $\delta > 0$, there is a positive integer N such that $\forall n \geq N$ and $\forall (x, r) \in B$, $T^n(x, r) \in Y_{n\delta}$. This implies that $T^n(B) \subset Y_{n\delta}$ and hence $\cup_{k=N}^n T^k(B) \subset Y_{n\delta}$. It follows that $\limsup_{n\to\infty}(1/n)\nu[\cup_{k=0}^{n-1}T^k(B)] \leq \delta$. Because $\delta > 0$ can be made arbitrarily small, $(1/n)\nu[\cup_{k=0}^{n-1}T^k(B)] \to 0$. However, because T^{-1} are ν-preserving and $T^k(B)$ are ν-disjoint, $\nu[\cup_{k=0}^{n-1}T^k(B)] = n\nu(B)$, which is impossible.

Let C be the set of the points $(x, r) \in Y_\varepsilon$ such that $T^k(x, r) \notin Y_\varepsilon$ for any $k > 0$. The invertibility of T implies that $T^k(C)$ for $k > 0$ are ν-disjoint; hence, $\nu(C) = 0$. Therefore, ν-almost all (x, r) in Y_ε will return to Y_ε in the sense that $T^k(x, r) \in Y_\varepsilon$ for some $k > 0$. Let $B = Y_\varepsilon - C$ (set difference). Because T is ν-preserving, it follows that, for ν-almost all $(x, r) \in B$, $T^k(x, r) \in B$ for some $k > 0$. This implies that ν-almost all (x, r) will return to B under T infinitely many times. $\qquad\square$

Corollary 7.1. *Let ξ_n be a real-valued stationary sequence with $E(|\xi_0|) < \infty$. If $\sum_{i=0}^n \xi_i \to \infty$ almost surely as $n \to \infty$, then $E(\xi_0) > 0$.*

Proof. This is an immediate consequence of Lemma 7.1 because $\xi_n = f(\tau^n \eta)$ for some random variable η taking value in a measurable space space S, a map $\tau: S \mapsto S$ preserving the distribution of η, and a real-valued measurable function f on S $\qquad\square$

By Proposition 7.3, the K-component k_t of a left Lévy process g_t in the Iwasawa decomposition $G = N^- AK$ is a Feller process in K with transition semigroup P_t^K. Recall (Appendix B.1) that a probability measure ν on K is called a stationary measure of k_t (as a Markov process) if $\nu P_t^K = \nu$. Because K is compact, there is a stationary measure of k_t. In fact, for any probability measure v' on K, any limiting point of $(1//t) \int_0^t v' P_s^K ds$ as $t \to \infty$ is a stationary measure of k_t.

Theorem 7.1. *Let g_t be a left Lévy process in G satisfying the stochastic integral equation (1.23), and let $g_t = \xi_t a_t^+ \eta_t$ and $g_t = n_t a_t k_t$ be respectively a Cartan decomposition and the Iwasawa decomposition $G = N^- AK$. Assume the hypotheses of Theorem 6.4 and the condition (7.7). Then P-almost surely the limit*

$$H^+ = \lim_{t \to \infty} \frac{1}{t} \log a_t$$

exists, is nonrandom, and is contained in $\mathfrak{a}_+$. Moreover, P-almost surely,

$$H^+ = \lim_{t \to \infty} (1/t) \log a_t^+.$$

Proof. By Proposition 6.2 and Theorem 6.4, $a_t^+ a_t^{-1}$ is P-almost surely bounded as $t \to \infty$. Therefore, if one of the limits, $\lim_{t \to \infty}(1/t) \log a_t$ or $\lim_{t \to \infty}(1/t) \log a_t^+$, exists, then both limits exist and are equal. For any $u \in G$, let $u = nbh$ be the Iwasawa decomposition $G = N^- AK$. By (7.4), $(ug_t)_A = b(hg_t)_A$. Since hg_t and g_t have the same $\overline{A_+}$-component in the Cartan decomposition, it follows that $\lim_{t \to \infty}(1/t) \log(ug_t)_A = \lim_{t \to \infty}(1/t) \log a_t$ if either of the two limits exists. Therefore, we can arbitrarily change g_0 without affecting the existence and the value of $\lim_{t \to \infty}(1/t) \log a_t$. In particular, we may and will assume that $g_0 = k_0 \in K$ and the distribution of k_0 is a stationary measure v of k_t. Then k_t is a stationary Markov process.

Since $a_0 = e$, (7.8) may be rewritten as

$$\log a_t = M_t + \int_0^t F(k_s) ds + \int_0^t \int_G J(k_{s-}, h) N(dsdh), \qquad (7.15)$$

where M_t is the martingale given by (7.9),

$$F(k) = \frac{1}{2} \sum_{i=1}^m [[\mathrm{Ad}(k)Y_i]_{\mathfrak{k}}, \mathrm{Ad}(k)Y_i]_{\mathfrak{a}} + [\mathrm{Ad}(k)Z_0]_{\mathfrak{a}}, \quad \text{and}$$
$$J(k, g) = \log[(kg)_A].$$

By Proposition 7.2, $\|J(k, g)\| \le \|g_{\mathfrak{p}}\|$.

Because the integrand in (7.9) is bounded, $M_t/t \to 0$ as $t \to \infty$. By the ergodic theory, $(1/t) \int_0^t F(k_s) ds$ converges almost surely to an $\mathfrak{a}$-valued random variable H' with $E(H') = \int_K F(k) v(dk)$.

To show the convergence of $(1/t) \int_0^t \int_G J(k_{s-}, g) N(ds dg)$, let us introduce two sequences of random variables, x_n and $\bar{x}_n$, for $n = 0, 1, 2, \ldots$, defined by

$$x_n = \int_n^{n+1} \int_G J(k_{s-}, g) N(ds dg) \quad \text{and} \quad \bar{x}_n = \int_n^{n+1} \int_G \| J(k_{s-}, g) \| N(ds dg).$$

Both are stationary sequences since k_t is stationary and N is a Poisson random measure associated to the natural filtration of the Lévy process g_t. Moreover, $\|x_n\| \leq \bar{x}_n$. Note that

$$E(\bar{x}_0) \leq E \int_0^1 \int_G \| g_\mathfrak{p} \| N(ds dg) = \int_G \| g_\mathfrak{p} \| \Pi(dg) < \infty.$$

It follows, by the ergodic theory, that $\sum_{i=0}^{n-1} x_i/n$ converges almost surely to some $\mathfrak{a}$-valued random variable H^* whose expectation is given by

$$E(H^*) = E(x_0) = E \left\{ \int_0^1 \int_G J(k_s, g) \Pi(dg) ds \right\} = \int_K \int_G J(k, g) \Pi(dg) \nu(dk).$$

However, $\sum_{i=0}^{n-1} \bar{x}_i/n$ also converges; hence, $\bar{x}_n/n \to 0$. This implies that

$$\frac{1}{t} \int_0^t \int_G J(k_{s-}, g) N(ds dg) \to H^*$$

as $t \to \infty$.

We have proved that, as $t \to \infty$, $(1/t) \log a_t$ converges almost surely to $H^+ = H' + H^*$ and

$$E(H^+) = \int_K F(k) \nu(dk) + \int_K \int_G J(k, g) \Pi(dg) \nu(dk).$$

We will show that H^+ is a constant element of $\mathfrak{a}$.

For $t > s$, $g_t = g_s(g_s^{-1} g_t) = g_s g'_{t-s}$, where $g'_u = g_s^{-1} g_{u+s}$ is a left Lévy process in G independent of $\sigma\{g_v; 0 \leq v \leq s\}$. Let $g'_t = \xi'_t a'_t \eta'_t$ be the Cartan decomposition. By (7.4), $a_t = a_s(k_s g'_{t-s})_A$. Then $\lim_{t \to \infty} (1/t)$ $\log(k_s g'_{t-s})_A = \lim_{t \to \infty} \log a_t = H^+$. This is also equal to $\lim_{t \to \infty} (1/t) \log a'_t$ because this limit is unchanged when a'_t is replaced by the $\overline{A_+}$-component of $k_s g'_{t-s}$ in the Cartan decomposition. It follows that H^+ is independent of $\sigma\{g_v; 0 \leq v \leq s\}$ for any $s > 0$; hence, it is independent of $\sigma\{g_t; t \in \mathbb{R}_+\}$. This implies that H^+ must be a constant.

To show $H^+ \in \mathfrak{a}_+$, it suffices to prove $\alpha(H^+) > 0$ for any positive root α. Recall that $g_0 = k_0$; hence, $a_1 = (g_1)_A = (k_0 g_0^{-1} g_1)_A$. Let $b_0 = a_1$ and inductively let $b_i = (k_i g_i^{-1} g_{i+1})_A$ for $i = 0, 1, 2, \ldots$. Then $\{b_i\}$ is a stationary sequence and

$$a_i = (g_i)_A = \left(g_{i-1} g_{i-1}^{-1} g_i\right)_A = a_{i-1} \left(k_{i-1} g_{i-1}^{-1} g_i\right)_A$$

$$= a_{i-1} b_{i-1} = a_{i-2} b_{i-2} b_{i-1} = \cdots = b_0 b_1 \cdots b_{i-1}.$$

Let $\xi_i = \alpha(\log b_i)$ for $i = 0, 1, 2, \ldots$. Then $\{\xi_i\}$ is a real-valued stationary sequence such that

$$\sum_{i=0}^{n-1} \xi_i = \sum_{i=0}^{n-1} \alpha(\log b_i) = \alpha(\log(b_0 b_1 \cdots b_{n-1})) = \alpha(\log a_n) \to \infty$$

almost surely as $n \to \infty$. By the note following Proposition 7.4, $E(|\xi_0|) = E[|\alpha(\log a_1)|] < \infty$. By Corollary 7.1, $E(\xi_0) > 0$. However,

$$\alpha(H^+) = \lim_{n \to \infty} \frac{1}{n} E[\alpha(\log a_n)] = \lim_{n \to \infty} \frac{1}{n} \sum_{i=0}^{n-1} E(\xi_i) = E(\xi_0).$$

This proves $H^+ \in \mathfrak{a}_+$. $\qquad\qquad\qquad\qquad\qquad\qquad\qquad\qquad\qquad\quad\square$

By Theorem 7.1, both a_t and a_t^+ converge to ∞ as $t \to \infty$ at nonrandom exponential rates equal in values to the components of H^+. Therefore, H^+ will be called the rate vector of the left Lévy process g_t.

From the proof of Theorem 7.1, it is easy to see that the rate vector is given by

$$H^+ = \int_K \left\{ \frac{1}{2} \sum_{i=1}^{m} [[Ad(k)Y_i]_\mathfrak{k}, Ad(k)Y_i]_\mathfrak{a} + [Ad(k)Z_0]_\mathfrak{a} \right\} \nu(dk)$$

$$+ \int_K \int_G \log[(kg)_A] \Pi(dg) \nu(dk), \qquad\qquad (7.16)$$

where ν is a stationary measure of the process k_t in K and Z_0 has the same meaning as in Proposition 7.4.

7.3. Haar Measure as Stationary Measure

Let g_t be a left Lévy process in G satisfying the hypotheses of Theorem 7.1 and let $g_t = n_t a_t k_t$ be the Iwasawa decomposition $G = N^- A K$. By Proposition 7.3, k_t is a Markov process in K. In this section, we obtain a useful formula for the rate vector H^+ of g_t when the normalized Haar measure on K, which will be denoted by dk in the following, is a stationary measure of k_t. Note that, by Proposition 7.3, if the left Lévy process g_t is right K-invariant, then k_t is a right Lévy process in K and, by the left invariance of the Haar measure, it is easy to show that dk is a stationary measure of k_t.

Lemma 7.2. *If $Y \in \mathfrak{p}$, then $\int_K dk \mathrm{Ad}(k)Y = 0$.*

Proof. By Proposition 5.15, there exist $H \in \mathfrak{a}$ and $h \in K$ such that $Y = \mathrm{Ad}(h)H$. Since $\int_K dk \mathrm{Ad}(k)Y = \int_K dk \mathrm{Ad}(kh)H = \int_K dk \mathrm{Ad}(k)H$, we may assume $Y \in \mathfrak{a}$. Recall $W = M'/M$ is the Weyl group introduced in Section 5.3. Let $\overline{Y} = \sum_{s \in W} s(Y)$. Then $s(\overline{Y}) = \overline{Y}$. Since W contains the reflection about the hyperplane $\alpha = 0$, for any root α, this implies $\alpha(\overline{Y}) = 0$; hence, $\overline{Y} = 0$. Let $|W|$ be the cardinality of W, and for $s \in W$, let $k_s \in M'$ represent s. Then

$$|W| \int_K dk \mathrm{Ad}(k)Y = \sum_{s \in W} \int_K dk \mathrm{Ad}(kk_s)Y = \int_K dk \mathrm{Ad}(k)\overline{Y} = 0. \qquad \square$$

Let α_j, H_j, X'_j, Y'_j, Z'_j, and H_ρ be defined as in Section 6.2.

Lemma 7.3. *For $X \in \mathfrak{g}$, let Y and Z be, respectively, its orthogonal projections to $\mathfrak{p}$ and $\mathfrak{k}$. Then*

$$[[\mathrm{Ad}(k)X]_\mathfrak{k}, \mathrm{Ad}(k)X]_\mathfrak{a} = [\mathrm{Ad}(k)[Z, Y]]_\mathfrak{a} + \sum_{j=1}^{r} \langle \mathrm{Ad}(k)Y, Y'_j \rangle^2 H_j.$$

Proof. Since $[\mathfrak{k}, \mathfrak{k}] \subset \mathfrak{k}$,

$$
\begin{aligned}
[[\mathrm{Ad}(k)X]_\mathfrak{k}, \mathrm{Ad}(k)X]_\mathfrak{a} &= [[\mathrm{Ad}(k)X]_\mathfrak{k}, \mathrm{Ad}(k)Y]_\mathfrak{a} \\
&= [\mathrm{Ad}(k)Z, \mathrm{Ad}(k)Y]_\mathfrak{a} + [[\mathrm{Ad}(k)Y]_\mathfrak{k}, \mathrm{Ad}(k)Y]_\mathfrak{a} \\
&= [\mathrm{Ad}(k)[Z, Y]]_\mathfrak{a} + [[\mathrm{Ad}(k)Y]_\mathfrak{k}, \mathrm{Ad}(k)Y]_\mathfrak{a}. \quad (7.17)
\end{aligned}
$$

Since $\mathrm{Ad}(k)Y = \sum_{j=1}^{r} \langle \mathrm{Ad}(k)Y, Y'_j \rangle Y'_j + H$ for some $H \in \mathfrak{a}$ and $(Y'_j)_\mathfrak{k} = Z'_j$, it follows that $[\mathrm{Ad}(k)Y]_\mathfrak{k} = \sum_{j=1}^{r} \langle \mathrm{Ad}(k)Y, Y'_j \rangle Z'_j$ and

$$
\begin{aligned}
[[\mathrm{Ad}(k)Y]_\mathfrak{k}, \mathrm{Ad}(k)Y]_\mathfrak{a} &= \sum_{i,j=1}^{r} \langle \mathrm{Ad}(k)Y, Y'_i \rangle \langle \mathrm{Ad}(k)Y, Y'_j \rangle [Z'_i, Y'_j]_\mathfrak{a} \\
&= \sum_{j=1}^{r} \langle \mathrm{Ad}(k)Y, Y'_j \rangle^2 H_j. \qquad \square
\end{aligned}
$$

The adjoint action $\text{Ad}(G)$ of G on $\mathfrak{g}$ given by $\mathfrak{g} \ni X \mapsto \text{Ad}(g)X \in \mathfrak{g}$ for $g \in G$ restricts to an action of K on $\mathfrak{p}$ given by $\mathfrak{p} \ni Y \mapsto \text{Ad}(k)Y \in \mathfrak{p}$ for $k \in K$, which will be denoted by $\text{Ad}_\mathfrak{p}(K)$. It is called irreducible if it has no nontrivial invariant subspace; that is, if V is a subspace of $\mathfrak{p}$ such that $\text{Ad}(k)V \subset V$ for all $k \in K$, then either $V = \mathfrak{p}$ or $V = \{0\}$.

Proposition 7.5. $\text{Ad}_\mathfrak{p}(K)$ *is irreducible in the following two cases:*

   (i) $G = SL(d, \mathbb{R})$ *and* $K = SO(d)$.
   (ii) $G = SO(1, d)$ *and* $K = \text{diag}\{1, SO(d)\}$.

Proof. In case (i), $\mathfrak{p}$ is the space of symmetric traceless matrices and two such matrices belong to the same orbit of $\text{Ad}_\mathfrak{p}(K)$ if and only if they have the same eigenvalues. Because any orbit contains a diagonal matrix and the Weyl group acts on $X \in \mathfrak{a}$ by permuting the diagonal elements of X, it suffices to prove the following elementary fact: Given any nonzero vector in the subspace of $\mathbb{R}^d$ determined by the equation $x_1 + \cdots + x_d = 0$, by permuting the components of this vector, we can get enough vectors to span the whole subspace.

In case (ii), by the discussion in Section 5.4, $\mathfrak{p} = \{\xi_y; y \in \mathbb{R}^d\}$, where ξ_y is defined by (5.23). Since $\text{Ad}(\text{diag}(1, C))\xi_y = \xi_{Cy}$ for $C \in SO(d)$ and $y \in \mathbb{R}^d$, it follows easily that $\text{Ad}_\mathfrak{p}(K)$ is irreducible in this case. $\qquad\square$

Lemma 7.4. *Assume* $\text{Ad}_\mathfrak{p}(K)$ *is irreducible. For any* $Y, Y' \in \mathfrak{p}$ *with* $\|Y\| = \|Y'\| = 1$, *we have* $\int_K dk \langle \text{Ad}(k)Y, Y'\rangle^2 = 1/p$, *where* $p = \dim(\mathfrak{p})$.

Proof. Let S be the unit sphere in $\mathfrak{p}$ and let $Y \in S$. Consider the function $\psi(W) = \int_K dk \langle \text{Ad}(k)Y, W\rangle^2$ defined on S. We will show that ψ is a constant on S. If not, let a and b be, respectively, its minimal and maximal values. Choose W_1 such that $\psi(W_1) = a$. Let $S' = \{X \in S; \langle X, W_1\rangle = 0\}$. Any $W \in S$ can be expressed as $W = xW_1 + yW_2$ for some $W_2 \in S'$ and $x^2 + y^2 = 1$. We have

$$0 = (d/dt)\psi(\sqrt{1 - t^2}W_1 + tW_2)|_{t=0} = 2\int_K dk \langle \text{Ad}(k)Y, W_1\rangle \langle \text{Ad}(k)Y, W_2\rangle.$$

It follows that

$$\psi(xW_1 + yW_2) = x^2\psi(W_1) + y^2\psi(W_2).$$

This is less than the maximal value b if $x \neq 0$. Therefore, b can only be obtained on S'. If $\psi(W_2) = b$, then by the invariance of the Haar measure dk, $\psi = b$ along the orbit of W_2 under $\text{Ad}_\mathfrak{p}(K)$. Hence, this orbit is orthogonal to W_1. The linear span of the orbit is a nontrivial invariant subspace of

$Ad_p(K)$, which contradicts the irreducibility of $Ad_p(K)$. This proves that ψ is a constant. Because $\langle Ad(k)Y, W \rangle = \langle Y, Ad(k^{-1})W \rangle$, it is easy to see that this constant is independent of $Y \in S$.

Let $W_1, \ldots, W_p$ be an orthonormal basis of $\mathfrak{p}$. Because, for $k \in K$, $Ad(k)$ is an isometry on $\mathfrak{p}$ with respect to the inner product $\langle \cdot, \cdot \rangle$, $\{\langle Ad(k)W_i, W_j \rangle\}$ is an orthogonal matrix. We have

$$p\psi = \sum_i \int_K dk \langle Ad(k)W_i, W_1 \rangle^2 = \int_K dk \sum_i \langle Ad(k)W_i, W_1 \rangle^2 = 1. \quad \square$$

For $X \in \mathfrak{g}$, let $X_\mathfrak{p}$ be its orthogonal projection to $\mathfrak{p}$. By (7.16) and Lemmas 7.2–7.4, we obtain (7.18) in the following theorem. If the left Lévy process g_t is right K-invariant, then, for any $k \in K$, the process kg_tk^{-1} has the same distribution as g_t. It follows that the Lévy measure Π is invariant under the map $g \mapsto kgk^{-1}$ for any $k \in K$. In this case,

$$\int_K \int_G \log[(kg)_A] \, dk \, \Pi(dg) = \int_K \int_G \log[(kgk^{-1})_A] \, dk \, \Pi(dg)$$

$$= \int_G \log(g_A) \Pi(dg).$$

We have proved the following theorem.

Theorem 7.2. *Let g_t be a left Lévy process in G determined by (1.23) and let $g_t = n_t a_t k_t$ be the Iwasawa decomposition $G = N^- AK$. Assume the hypotheses of Theorem 7.1, the irreducibility of $Ad_p(K)$, and that the normalized Haar measure dk on K is a stationary measure of k_t. Then the rate vector H^+ of the process g_t is given by*

$$H^+ = \left[\frac{1}{p} \sum_{i=1}^m \|(Y_i)_\mathfrak{p}\|^2 \right] H_\rho + \int_K \int_G \log[(kg)_A] \, dk \, \Pi(dg), \qquad (7.18)$$

where $p = \dim(\mathfrak{p})$, and $Y_1, Y_2, \ldots, Y_m \in \mathfrak{g}$ are in (1.23). Note that the integral term on the right-hand side of this equation vanishes if g_t is continuous or more generally if Π is supported by K. Moreover, if g_t is right K-invariant, then this term is equal to $\int_G \log(g_A) \Pi(dg)$.

Let $\{a_{ij}\}$ be the symmetric matrix in the generator L of g_t given by (1.7) with respect to a basis $\{X_1, X_2, \ldots, X_d\}$ of $\mathfrak{g}$. By (1.22), there are constants σ_{ij} such that

$$Y_i = \sum_{j=1}^d \sigma_{ij} X_j \qquad \text{for} \quad 1 \le i \le m.$$

Then $a_{ij} = \sum_{k=1}^{m} \sigma_{ki}\sigma_{kj}$ and

$$\sum_{k=1}^{m} \|(Y_k)_{\mathfrak{p}}\|^2 = \sum_{k=1}^{m} \sum_{i,j=1}^{d} \sigma_{ki}\sigma_{kj} \langle (X_i)_{\mathfrak{p}}, (X_j)_{\mathfrak{p}} \rangle = \sum_{i,j=1}^{d} a_{ij} \langle (X_i)_{\mathfrak{p}}, (X_j)_{\mathfrak{p}} \rangle.$$

Assume the basis $\{X_1, \ldots, X_d\}$ is orthonormal such that $X_1, \ldots, X_p$ form a basis of $\mathfrak{p}$ and $X_{p+1}, \ldots, X_d$ form a basis of $\mathfrak{k}$. Then $\sum_{k=1}^{m} \|(Y_k)_{\mathfrak{p}}\|^2 = \sum_{i=1}^{p} a_{ii}$. It follows that the rate vector given in (7.18) can also be written as

$$H^+ = \left[\frac{1}{p} \sum_{i=1}^{p} a_{ii} \right] H_\rho + \int_K \int_G \log[(kg)_A] \, dk \, \Pi(dg). \tag{7.19}$$

In the rest of this section, we derive a necessary and sufficient condition for dk to be a stationary measure of k_t for a continuous left Lévy process g_t. We also obtain an example of g_t that satisfies this condition but is not right K-invariant.

A continuous left Lévy process g_t is the solution of the stochastic differential equation

$$dg_t = \sum_{i=1}^{m} Y_i^l(g_t) \circ dW_t^i + Y_0^l(g_t) dt \tag{7.20}$$

on G, where $Y_0, Y_1, \ldots, Y_m \in \mathfrak{g}$ and $W_t = (W_t^1, \ldots, W_t^m)$ is an m-dimensional standard Brownian motion.

Let L_K be the generator of k_t. By (7.14) with $N = 0$,

$$L_K = (1/2) \sum_{i=1}^{m} U_i U_i + U_0,$$

where U_i is the vector field on K given by $U_i(k) = [\text{Ad}(k)Y_i]_{\mathfrak{k}}^r$ for $k \in K$. Let L_K^* be the adjoint of L_K; that is, $(L_K\phi, \psi) = (\phi, L_K^*\psi)$ for $\phi, \psi \in C^2(K)$, where $(\phi, \psi) = \int_K dk \phi(k) \psi(k)$. We have

$$L_K^* = (1/2) \sum_{i=1}^{m} U_i^* U_i^* + U_0^*.$$

To obtain an explicit expression for L_K^*, we will write U and Y for U_i and Y_i for fixed i. Let Y_j' and Z_j' for $j = 1, \ldots, r$ be defined as before. Recall that $[Z_j', Y_j'] = H_j$ and $H_\rho = (1/2) \sum_{j=1}^{r} H_j$. We can choose $Y_{r+1}', \ldots, Y_p'$ in $\mathfrak{a}$

and $Z'_{r+1}, \ldots, Z'_q$ in the Lie algebra $\mathfrak{m}$ of M such that $Y'_1, \ldots, Y'_p, Z_1, \ldots, Z'_q$ form an orthonormal basis of $\mathfrak{g}$ with respect to $\langle \cdot, \cdot \rangle$. We have

$$
\mathrm{Ad}(k)Y = \sum_{j=1}^{p} \langle \mathrm{Ad}(k)Y, Y'_j \rangle Y'_j + \sum_{j=1}^{q} \langle \mathrm{Ad}(k)Y, Z'_j \rangle Z'_j.
$$

Hence, $[\mathrm{Ad}(k)Y]_{\mathfrak{k}} = \sum_{j=1}^{r} \langle \mathrm{Ad}(k)Y, Y'_j + Z'_j \rangle Z'_j + \sum_{j>r} \langle \mathrm{Ad}(k)Y, Z'_j \rangle Z'_j$ and

$$
U\phi(k) = \left[\sum_{j=1}^{r} \langle \mathrm{Ad}(k)Y, Y'_j + Z'_j \rangle \frac{d}{ds} \phi\left(e^{sZ'_j}k\right) \right.
$$

$$
\left. + \sum_{j>r} \langle \mathrm{Ad}(k)Y, Z'_j \rangle \frac{d}{ds} \phi\left(e^{sZ'_j}k\right) \right]\Big|_{s=0}.
$$

By the invariance of the Haar measure dk, we have

$$
\int_K dk \psi(k) \sum_{j=1}^{r} \langle \mathrm{Ad}(k)Y, Y'_j + Z'_j \rangle \frac{d}{ds} \phi\left(e^{sZ'_j}k\right)\big|_{s=0}
$$

$$
= \int_K dk \phi(k) \sum_{j=1}^{r} \frac{d}{ds} \psi\left(e^{-sZ'_j}k\right) \langle \mathrm{Ad}\left(e^{-sZ'_j}k\right)Y, Y'_j + Z'_j \rangle\big|_{s=0}
$$

$$
= \int_K dk \phi(k) \sum_{j=1}^{r} \left[-\langle \mathrm{Ad}(k)Y, Y'_j + Z'_j \rangle \frac{d}{ds} \psi\left(e^{sZ'_j}k\right)\big|_{s=0} \right.
$$

$$
\left. + \psi(k) \langle \mathrm{Ad}(k)Y, [Z'_j, Y'_j + Z'_j] \rangle \right]
$$

$$
= \int_K dk \phi(k) \left[-\sum_{j=1}^{r} \langle \mathrm{Ad}(k)Y, Y'_j + Z'_j \rangle \frac{d}{ds} \psi\left(e^{sZ'_j}k\right)\big|_{s=0} \right.
$$

$$
\left. + 2\psi(k) \langle \mathrm{Ad}(k)Y, H_\rho \rangle \right].
$$

A similar computation yields

$$
\int_K dk \psi(k) \sum_{j>r} \langle \mathrm{Ad}(k)Y, Z'_j \rangle \frac{d}{ds} \phi\left(e^{sZ'_j}k\right)\big|_{s=0}
$$

$$
= -\int_K dk \phi(k) \sum_{j>r} \langle \mathrm{Ad}(k)Y, Z'_j \rangle \frac{d}{ds} \psi\left(e^{sZ'_j}k\right)\big|_{s=0}.
$$

It follows that

$$
\int_K dk \psi(k) U\phi(k) = \int_K dk \phi(k)[-U\psi(k) + 2\psi(k)\langle \mathrm{Ad}(k)Y, H_\rho \rangle].
$$

Hence,

$$U^*\psi(k) = -U\psi(k) + 2\langle\mathrm{Ad}(k)Y, H_\rho\rangle\psi(k)$$

and

$$
\begin{aligned}
U^*U^*\psi(k) \\
&= UU\psi(k) - 4\langle\mathrm{Ad}(k)Y, H_\rho\rangle U\psi(k) \\
&\quad + [-2U\langle\mathrm{Ad}(k)Y, H_\rho\rangle + 4\langle\mathrm{Ad}(k)Y, H_\rho\rangle^2]\psi(k) \\
&= UU\psi(k) - 4\langle\mathrm{Ad}(k)Y, H_\rho\rangle U\psi(k) \\
&\quad + [-2\langle[[\mathrm{Ad}(k)Y]_\mathfrak{k}, \mathrm{Ad}(k)Y], H_\rho\rangle + 4\langle\mathrm{Ad}(k)Y, H_\rho\rangle^2]\psi(k).
\end{aligned}
$$

This gives us an explicit expression for $L_K^* = (1/2)\sum_{i=1}^m U_i^* U_i^* + U_0^*$. This expression contains three different types of terms: a ψ term, the terms involving its first-order derivatives, and those involving its second-order derivatives. The coefficient of the ψ term in $L_K^*\psi$ is

$$\sum_{i=1}^m \{-\langle[[\mathrm{Ad}(k)Y_i]_\mathfrak{k}, \mathrm{Ad}(k)Y_i], H_\rho\rangle + 2\langle\mathrm{Ad}(k)Y_i, H_\rho\rangle^2\} + \langle\mathrm{Ad}(k)Y_0, H_\rho\rangle. \tag{7.21}$$

Let μ be a measure on K with a smooth density ψ with respect to the Haar measure and let P_t^K be the transition semigroup of k_t. If μ is a stationary measure of k_t, then $(\psi, P_t^K f) = \mu(P_t^K f) = \mu(f)$ for any $f \in C^\infty(K)$. Therefore, $(L_K^*\psi, f) = (\psi, L_K f) = (d/dt)(\psi, P_t^K f)|_{t=0} = 0$. From this, it is easy to see that μ is a stationary measure of k_t if and only if $L_K^*\psi = 0$. It follows that dk is a stationary measure of k_t if and only if (7.21) vanishes. We obtain the following proposition:

Proposition 7.6. *Let g_t be a left invariant diffusion process in G determined by (7.20) and let k_t be its K-component in the Iwasawa decomposition $G = N^-AK$. Then the normalized Haar measure on K is a stationary measure of k_t if and only if the expression in (7.21) vanishes for all $k \in K$.*

By Proposition 7.3, if g_t is right K-invariant, then its K-component k_t in the Iwasawa decomposition $G = N^-AK$ is a right Lévy process in K; hence, the normalized Haar measure dk on K is a stationary measure of k_t. We can construct an example that is not right K-invariant but for which dk is still a stationary measure of k_t.

Let $\{Y_1, \ldots, Y_p\}$ be an orthonormal basis of $\mathfrak{p}$ and let $Y_0 = 0$. Choose arbitrary elements $Y_{p+1}, \ldots, Y_m \in \mathfrak{k}$ so that the continuous left Lévy process g_t determined by (7.20) is not right K-invariant. For any $k \in K$,

$\text{Ad}(k)Y_i = \sum_{j=1}^{p} a_{ij}(k)Y_j$, where $\{a_{ij}(k)\}$ is an orthogonal matrix. It follows that $\sum_{i=1}^{m}[[\text{Ad}(k)Y_i]_{\mathfrak{k}}, \text{Ad}(k)Y_i]_{\mathfrak{a}} = \sum_{i=1}^{p}[(Y_i)_{\mathfrak{k}}, Y_i]_{\mathfrak{a}}$. Let $Y_1', \ldots, Y_r'$, $Z_1', \ldots, Z_r'$, and $H_1, \ldots, H_r$ be defined as in Section 6.2. We may assume $Y_i = Y_i'$ for $1 \le i \le r$ and $Y_{r+1}, \ldots, Y_p \in \mathfrak{a}$. Then $\sum_{i=1}^{p}[(Y_i)_{\mathfrak{k}}, Y_i]_{\mathfrak{a}} = \sum_{i=1}^{r}[Z_i', Y_i']_{\mathfrak{a}} = \sum_{i=1}^{r} H_i = 2H_\rho$. Therefore, the first term in (7.21), $-\sum_{i=1}^{m}\langle[[\text{Ad}(k)Y_i]_{\mathfrak{k}}, \text{Ad}(k)Y_i], H_\rho\rangle$, is equal to $-2\langle H_\rho, H_\rho\rangle$. The second term is

$$\sum_{i=1}^{m} 2\langle \text{Ad}(k)Y_i, H_\rho\rangle^2 = 2 \sum_{i,u,v=1}^{p} a_{iu}(k)a_{iv}(k)\langle Y_u, H_\rho\rangle\langle Y_v, H_\rho\rangle$$
$$= 2 \sum_{i=r+1}^{p} \langle Y_i, H_\rho\rangle^2 = 2\langle H_\rho, H_\rho\rangle,$$

where the last equality is the Parseval identity. This shows that the expression in (7.21) vanishes for all $k \in K$; hence, dk is a stationary measure of k_t.

Note that the expression (7.21) should vanish whenever g_t is right K-invariant, but it is not completely trivial to prove this directly from (7.21).

7.4. Rate of Convergence of the Nilpotent Component

As in the previous sections, let g_t be a left Lévy process in G and let $g_t = n_t a_t k_t$ be the Iwasawa decomposition $G = N^- A K$. We know that if g_t satisfies the hypotheses of Theorem 6.4, then, almost surely, $n_t \to n_\infty$, or equivalently, $n_t^{-1} n_\infty \to e$, as $t \to \infty$. In this section, we investigate how fast this convergence takes place. The result of this section together with the existence of the rate vector H^+ will play an important role in the study of the dynamical behavior of the Lévy processes.

Let $\{X_1, \ldots, X_d\}$ be an orthonormal basis of $\mathfrak{g}$ with respect to the inner product $\langle \cdot, \cdot \rangle$ such that each X_i is contained in $\mathfrak{g}_\alpha$ for some root α or $\alpha = 0$, which will be denoted by α_i. Then $\alpha_1, \ldots, \alpha_d$ form a complete set of all the roots, including zero, each of which is repeated as many times as its multiplicity.

For $g \in G$, let $\tilde{g}$ be the matrix representing $\text{Ad}(g)$ under the basis $\{X_i\}$; that is, $\text{Ad}(g)X_j = \sum_{i=1}^{d} \tilde{g}_{ij} X_i$. Then $\text{Ad}(gh) = \tilde{g}\tilde{h}$ and $\text{Ad}(g^{-1}) = \tilde{g}^{-1}$ for $g, h \in G$. Any $d \times d$ real matrix may be regarded as a point in the Euclidean space $\mathbb{R}^{d^2}$, and vice versa. For $X \in \mathbb{R}^{d^2}$, its Euclidean norm $|X| = (\sum_{i,j} X_{ij}^2)^{1/2}$ satisfies the product inequality $|XY| \le |X||Y|$ for any $X, Y \in \mathbb{R}^{d^2}$, where XY is the matrix product.

Let $\{h_{ij}(t)\}$ be the matrix representing the random linear map $\mathrm{Ad}(n_t^{-1}n_\infty)$: $\mathfrak{g} \to \mathfrak{g}$ under the basis $\{X_1, \ldots, X_d\}$; that is,

$$\mathrm{Ad}\left(n_t^{-1}n_\infty\right) X_j = \sum_{i=1}^{d} h_{ij}(t)X_i. \qquad (7.22)$$

Note that $h_{ij}(t)$ depends also on $\omega \in \Omega$, so it may be written as $h_{ij}(t, \omega)$ if this dependence needs to be indicated explicitly.

Let Δ_+ be the set of all the positive roots and let $[\Delta_+]$ be the set of all the nontrivial linear combinations of positive roots with integer coefficients, that is,

$$[\Delta_+] = \left\{ \sum_{\alpha \in \Delta_+} c_\alpha \alpha; \ c_\alpha \geq 0 \text{ are integers with } c_\alpha > 0 \text{ for some } \alpha \right\}.$$

Since $[\mathfrak{g}_\alpha, \mathfrak{g}_\beta] \subset \mathfrak{g}_{\alpha+\beta}$, we see that, for $n \in N^-$ and $X \in \mathfrak{g}_\alpha$,

$$\mathrm{Ad}(n)X = e^{\mathrm{ad}(\log n)} X = X + Y \text{ with } Y \in \sum_{\beta \in [\Delta_+]} \mathfrak{g}_{\alpha-\beta}. \qquad (7.23)$$

It follows that $h_{ii}(t) = 1$, and $h_{ij}(t) = 0$ if $i \neq j$ with $\alpha_i \notin \{\alpha_j - \beta; \ \beta \in [\Delta_+]\}$. In particular,

$$h_{ij}(t) = 0 \quad \text{if } i \neq j \quad \text{and} \quad \alpha_j(H^+) \leq \alpha_i(H^+). \qquad (7.24)$$

Because $n_t^{-1}n_\infty \to e$, the matrix $\{h_{ij}(t)\}$ converges to the identity matrix I_d almost surely as $t \to \infty$. The inequality (7.27) in the following theorem together with (7.24) says that the nonzero off-diagonal elements of this matrix converge to zero exponentially. We may identify $\mathrm{Ad}(g)$ with its matrix representation, thus writing $\tilde{g} = \mathrm{Ad}(g)$. For typographical convenience, $\tilde{g} = \mathrm{Ad}(g)$ may be written as $g\tilde{\ }$. In particular, $g_{\tilde{N}}$ denotes $\mathrm{Ad}(g_N)$.

Theorem 7.3. *Assume the hypotheses of Theorem 7.1. Let $H^+ \in \mathfrak{a}_+$ be the rate vector given in Theorem 7.1. Then, for P-almost all ω,*

$$\lim_{t\to\infty} \frac{1}{t} \log a_t(\omega) = H^+ \quad \text{and} \quad n_\infty(\omega) = \lim_{t\to\infty} n_t(\omega), \qquad (7.25)$$

where n_∞ is an N^--valued random variable, and

$$\forall \varepsilon > 0, \quad \sup_{k \in K} \left| \left[k g_t^{-1}(\omega) g_{t+1}(\omega) \right]_{\tilde{N}} \right| \leq e^{\varepsilon t} \text{ for sufficiently large } t > 0. \qquad (7.26)$$

Moreover, if ω satisfies (7.25) and (7.26), then, for any $\varepsilon > 0$,

$$|h_{ij}(t, \omega)| \leq \exp\{-t[\alpha_j(H^+) - \alpha_i(H^+) - \varepsilon]\} \text{ for sufficiently large } t > 0.$$
$$(7.27)$$

The rest of this section is devoted to the proof of Theorem 7.3. We already know from Theorem 7.1 that (7.25) holds for P-almost all ω. Without loss of generality, we will assume that $g_0 = e$.

Let $\tilde{G}$ be the group of all the linear automorphisms on $\mathfrak{g}$. It may be identified with the general linear group $GL(d, \mathbb{R})$. Its Lie algebra $\tilde{\mathfrak{g}}$ may be identified with $\mathfrak{gl}(d, \mathbb{R})$. As in Section 1.5, $\tilde{\mathfrak{g}}$ may be regarded as the Euclidean space $\mathbb{R}^{d^2}$ and $\tilde{G}$ as a dense open subset of $\mathbb{R}^{d^2}$.

The left Lévy process g_t in G induces a left Lévy process $\tilde{g}_t$ in $\tilde{G}$ given by $\tilde{g}_t = \mathrm{Ad}(g_t)$. Suppose g_t satisfies the stochastic integral equation (1.23). Note that, for $g \in G, Y \in \mathfrak{g}$, and $f \in C^2(\tilde{G}), Y^l(f \circ \mathrm{Ad})(g) = \tilde{Y}^l f(\tilde{g})$, where $\tilde{g} = \mathrm{Ad}(g)$ and $\tilde{Y} = \mathrm{ad}(Y) \in \tilde{\mathfrak{g}}$. Replacing g_t, Y_i, Z, X_j, Π, and N in (1.23) by $\tilde{g}_t, \tilde{Y}_i, \tilde{Z}, \tilde{X}_j, \tilde{\Pi} = \mathrm{Ad}\,\Pi$, and $(\mathrm{id}_{\mathbb{R}_+} \times \mathrm{Ad})N$, respectively, we obtain a stochastic integral equation satisfied by $\tilde{g}_t$ for any $f \in C_b(\tilde{G}) \cap C^2(\tilde{G})$. By Theorem 1.3, $\tilde{g}_t$ satisfies a stochastic integral equation of the form (1.30) if $\tilde{\Pi}$ has a compact support.

By Proposition 5.1(a), for any $k \in K, \tilde{k}$ is an orthogonal matrix. Because G is unimodular, $\det(\tilde{g}) = 1$ for any $g \in G$.

Lemma 7.5. *For any $t \in \mathbb{R}_+$,*

$$E\left[\sup_{0 \leq s \leq t} \log |\tilde{g}_s|\right] < \infty.$$

The same inequality holds when $\tilde{g}_s$ is replaced by $\tilde{g}_s^{-1}$.

Proof. If $\tilde{\Pi}$ is compactly supported, then by Theorem 1.3, $E[\sup_{0 \leq s \leq t} |g_s|^2] < \infty$. This is stronger than the desired inequality. Since this result also holds for the right Lévy process $\tilde{g}_t^{-1}$, $E[\sup_{0 \leq s \leq t} |\log \tilde{g}_s^{-1}|^2] < \infty$. The lemma is proved in the case when $\tilde{\Pi}$ has a compact support.

In general, note that any $g \in G$ may be written as $g = ke^Y$ with $k \in K$ and $Y = g_{\mathfrak{p}}$. Since $\tilde{k}$ is an orthogonal matrix, $|\tilde{k}| = \sqrt{d}$. We have

$$|\tilde{g}| \leq \sqrt{d}\,|\mathrm{Ad}(e^Y)| = \sqrt{d}\,|e^{\mathrm{ad}(Y)}| \leq \sqrt{d}\,e^{c\|Y\|} = \sqrt{d}\,\exp(c\|g_{\mathfrak{p}}\|) \quad (7.28)$$

for some constant $c > 0$.

Let $\tilde{\Pi} = \tilde{\Pi}_1 + \tilde{\Pi}_2$, where $\tilde{\Pi}_1$ is supported by a compact subset of $\tilde{G}$ and $\tilde{\Pi}_2$ is finite. By Proposition 1.5, we have the following independent objects: exponentially distributed random variables $\tau_1, \tau_2, \tau_3, \ldots$ of a common rate $\lambda = \tilde{\Pi}_2(\tilde{G})$; $\tilde{G}$-valued random variables $\sigma_1, \sigma_2, \sigma_3, \ldots$ with common distribution $\tilde{\Pi}_2/\lambda$; and a left Lévy process h_t that is the solution of a stochastic integral equation of the form (1.30) with Lévy measure $\tilde{\Pi}_1$ satisfying the initial condition $h_0 = I_d$ such that, with $T_i = \sum_{j=1}^{i} \tau_j$,

$$\tilde{g}_0 = I_d, \quad \tilde{g}_t = h_t \text{ for } 0 \leq t < T_1, \text{ and inductively}$$
$$\tilde{g}_t = h(T_1-)\sigma_1 h(T_1)^{-1} h(T_2-)\sigma_2 \cdots h(T_{i-1})^{-1} h(T_i-)\sigma_i \, h(T_i)^{-1} \, h_t \text{ for}$$
$$T_i \leq t < T_{i+1} \text{ and } i \geq 1.$$

On $A_i = [T_i \leq t < T_{i+1}]$,

$$\log |\tilde{g}_t| \leq \sum_{j=1}^{i} \log |h_{T_j-}| + \sum_{j=1}^{i} \log |h_{T_j}^{-1}| + \log |h_t| + \sum_{j=1}^{i} \log |\sigma_j|.$$

Let

$$C_t = E \left\{ \sup_{0 \leq s \leq t} \left[\log |h_s| + \log \left| h_s^{-1} \right| \right] \right\}.$$

By (7.28) and (7.7), the constant c_1 defined by

$$c_1 = \int_{\tilde{G}} \log |\tilde{g}| \, \tilde{\Pi}_2(d\tilde{g})$$

is finite. Since the process h_t and σ_i are independent of A_i, and $E[\log |\sigma_i|] = c_1/\lambda$, we have

$$E \left[\sup_{0 \leq s \leq t} \log |\tilde{g}_s| \right] = \sum_{i=0}^{\infty} E \left[\sup_{0 \leq s \leq t} \log |\tilde{g}_s|; A_i \right]$$
$$\leq \sum_{i=0}^{\infty} [(i+1)C_t + i(c_1/\lambda)] P(A_i) = \sum_{i=0}^{\infty} [(i+1)C_t + i(c/\lambda)] e^{-\lambda} \frac{(\lambda t)^i}{i!} < \infty.$$

Since $\tilde{g}_t^{-1}$ is a right Lévy process in $\tilde{G}$ and $\|[g^{-1}]_p\| = \|[g]_p\|$, the argument here can be easily modified to prove the same conclusion for $\tilde{g}_t^{-1}$. □

We now prove (7.26). Note that $g_{t+s} = g_t g'_s$, where the process $g'_s = g_t^{-1} g_{t+s}$ is identical in distribution to g_s. Let i be a positive integer. For $i \leq t \leq i+1$, $g_t^{-1} g_{t+1} = g'^{-1}_{t-i} g_i^{-1} g_{i+1} g''_{t-i}$, where g_t, g'_t, and g''_t are three

Lévy processes of the same distribution. Since $g_i^{-1} g_{i+1}$ is identical in distribution to g_1, by Lemma 7.5, we have

$$E\left[\sup_{i \le t \le i+1} \log\left|(g_t^{-1} g_{t+1})^{\sim}\right|\right]$$

$$\le E\left\{\sup_{i \le t \le i+1} \log\left[\left|(g_{t-i}'^{-1})^{\sim}\right| \left|(g_i^{-1} g_{i+1})^{\sim}\right| \left|(g_{t-i}'')^{\sim}\right|\right]\right\}$$

$$\le E\left[\sup_{0 \le s \le 1} \log\left|(g_s'^{-1})^{\sim}\right|\right] + E\left[\log\left|(g_i^{-1} g_{i+1})^{\sim}\right|\right] + E\left[\sup_{0 \le s \le 1} \log\left|(g_s'')^{\sim}\right|\right]$$

$$< \infty.$$

We now assume that the α_is are ordered such that

$$\alpha_1(H^+) \ge \alpha_2(H_+) \ge \cdots \ge \alpha_d(H^+). \tag{7.29}$$

Although Theorem 7.3 will be proved under this additional assumption, it is easy to see that the statement of this theorem is not affected by this assumption. By (7.24) and (7.29), for $n \in N^-$, $\tilde{n}$ is a lower triangular matrix with all diagonal elements equal to 1.

If $a \in A$, then $\tilde{a}$ is a diagonal matrix such that the product of its diagonal elements is equal to 1. This implies that $|\tilde{a}^{-1}| \le C|\tilde{a}|^{d-1}$ for some constant C.

For $g \in G$ and $k \in K$, $kg = (kg)_N (kg)_A k'$ for some $k' \in K$; hence, $(kg)_N (kg)_A = kgk'^{-1}$. Since $(kg)_{\tilde{A}}$ and $(kg)_{\tilde{N}}(kg)_{\tilde{A}}$ have the same diagonal, and $\tilde{k}$ as an orthogonal matrix has Euclidean norm equal to $\sqrt{d}$, we see that

$$|(kg)_{\tilde{A}}| \le |\tilde{k}\tilde{g}\tilde{k}'^{-1}| \le d|\tilde{g}|.$$

Therefore,

$$|(kg)_{\tilde{N}}| = |\tilde{k}\tilde{g}\tilde{k}'^{-1}[(kg)_{\tilde{A}}]^{-1}| \le d|\tilde{g}| \cdot |[(kg)_{\tilde{A}}]^{-1}| \le Cd^d |\tilde{g}|^d.$$

We have

$$E\left[\sup_{i \le t \le i+1} \sup_{k \in K} \log\left|(kg_t^{-1} g_{t+1})_{\tilde{N}}\right|\right]$$

$$\le E\left[d \sup_{i \le t \le i+1} \log\left|(g_t^{-1} g_{t+1})^{\sim}\right| + \log C + d \log d\right] < \infty.$$

For $i = 1, 2, 3, \ldots$, $u_i = \sup_{i \le t \le i+1} \sup_{k \in K} \log|(kg_t^{-1} g_{t+1})_{\tilde{N}}|$ are iid random variables with finite expectation. As a consequence of the strong law of large numbers, $(1/i)u_i \to 0$ as $i \to \infty$. This implies (7.26).

We will omit writing the fixed ω that satisfies (7.25) and (7.26) in the proof of (7.27). Then $g_t^{-1} g_{t+1} = k_t^{-1} a_t^{-1} n_t^{-1} n_{t+1} a_{t+1} k_{t+1}$ and

$$k_t g_t^{-1} g_{t+1} = \left(a_t^{-1} n_t^{-1} n_{t+1} a_t \right) a_t^{-1} a_{t+1} k_{t+1}.$$

It follows that $n_t^{-1} n_{t+1} = a_t (k_t g_t^{-1} g_{t+1}) _N a_t^{-1}$. Let $n_t' = \mathrm{Ad}((k_t g_t^{-1} g_{t+1})_N)$ and

$$n_t' X_j = \sum_i c_{ij}(t) X_i$$

for some $c_{ij}(t)$. Because $\mathrm{Ad}(a) X = e^{\mathrm{ad}(\log a)} X = e^{\alpha(\log a)} X$ for $a \in A$ and $X \in \mathfrak{g}_\alpha$,

$$\mathrm{Ad}\left(n_t^{-1} n_{t+1} \right) X_j = \tilde{a}_t n_t' \tilde{a}_t^{-1} X_j = e^{-\alpha_j (\log a_t)} \tilde{a}_t n_t' X_j$$

$$= \sum_i e^{-(\alpha_j - \alpha_i)(\log a_t)} c_{ij}(t) X_i. \qquad (7.30)$$

The coefficients $c_{ij}(t)$ have the same properties as those satisfied by $h_{ij}(t)$. In particular, $c_{ii}(t) = 1$, and $c_{ij}(t) = 0$ if $i \neq j$ and $\alpha_j(H^+) \leq \alpha_i(H^+)$. This implies that $\{c_{ij}(t)\}$ is a lower triangular matrix.

Fix $\delta > 0$. By (7.26), the norm of the matrix $\{c_{ij}(t)\}$ is $\leq e^{c\delta t/2}$ for sufficiently large $t > 0$, where $c > 0$ is given by

$$c = \min\{(\alpha_j - \alpha_i)(H^+); \ (\alpha_j - \alpha_i)(H^+) > 0\}.$$

Let

$$b_{ij}(t) = c_{ij}(t) e^{-\delta t (\alpha_j - \alpha_i)(H^+) + (\alpha_j - \alpha_i)(t H^+ - \log a_t)}.$$

Then $\{b_{ij}(t)\}$ is a lower triangular matrix with all its diagonal elements equal to 1 and

$$e^{-(\alpha_j - \alpha_i)(\log a_t)} c_{ij}(t) = e^{-t(1-\delta)(\alpha_j - \alpha_i)(H^+)} b_{ij}(t).$$

Since $(1/t) \log a_t \to H^+$ as $t \to \infty$, if $i \neq j$, then

$$|b_{ij}(t)| \leq e^{c\delta t/2} e^{-\delta t(\alpha_j - \alpha_i)(H^+) + t(\alpha_j - \alpha_i)(H^+ - (1/t)\log a_t)} \leq e^{-c\delta t/3}$$

for sufficiently large $t > 0$.

We have $\mathrm{Ad}(n_t^{-1} n_\infty) X_j = \lim_{k \to \infty} \mathrm{Ad}(n_t^{-1} n_{t+1} n_{t+1}^{-1} n_{t+2} \cdots n_{t+k}^{-1} n_{t+k+1}) X_j$ and, by (7.30),

$$\mathrm{Ad}\left(n_t^{-1} n_{t+1} n_{t+1}^{-1} n_{t+2} \cdots n_{t+k}^{-1} n_{t+k+1} \right) X_j$$

$$= \mathrm{Ad}\left(n_t^{-1} n_{t+1} \right) \mathrm{Ad}\left(n_{t+1}^{-1} n_{t+2} \right) \cdots \mathrm{Ad}\left(n_{t+k-1}^{-1} n_{t+k} \right) \mathrm{Ad}\left(n_{t+k}^{-1} n_{t+k+1} \right) X_j$$

$$= \sum_{i, i_1, \dots, i_k} e^{-(\alpha_j - \alpha_{i_1})(\log a_{t+k})} c_{i_1 j}(t+k) e^{-(\alpha_{i_1} - \alpha_{i_2})(\log a_{t+k-1})} c_{i_2 i_1}(t+k-1)$$

$$\times \cdots e^{-(\alpha_{i_{k-1}} - \alpha_{i_k})(\log a_{t+1})} c_{i_k i_{k-1}}(t+1) e^{-(\alpha_{i_k} - \alpha_i)(\log a_t)} c_{i i_k}(t) X_i$$

$$= \sum_{i,i_1,\dots,i_k} e^{-(t+k)(1-\delta)(\alpha_j - \alpha_{i_1})(H^+)} b_{i_1 j}(t+k) e^{-(t+k-1)(1-\delta)(\alpha_{i_1} - \alpha_{i_2})(H^+)}$$

$$\times \, b_{i_2 i_1}(t+k-1) \cdots e^{-(t+1)(1-\delta)(\alpha_{i_{k-1}} - \alpha_{i_k})(H^+)} b_{i_k i_{k-1}}(t+1)$$

$$\times \, e^{-t(1-\delta)(\alpha_{i_k} - \alpha_i)(H^+)} b_{i i_k}(t) X_i$$

$$= \sum_j e^{-t(1-\delta)(\alpha_j - \alpha_i)(H^+)} C_{ij}(t,k) X_i,$$

where the matrix $\{C_{ij}(t,k)\}$ is the matrix product $b^{(0)} b^{(1)} \cdots b^{(k)}$ with

$$b_{ij}^{(h)} = e^{-h(1-\delta)(\alpha_j - \alpha_i)(H^+)} b_{ij}(t+h)$$

for $h = 0, 1, 2, \dots, k$. When $t > 0$ is sufficiently large, $|b_{ij}(t)| \le 1$. Note that $b^{(h)}$ is a lower triangular matrix with all the diagonal elements equal to 1. Let $\eta = (1 - \delta)c$. It is easy to see that any off-diagonal element of $b^{(0)} b^{(1)}$ has an absolute value bounded by $1 + de^{-\eta}$, any off-diagonal element of $b^{(0)} b^{(1)} b^{(2)}$ has an absolute value bounded by

$$(1 + de^{-\eta}) + de^{-2\eta}(1 + de^{-\eta}) = (1 + de^{-\eta})(1 + de^{-2\eta}),$$

..., and any off-diagonal element of $b^{(0)} b^{(1)} \cdots b^{(k)}$ has an absolute value bounded by

$$(1 + de^{-\eta})(1 + de^{-2\eta}) \cdots (1 + de^{-k\eta}).$$

Because the infinite product $\prod_{h=1}^{\infty} (1 + de^{-h\eta})$ converges, it follows that there is a constant $C > 0$ such that $|C_{ij}(t,k)| \le C$ for all $k > 0$. Since $\mathrm{Ad}(n_t^{-1} n_\infty) X_j = \sum_i h_{ij}(t) X_i$, the argument here implies

$$h_{ij}(t) = e^{-t(1-\delta)(\alpha_j - \alpha_i)(H^+)} \lim_{k \to \infty} C_{ij}(t,k).$$

Letting $\varepsilon = \delta \max_{j>i} (\alpha_j - \alpha_i)(H^+)$ yields (7.27). This completes the proof of Theorem 7.3.

7.5. Right Lévy Processes

In the previous sections, we have concentrated on left Lévy processes. We now convert the main results proved for left Lévy processes to the corresponding results for right Lévy processes by the transformation $g \mapsto \Theta(g)^{-1}$. This transformation has been used in Section 6.5. Although the conversion is quite natural, it is not completely trivial. It is also a good idea to have these results stated explicitly for easy reference.

Let g_t be a right Lévy process in G. Let $g_t = k_t a_t n_t$ be the Iwasawa decomposition $G = KAN^+$. Proposition 7.3 has a corresponding version for right Lévy processes, which says that k_t is a Feller process in K, and if g_t is left K-invariant, then k_t is a left Lévy process in K. Let $g_t' = \Theta(g_t)^{-1}$. Then g_t' is a left Lévy process and $g_t' = n_t' a_t k_t^{-1}$, where $n_t' = \Theta(n_t)^{-1}$, is the Iwasawa decomposition $G = N^- A K$. Note that the A-component of g_t under the decomposition $G = KAN^+$ coincides with the A-component of g_t' under the decomposition $G = N^- A K$. Note also that the transformation $\Theta(\cdot)^{-1}$ fixes the $\overline{A_+}$-component in the Cartan decomposition $G = K\overline{A_+}K$. The following result follows immediately from Theorem 7.1.

Theorem 7.4. *Let g_t be a right Lévy process in G, and let $g_t = \xi_t a_t^+ \eta_t$ and $g_t = k_t a_t n_t$ be respectively a Cartan decomposition and the Iwasawa decomposition $G = KAN^+$. Assume the hypotheses of Theorem 6.5 and the condition (7.7). Then P-almost surely the limit $H^+ = \lim_{t \to \infty}(1/t) \log a_t$ exists, is nonrandom, and is contained in $\mathfrak{a}_+$. Moreover, $H^+ = \lim_{t \to \infty}(1/t) \log a_t^+$ P-almost surely.*

As for a left Lévy process, H^+ is called the rate vector of the right Lévy process g_t.

The right Lévy process g_t satisfies a stochastic integral equation of the form (1.23) for $f \in C_b(G) \cap C^2(G)$, with Y_i^l, Z^l, X_j^l, and g_{s-h} replaced by Y_i^r, Z^r, X_j^r, and hg_{s-}, respectively. In this equation, if we replace f by $f \circ \Theta(\cdot)^{-1}$, then using the easily proved identify

$$X^r[f \circ \Theta(\cdot)^{-1}] = [-\theta(X)^l f] \circ \Theta(\cdot)^{-1}$$

for $X \in \mathfrak{g}$ and $f \in C^1(G)$, we will see that the left Lévy process g_t' satisfies Equation (1.23) with Y_i, Z, X_j, x_j, N, and Π replaced by $-\theta(Y_i)$, $-\theta(Z)$, $-\theta(X_j)$, $x_j \circ \Theta(\cdot)^{-1}$, $[\mathrm{id}_{\mathbb{R}_+} \times \Theta(\cdot)^{-1}]N$, and $\Theta(\cdot)^{-1}\Pi$, respectively. Note that $x_i \circ \Theta(\cdot)^{-1}$ are the coordinate functions associated to the basis $\{-\theta(X_1), \ldots, -\theta(X_d)\}$ and dk is invariant under the inverse map on K. We can now easily convert Theorem 7.2 to right Lévy processes.

Theorem 7.5. *Let g_t be a right Lévy process in G and let $g_t = k_t a_t n_t$ be the Iwasawa decomposition $G = KAN^+$. Assume the hypotheses of Theorem 7.4, the irreducibility of $\mathrm{Ad}_\mathfrak{p}(K)$, and that the normalized Haar measure dk on K is a stationary measure of k_t. Then the rate vector H^+ of the process g_t is given by*

$$H^+ = \left[\frac{1}{p} \sum_{i=1}^m \|(Y_i)_\mathfrak{p}\|^2 \right] H_\rho + \int_K \int_G \log[(gk)_A] \, dk \Pi(dg). \tag{7.31}$$

Note that the integral term on the right-hand side of this formula vanishes if g_t is continuous or more generally if Π is supported by K. Moreover, if g_t is left K-invariant, then this term is equal to $\int_G \log(g_A)\Pi(dg)$.

In Section 7.1, we defined $g = g_N g_A g_K$ and $X = X_n + X_a + X_\mathfrak{k}$ as the Iwasawa decomposition $G = N^- A K$ and the corresponding direct sum decomposition $\mathfrak{g} = \mathfrak{n}^- \oplus \mathfrak{a} \oplus \mathfrak{k}$ at the Lie algebra level (see (7.2) and (7.3)). When dealing with right Lévy processes, it is convenient to work under a different decomposition. Therefore, for any $g \in G$ and $X \in \mathfrak{g}$, we now let

$$g = (g_K)(g_A)(g_N) \quad \text{be the decomposition } G = KAN^+ \qquad (7.32)$$

and let

$$X = X_\mathfrak{k} + X_a + X_n \quad \text{be the direct sum decomposition } \mathfrak{g} = \mathfrak{k} \oplus \mathfrak{a} \oplus \mathfrak{n}^+.$$
$$(7.33)$$

Note that g_K, g_A, g_N, $X_\mathfrak{k}$, and X_n defined here have different meanings from those defined by (7.2) and (7.3) in Section 7.1, but X_a remains unchanged because it is still equal to the orthogonal projection of X into $\mathfrak{a}$.

For $g \in G$ and $X \in \mathfrak{g}$, let $g = (g'_N)(g'_A)(g'_K)$ and $X = X'_n + X'_a + X'_\mathfrak{k}$ be, respectively, the decompositions given by (7.2) and (7.3). Then it is easy to show that

$$[\Theta(g)^{-1}]'_A = g_A, \quad [\Theta(g)^{-1}]'_K = (g_K)^{-1}, \quad [\theta(X)]'_a = -X_a, \quad \text{and}$$
$$[(\theta X)]'_\mathfrak{k} = X_\mathfrak{k}. \qquad (7.34)$$

For any $g \in G$, $\Theta \circ c_g = c_{\Theta(g)} \circ \Theta$, where c_g is the conjugation map on G. Taking the differential maps, we obtain

$$\theta \circ \mathrm{Ad}(g) = \mathrm{Ad}(\Theta(g)) \circ \theta. \qquad (7.35)$$

As in Section 7.4, let $\{X_1, \ldots, X_d\}$ be an orthonormal basis of $\mathfrak{g}$ such that each X_i is contained in $\mathfrak{g}_\alpha$ for some root α or $\alpha = 0$, which will be denoted by α_i. By (7.35), it is easy to show that, for any $n \in N^+$, $-\theta[\mathrm{Ad}(n)X_j] = \mathrm{Ad}(n'^{-1})X'_j$, where $n' = \Theta(n)^{-1} \in N^-$ and $X'_j = -\theta(X_j) \in \mathfrak{g}_{-\alpha_j}$. Let $\{h_{ij}(t)\} = \{h_{ij}(t, \omega)\}$ be the matrix representing the random linear map $\mathrm{Ad}(n_t n_\infty^{-1}): \mathfrak{g} \to \mathfrak{g}$ under the basis $\{X_1, \ldots, X_d\}$, that is,

$$\mathrm{Ad}\left(n_t n_\infty^{-1}\right) X_j = \sum_{i=1}^d h_{ij}(t) X_i. \qquad (7.36)$$

Note that the $h_{ij}(t)$ defined here are different from those defined by (7.22) in Section 7.4. Because $n_t n_\infty^{-1} \in N^+$, we see that $h_{ii}(t) = 1$, and $h_{ij}(t) = 0$ if $i \neq j$ and $\alpha_i(H^+) \leq \alpha_j(H^+)$.

Applying $-\theta$ to (7.36), we obtain

$$\mathrm{Ad}\big(n_t'^{-1}n_\infty'\big)X_j' = \sum_{i=1}^d h_{ij}(t)X_i'.$$

The next theorem now follows directly from Theorem 7.3. As in Section 7.4, for $g \in G$, $\tilde{g} = \mathrm{Ad}(g)$ is regarded as a matrix $\{\tilde{g}_{ij}\}$ determined by $\tilde{g}X_j = \sum_{i=1}^d \tilde{g}_{ij}X_i$ with the Euclidean norm $|\tilde{g}| = (\sum_{i,j} \tilde{g}_{ij}^2)^{1/2}$ and $\mathrm{Ad}(g_N)$ with g_N given by (7.32) is written as $g_{\tilde{N}}$.

Theorem 7.6. *Assume the hypotheses of Theorem 7.4. Let $H^+ \in \mathfrak{a}_+$ be the rate vector given in Theorem 7.4 and let $h_{ij}(t)$ be defined by (7.36). Then, for P-almost all ω,*

$$\lim_{t\to\infty} \frac{1}{t}\log a_t(\omega) = H^+ \quad and \quad n_\infty(\omega) = \lim_{t\to\infty} n_t(\omega), \tag{7.37}$$

where n_∞ is an N^+-valued random variable, and

$$\forall \varepsilon > 0, \quad \sup_{k\in K}\big|\big[g_{t+1}(\omega)g_t^{-1}(\omega)k\big]_{\tilde{N}}\big| \le e^{\varepsilon t} \ \text{for sufficiently large } t > 0. \tag{7.38}$$

Moreover, if ω satisfies (7.37) and (7.38), then, for any $\varepsilon > 0$,

$$|h_{ij}(t,\omega)| \le \exp\{-t[\alpha_i(H^+) - \alpha_j(H^+) - \varepsilon]\} \ \text{for sufficiently large } t > 0. \tag{7.39}$$

8

Lévy Processes as Stochastic Flows

We define a stochastic flow on a manifold M as a right Lévy process in Diff(M). In this chapter, we look at the dynamical aspects of a right Lévy process regarded as a stochastic flow. The first section contains some basic definitions and facts about a general stochastic flow. Although these facts will not be used to prove anything, they provide a general setting under which one may gain a better understanding of the results to be proved. In the rest of the chapter, the limiting properties of Lévy processes are applied to study the asymptotic stability of the induced stochastic flows on certain compact homogeneous spaces. In Section 8.2, the properties of the Lévy process are transformed to a form more suitable for the study of its dynamical behavior, in which the dependence on ω and the initial point g is made explicit. In Section 8.3, the explicit formulas, in terms of the group structure, for the Lyapunov exponents and the associated stable manifolds are obtained. A clustering property of the stochastic flow related to the rate vector of the Lévy process is studied in Section 8.4. Some explicit results for $SL(d, \mathbb{R})$-flows and $SO(1, d)$-flows on $SO(d)$ and S^{d-1} are presented in the last three sections. The main results of this chapter are taken from Liao [40, 41, 42].

8.1. Introduction to Stochastic Flows

Let M be a manifold and let Diff(M) be the group of all the diffeomorphisms $M \to M$ with the composition as multiplication and the identity map id$_M$ as the group identity element. We will equip Diff(M) with the compact-open topology under which a set is open if and only if it is a union of the sets of the following form:

$$\{f \in \text{Diff}(M); \ f(C_1) \subset O_1, \ f(C_2) \subset O_2, \ldots, f(C_k) \subset O_k\},$$

where $C_1, C_2, \ldots, C_k$ are compact and $O_1, O_2, \ldots, O_k$ are open subsets of M, and k is an arbitrary positive integer. Under this topology, Diff(M)

becomes a topological group that acts continuously on M in the sense that the map $\text{Diff}(M) \times M \ni (\phi, x) \mapsto \phi x \in M$ is continuous.

Let ϕ_t be a right Lévy process in $\text{Diff}(M)$ with $\phi_0 = \text{id}_M$. It is easy to see that Proposition 2.1 holds also when G is a topological group acting continuously on a manifold M. Therefore, for $x \in M$, the one-point motion of ϕ_t from x, $x_t = \phi_t x$, is a Markov process in M with the Feller transition semigroup P_t^M given by $P_t^M f(z) = E[f(\phi_t z)]$ for $f \in \mathcal{B}(M)_+$ and $z \in M$.

For any right Lévy process ϕ_t in a topological group G starting at the group identity element e, by properly choosing an underlying probability space $(\Omega, \mathcal{F}, P)$, a family of P-preserving maps $\theta_t \colon \Omega \to \Omega, t \in \mathbb{R}_+$, called the time-shift operators, may be defined such that it is a semigroup in t in the sense that $\theta_{t+s} = \theta_t \circ \theta_s$ for $s, t \in \mathbb{R}_+$ and $\theta_0 = \text{id}_\Omega$. Moreover, the process ϕ_t satisfies the following so-called cocycle property.

$$\forall s, t \in \mathbb{R}_+ \text{ and } \omega \in \Omega, \quad \phi_{s+t}(\omega) = \phi_s(\theta_t \omega)\phi_t(\omega). \tag{8.1}$$

For example, we may let $(\Omega, \mathcal{F}, P)$ be the canonical sample space of the process ϕ_t mentioned in Appendix B.1; that is, Ω is the space of all the càdlàg maps $\omega \colon \mathbb{R}_+ \to G$ with $\omega(0) = e$, $\mathcal{F}$ is the σ-algebra generated by the maps $\Omega \ni \omega \mapsto \omega(t) \in G$ for $t \in \mathbb{R}_+$, and $\phi_t(\omega) = \omega(t)$. For $t \in \mathbb{R}_+$, define $\theta_t \colon \Omega \to \Omega$ by

$$\forall s \in \mathbb{R}_+, \quad \theta_t(\omega)(s) = \omega(s+t)\omega(t)^{-1}. \tag{8.2}$$

It is easy to see that $\{\theta_t; \ t \in \mathbb{R}_+\}$ is a semigroup in t and the cocycle property (8.1) holds. Moreover, θ_t is P-preserving because, for any $0 = t_0 < t_1 < t_2 < \cdots < t_k$, the joint distribution of the increments $\phi_{t_1}\phi_{t_0}^{-1}, \phi_{t_2}\phi_{t_1}^{-1}, \ldots, \phi_{t_k}\phi_{t_{k-1}}^{-1}$ is clearly independent of the time shift by t. Note that Ω may be extended to be a product space to accommodate additional random variables that are independent of the process ϕ_t.

Given a semigroup of time-shift operators $\{\theta_t; t \in \mathbb{R}_+\}$, a stochastic process ϕ_t in $\text{Diff}(M)$ with $\phi_0 = \text{id}_M$ that satisfies the cocycle property (8.1) is called a random dynamical system (see Arnold [4]) or a stochastic flow on M. The preceding discussion shows that a right Lévy process ϕ_t in $\text{Diff}(M)$ with $\phi_0 = \text{id}_M$ is a stochastic flow on M. In the rest of this work, a stochastic flow on M will always mean a right Lévy process ϕ_t in $\text{Diff}(M)$ with $\phi_0 = \text{id}_M$.

A stochastic flow may be obtained by solving a stochastic differential equation on M of the following form:

$$dx_t = \sum_{i=1}^m X_i(x_t) \circ dW_t^i + X_0(x_t)dt, \tag{8.3}$$

where $X_0, X_1, \ldots, X_m$ are vector fields on M, $W_t = (W_t^1, \ldots, W_t^m)$ is a standard m-dimensional Brownian motion, and m is some positive integer. A stochastic flow on M is said to be generated by the stochastic differential equation (8.3) if its one-point motions are the solutions of this equation. In the special case when $X_1 = X_2 = \cdots = X_m = 0$, the stochastic flow becomes the deterministic solution flow of the ordinary differential equation $\frac{d}{dt}x(t) = X_0(x(t))$ on M, which will be called the flow of the vector field X_0.

It is known that the stochastic flow generated by the stochastic differential equation (8.3) exists uniquely and is continuous in t under either of the following two conditions:

(i) M is a Euclidean space and the vector fields $X_0, X_1, \ldots, X_m$ are uniformly Lipschitz continuous on M.

(ii) M is compact.

The existence under the condition (i) is given, for example, by Kunita [36, theorem 4.6.5]. From this, the existence under (ii) follows because a compact manifold can be embedded in a Euclidean space and the vector fields on this manifold may be extended to be vector fields on the Euclidean space with compact supports.

In fact, any stochastic flow on a compact manifold M that is continuous in time t is generated by a more general type of stochastic differential equation on M, possibly containing infinitely many vector fields (see Baxendale [7]).

Let G be a Lie group that acts on M. If the action of G on M is effective, that is, if $gx = x$ for all $x \in M$ implies $g = e$, then G may be regarded a subgroup of $\text{Diff}(M)$. In this case, any right Lévy process ϕ_t in G with $\phi_0 = e$ may be regarded as a stochastic flow on M. In general, $H = \{g \in G; gx = x$ for any $x \in M\}$ is a closed normal subgroup of G and the quotient group G/H acts on M effectively by $M \ni x \mapsto (gH)x = gx \in M$ for $gH \in G/H$. If ϕ_t is a right Lévy process in G with $\phi_0 = e$, then $\phi_t H$ is a right Lévy process in G/H and thus may be regarded as a stochastic flow on M. Such a stochastic flow will be said to be induced by the right Lévy process ϕ_t in G and will be called a G-flow on M. For simplicity, ϕ_t itself will also be regarded as a stochastic flow and will be called a G-flow on M.

Let $\mathfrak{g}$ be the Lie algebra of G and let $Y_0, Y_1, \ldots, Y_m \in \mathfrak{g}$ be such that $X_i = Y_i^*$ for $i = 0, 1, \ldots, m$, where Y^* is the vector field on M induced by $Y \in \mathfrak{g}$ defined by (2.3). If ϕ_t is a right invariant diffusion process in G (that is, a continuous right Lévy process in G) with $\phi_0 = e$ satisfying the stochastic differential equation

$$d\phi_t = \sum_{i=1}^{m} Y_i^r(\phi_t) \circ dW_t^i + Y_0^r(\phi_t)dt, \tag{8.4}$$

then its one-point motion $x_t = \phi_t x$, $x \in M$, is a solution of (8.3); therefore, ϕ_t is a stochastic flow generated by (8.3). In this case, the Lie algebra $\text{Lie}(X_0, X_1, \ldots, X_m)$ is finite dimensional and is contained in $\mathfrak{g}^* = \{Y^*; Y \in \mathfrak{g}\}$. Conversely, if $X_0, X_1, \ldots, X_m$ are complete vector fields and $\text{Lie}(X_0, X_1, \ldots, X_m)$ is finite dimensional, then it can be shown that the stochastic flow ϕ_t generated by (8.3) is a right invariant diffusion process in a Lie group G that acts on M effectively. In fact, by Palais's result mentioned in Appendix A.2, there is such a Lie group G with Lie algebra $\mathfrak{g}$ such that $X_i = Y_i^*$ for some $Y_i \in \mathfrak{g}$ and ϕ_t satisfies (8.4). See also [36, theorem 4.8.7].

We now state some basic definitions and facts about ϕ_t regarded as a stochastic flow. Although these facts will not be needed in the proofs in the rest of this chapter, they may help us to gain a better understanding of the results to be presented.

For $t \in \mathbb{R}_+$, let $\Phi_t: M \times \Omega \to M \times \Omega$ be the map defined by $\Phi_t(x, \omega) = (\phi_t(\omega)x, \theta_t\omega)$. This is a semigroup in t, that is, $\Phi_0 = \text{id}_{M \times \Omega}$ and $\Phi_{t+s} = \Phi_t \Phi_s$ for any $s, t \in \mathbb{R}_+$, and is called the skew-product flow associated to the stochastic flow ϕ_t. Note that the stochastic flow ϕ_t itself is not a semigroup in t.

Recall that a stationary measure ρ of the one-point motion x_t of ϕ_t as a Markov process in M is a probability measure on M such that $\rho P_t^M = \rho$, where P_t^M is the transition semigroup of x_t. Such a measure will also be called a stationary measure of the stochastic flow ϕ_t on M and can be characterized by $\rho = E(\phi_t \rho)$ for $t \geq 0$, where $E(\phi_t \rho)$ denotes the probability measure on M defined by $E(\phi_t \rho)(f) = E[\phi_t \rho(f)] = E[\rho(f \circ \phi_t)]$ for any $f \in \mathcal{B}(M)_+$. Note the difference between this measure and the stationary measure of ϕ_t as a Markov process in $\text{Diff}(M)$. It is easy to show that if ρ is a stationary measure of ϕ_t on M, then the skew-product flow Φ_t preserves the measure $\rho \times P$ on $M \times \Omega$.

A subset B of M is called invariant (under the transition semigroup P_t^M and the stationary measure ρ) if $B \in \mathcal{B}(M)$ and $P_t^M 1_B = 1_B$ ρ-almost surely. The stationary measure ρ is called ergodic if any invariant subset of M has ρ-measure 0 or 1. It can be shown that ρ is ergodic if and only if any $\{\Phi_t\}$-invariant subset of $M \times \Omega$ has $\rho \times P$-measure 0 or 1 (see [13, appendix B]). Here, a subset Γ of $M \times \Omega$ is called $\{\Phi_t\}$-invariant if $\Gamma \in \mathcal{B}(M) \times \mathcal{F}$ and $\Phi_t^{-1}(\Gamma) = \Gamma$ for any $t \in \mathbb{R}_+$.

We will assume M is equipped with a Riemannian metric. Let $\| \cdot \|_x$ denote the Riemannian norm on the tangent space $T_x M$ of M at $x \in M$ and let $\text{dist}(x, y)$ denote the Riemannian distance between two points x and y in M.

Let ρ be an ergodic stationary measure of ϕ_t and, for $u > 0$, let $\log^+ u = \max(\log u, 0)$. Assume

$$\int_M \rho(dx) E \left[\sup_{0 \leq t \leq 1} \log^+ |D\phi_t(x)| \right] < \infty \quad \text{and}$$

$$\int_M \rho(dx) E \left[\sup_{0 \leq t \leq 1} \log^+ |D\phi_t(x)^{-1}| \right] < \infty, \tag{8.5}$$

where, for any linear map $L \colon T_x M \to T_y M$, $|L| = \sqrt{\sum_{i,j} L_{ij}^2}$ is the Euclidean norm of the matrix $\{L_{ij}\}$ representing L under orthonormal bases in $T_x M$ and $T_y M$. Note that $|L|$ is independent of the choice of the orthonormal bases.

By Theorem 4.2.6 in Arnold [4] (see also Carverhill [13] in the case when ϕ_t is generated by (8.3)), there exist a $\{\Phi_t\}$-invariant subset Γ of $M \times \Omega$ of full $\rho \times P$-measure, constants $\lambda_1 > \lambda_2 > \cdots > \lambda_l$, and, for any $(x, \omega) \in \Gamma$, subspaces

$$T_x M = V_1(x, \omega) \underset{\neq}{\supset} V_2(x, \omega) \underset{\neq}{\supset} \cdots \underset{\neq}{\supset} V_l(x, \omega) \underset{\neq}{\supset} V_{l+1}(x, \omega) = \{0\},$$

such that

$$\forall v \in [V_i(x, \omega) - V_{i+1}(x, \omega)], \quad \lim_{t \to \infty} \frac{1}{t} \log \|D\phi_t(\omega)v\|_{\phi_t(x)} = \lambda_i \tag{8.6}$$

for $i = 1, 2, \ldots, l$, where $V_i(x, \omega) - V_{i+1}(x, \omega)$ is the set difference. Moreover, $c_i = \dim V_i(x, \omega)$ is independent of (x, ω). The numbers λ_i are called the Lyapunov exponents, $V_i(x, \omega)$ is called the subspace of $T_x M$ associated to the exponent λ_i and $d_i = c_i - c_{i+1}$ is called the multiplicity of λ_i. If $d_i = 1$, then the exponent λ_i is called simple. The Lyapunov exponents are the limiting exponential rates at which the lengths of tangent vectors grow or decay under the stochastic flow ϕ_t. The complete set of the Lyapunov exponents together with their respective multiplicities is called the Lyapunov spectrum of the stochastic flow ϕ_t.

Two Riemannian metrics $\{\| \cdot \|_x; \ x \in M\}$ and $\{\| \cdot \|'_x; \ x \in M\}$ on M are called equivalent if there is a constant $c > 0$ such that $c^{-1} \|v\|_x \leq \|v\|'_x \leq c \|v\|_x$ for any $x \in M$ and $v \in T_x M$. If M is compact, then any two Riemannian metrics are equivalent. It is clear that the Lyapunov spectrum does not depend on the choice of equivalent Riemannian metrics.

Let $\lambda_i < 0$ be a negative Lyapunov exponent and $(x, \omega) \in \Gamma$. A connected submanifold M' of M containing x is called a stable manifold of λ_i at (x, ω) if $M' \subset \{y \in M; \quad (y, w) \in \Gamma\}$,

$$T_x M' = V_i(x, \omega) \tag{8.7}$$

and, for any constant λ with $\lambda_i < \lambda < 0$ and any compact subset C of M', there exists $c > 0$ such that,

$$\forall y \in C \text{ and } t \in \mathbb{R}_+, \quad \text{dist}(\phi_t(\omega)x, \phi_t(\omega)y) \le ce^{\lambda t}. \tag{8.8}$$

Note that, for a fixed $y \in M$, the inequality in (8.8) holds for any $\lambda > \lambda_i$ if and only if

$$\limsup_{t \to \infty} \frac{1}{t} \log \text{dist}(\phi_t(\omega)x, \phi_t(\omega)y) \le \lambda_i. \tag{8.9}$$

A stable manifold of λ_i at (x, ω) is called maximal if it contains any stable manifold of λ_i at (x, ω). It is easy to see that if M' is a stable manifold of λ_i at (x, ω) and if it contains any $y \in M$ for which (8.9) holds, then it is maximal.

For the stochastic flow generated by the stochastic differential equation (8.3), the local existence of the stable manifolds for a negative Lyapunov exponent can be found in Carverhill [13]. More precisely, it is shown that the set M' defined in the last paragraph is locally a submanifold and is tangent to $V_i(x, \omega)$ at x. Intuitively, one would expect that the maximal stable manifolds of a negative Lyapunov exponents form a foliation of a dense open subset of M. However, such a global theory of stable manifolds under a general setting can be quite complicated (see chapter 7 in Arnold [4]).

8.2. Lévy Processes as Dynamical Systems

Throughout the rest of this chapter, let G be a connected semi-simple Lie group of noncompact type and of a finite center with Lie algebra $\mathfrak{g}$ and let ϕ_t be a right Lévy process in G with $\phi_0 = e$.

We will continue to use the standard notations and definitions on semi-simple Lie groups and Lévy processes introduced in the earlier chapters, but with the following exception: The centralizer of A in K and its Lie algebra, which are denoted by M and $\mathfrak{m}$ in the earlier chapters, will now be denoted by U and $\mathfrak{u}$. The reason for this is that we would like to reserve the letter M to denote the manifold on which G acts. Similarly, the normalizer M' of A in K will now be denoted by U'.

Applying the transformation $g \mapsto \Theta(g)^{-1}$ to Proposition 6.2, we obtain the following result:

Lemma 8.1. *Let $\{g_j\}$ be a sequence in G. Then the following two statements are equivalent:*

(a) *There is a Cartan decomposition $g_j = \xi_j a_j^+ \eta_j$ such that a_j^+ is positive contracting and $U\eta_j \to U\eta_\infty$ for some $\eta_\infty \in K \cap (N^- U A N^+)$.*

(b) *Let $g_j = k_j a_j n_j$ be the Iwasawa decomposition $G = KAN^+$. Then a_j is positive contracting and $n_j \to n_\infty$ for some $n_\infty \in N^+$.*

Moreover, these statements imply that $(N^- A U)\eta_\infty = (N^- A U)n_\infty$, $(a_j^+)^{-1}a_j$ converges in A, and $k_j = \xi_j \xi_j'$, where $K \ni \xi_j' \to u$ for some $u \in U$.

We now introduce the following hypothesis:

(H). G_μ is totally right irreducible and T_μ contains a contracting sequence in G, and

$$\int_G \|g_\mathfrak{p}\| \, \Pi(dg) < \infty,$$

where $\| \cdot \|$ is the norm on $\mathfrak{g}$ induced by the Killing form, G_μ and T_μ are respectively the group and the semigroup generated by the right Lévy process ϕ_t, and Π is the Lévy measure of ϕ_t. Recall that, for $g \in G$, $g_\mathfrak{p}$ is the unique element of $\mathfrak{p}$ such that $g = ke^{g_\mathfrak{p}}$ for some $k \in K$.

See Section 6.6 for some sufficient conditions that guarantee conditions for G_μ and T_μ stated in the hypothesis (H). Under this hypothesis, by Theorem 7.4, the rate vector $H^+ = \lim_{t\to\infty}(1/t)\log a_t$ exists, is nonrandom, and is contained in $\mathfrak{a}_+$, where a_t is the A-component of ϕ_t in the Iwasawa decomposition $G = KAN^+$.

Denote by $D(\mathbb{R}_+, G)$ the space of all the càdlàg functions $\mathbb{R}_+ \to G$. Define

$$\Lambda = \{g(\cdot) \in D(\mathbb{R}_+, G); \lim_{t\to\infty}\frac{1}{t}\log a(t) = H^+ \text{ and } n(\infty) = \lim_{t\to\infty} n(t) \text{ exists},$$
$$\text{where } g(t) = k(t)a(t)n(t) \text{ is the Iwasawa decomposition } G = KAN^+\}.$$
$$(8.10)$$

By Lemma 8.1, if $g(\cdot) \in \Lambda$ with Cartan decomposition $g(t) = \xi(t)a^+(t)\eta(t)$, then $(1/t)\log a^+(t) \to H^+$ and $(N^- U A)\eta(t) \to (N^- U A)\eta(\infty) = (N^- U A)n(\infty)$ for some $\eta(\infty) \in K$ as $t \to \infty$. By properly choosing the two K-components in the Cartan decomposition, we may assume $\eta(t) \to \eta(\infty)$.

By Theorems 6.5 and 7.4, and noting that $\phi_t g$ is a right Lévy process starting at $g \in G$, we obtain the following result:

Proposition 8.1. *Assume the hypothesis (H). For any $g \in G$, $\phi_\cdot(\omega)g \in \Lambda$ for P-almost all ω, where $\phi_\cdot(\omega)$ denotes the map $t \to \phi_t(\omega)$.*

Lemma 8.2. *Let $g \in G$ and $g(\cdot) \in \Lambda$ with Iwasawa decomposition $g(t) = k(t)a(t)n(t)$ of $G = KAN^+$. Then $g(\cdot)g \in \Lambda$ if and only if $g \in n(\infty)^{-1}(N^- U A N^+)$.*

Proof. Let $\zeta = n(\infty)g$ and $\zeta(t) = n(t)g$. Then $\zeta(t) \to \zeta$. Let $\zeta = k'a'n'$ and $\zeta(t) = k'(t)a'(t)n'(t)$ be the Iwasawa decomposition $G = KAN^+$. Then $(k'(t), a'(t), n'(t)) \to (k', a', n')$ and $g(t)g = k(t)a(t)\zeta(t) = k(t)a(t)k'(t)a'(t)n'(t)$. Since, for large $t > 0$, $\log a(t) \in \mathfrak{a}_+$, we may regard $k(t)a(t)k(t)'$ as a Cartan decomposition. Let $k(t)a(t)k'(t) = \tilde{k}(t)\tilde{a}(t)\tilde{n}(t)$, where the right-hand side is the Iwasawa decomposition $G = KAN^+$. Then

$$g(t)g = \tilde{k}(t)\tilde{a}(t)\tilde{n}(t)a'(t)n'(t) = \tilde{k}(t)[\tilde{a}(t)a'(t)][a'(t)^{-1}\tilde{n}(t)a'(t)n'(t)].$$

Let $g(t)g = k^g(t)a^g(t)n^g(t)$ be the Iwasawa decomposition $G = KAN^+$. Then $k^g(t) = \tilde{k}(t)$, $a^g(t) = \tilde{a}(t)a'(t)$, and $n^g(t) = a'(t)^{-1}\tilde{n}(t)a'(t)n'(t)$. If $g \in n(\infty)^{-1}(N^- U A N^+)$, then $\zeta = n(\infty)g \in N^- U A N^+$ and so $k' \in N^- U A N^+$. Since $k'(t) \to k'$, applying Lemma 8.1 to $k(t)a(t)k'(t) = \tilde{k}(t)\tilde{a}(t)\tilde{n}(t)$ yields $\lim_{t\to\infty}(1/t)\log\tilde{a}(t) = H^+$ and the existence of $\lim_{t\to\infty}\tilde{n}(t)$. Therefore,

$$\lim_{t\to\infty} \frac{1}{t}\log a^g(t) = \lim \frac{1}{t}\log[\tilde{a}(t)a'(t)] = H^+$$

and $\lim_t n^g(t) = \lim_t[a'(t)^{-1}\tilde{n}(t)a'(t)n'(t)]$ exists; that is, $g(\cdot)g \in \Lambda$. However, if $g(\cdot)g \in \Lambda$, then from $\lim_{t\to\infty}(1/t)\log a^g(t) = H^+$ and the existence of $\lim_{t\to\infty} n^g(t)$, we get $\lim_t(1/t)\log\tilde{a}(t) = H^+$ and the existence of $\lim_t \tilde{n}(t)$. Applying Lemma 8.1 again to $k(t)a(t)k'(t) = \tilde{k}(t)\tilde{a}(t)\tilde{n}(t)$, we see that there exist $u, u(t) \in U$ such that $u(t)k'(t) \to uk' \in N^- U A N^+$; hence, $k' \in N^- U A N^+$. Since $\zeta = k'a'n'$, $\zeta \in N^- U A N^+$ and

$$g = n(\infty)^{-1}\zeta \in n(\infty)^{-1}(N^- U A N^+). \qquad \square$$

As in Section 7.5, let $g = g_K g_A g_N$ be the Iwasawa decomposition $G = KAN^+$ and let $X = X_{\mathfrak{k}} + X_{\mathfrak{a}} + X_{\mathfrak{n}}$ be the decomposition $\mathfrak{g} = \mathfrak{k} \oplus \mathfrak{a} \oplus \mathfrak{n}^+$. Let Ω' be the set of $\omega \in \Omega$ such that

$$\forall \varepsilon > 0, \quad \sup_{k\in K} \left| \left[\phi_{t+1}(\omega)\phi_t^{-1}(\omega)k \right]_{\tilde{N}} \right| \le e^{\varepsilon t} \text{ for sufficiently large } t > 0,$$

$$(8.11)$$

where, for any $g \in G$, $g_{\tilde{N}}$ is the matrix representing the linear map Ad(g_N): $\mathfrak{g} \to \mathfrak{g}$ with respect to an orthonormal basis of $\mathfrak{g}$ and $|g_{\tilde{N}}| = [\sum_{i,j}(g_{\tilde{N}})_{ij}^2]^{1/2}$ is its Euclidean norm that is independent of the choice for the basis. Let $\{\theta_t\}$ be the semigroup of time-shift operators associated to the Lévy process ϕ_t defined in the previous section. By the cocycle property (8.1), it is easy to show that Ω' is $\{\theta_t\}$-invariant in the sense that $\theta_t^{-1}\Omega' = \Omega'$ for any $t \in \mathbb{R}_+$.

Define

$$\Gamma_0 = \{(g, \omega) \in G \times \Omega'; \ \phi_\cdot(\omega)g \in \Lambda\}. \tag{8.12}$$

Let Φ_t^0 be the skew-product flow associated to ϕ_t regarded as a stochastic flow on $G = G/\{e\}$ given by $\Phi_t^0(g, \omega) = (\phi_t(\omega)g, \theta_t\omega)$. For $g \in G$, let $\Gamma_0(g) = \{\omega \in \Omega; (g, \omega) \in \Gamma_0\}$, and call it the g-section of Γ_0. For $\omega \in \Omega$, let $\Gamma_0(\omega) = \{g \in G; (g, \omega) \in \Gamma_0\}$, and call it the ω-section of Γ_0. The projection of Γ_0 to Ω is

$$\Omega_0 = \{\omega \in \Omega; \ \Gamma_0(\omega) \text{ is nonempty}\}. \tag{8.13}$$

For $g \in G$, let n_t^g be the N^+-component of $\phi_t g$ in the Iwasawa decomposition $G = KAN^+$. Note that $n_\infty^g(\omega) = \lim_{t\to\infty} n_t^g(\omega)$ exists for $\omega \in \Gamma_0(g)$.

Proposition 8.2. *Γ_0 is $\{\Phi_t^0\}$-invariant, Ω_0 is $\{\theta_t\}$-invariant and, for any $\omega \in \Omega_0$,*

$$\phi_t(\omega)\Gamma_0(\omega) = \Gamma_0(\theta_t\omega). \tag{8.14}$$

Moreover, under the hypothesis (H), $P[\Gamma_0(g)] = 1$ for any $g \in G$, and, for $(g, \omega) \in \Gamma_0$,

$$\Gamma_0(\omega) = g\, n_\infty^g(\omega)^{-1} N^- U A N^+. \tag{8.15}$$

Proof. The $\{\Phi_t^0\}$-invariance of Γ_0 follows from the $\{\theta_t\}$-invariance of Ω' and the cocycle property (8.1). From this it is easy to show the $\{\theta_t\}$-invariance of Ω_0 and (8.14). By Proposition 8.1, $P[\Gamma_0(g)] = 1$ for any $g \in G$. By Lemma 8.2, for any $h \in G$, $(gh, \omega) \in \Gamma_0$ if and only if $h \in n_\infty^g(\omega)^{-1}N^-UAN^+$; hence, $gh \in \Gamma_0(\omega)$ if and only if $gh \in g\, n_\infty^g(\omega)^{-1}N^-UAN^+$. This proves (8.15). $\square$

Note that $P[\Gamma_0(g)] = 1$ implies $P(\Omega_0) = 1$. By (8.15), for $\omega \in \Omega_0$, $\Gamma_0(\omega)$ is an open subset of G with a lower dimensional complement. The expression of $\Gamma_0(\omega)$ given in (8.15) should be independent of the choice of $g \in \Gamma_0(\omega)$. This can also be verified directly by using Lemma 8.3 that follows.

Let $\phi_t = \xi_t a_t^+ \eta_t$ and $\phi_t = k_t a_t n_t$ be respectively a Cartan decomposition and the Iwasawa decomposition $G = KAN^+$. For any $g \in G$, let $\phi_t g = \xi_t^g a_t^{g+} \eta_t^g = k_t^g a_t^g n_t^g$ be the corresponding decompositions for $\phi_t g$.

Lemma 8.3. *Let $g, h \in G$ and $\omega \in \Gamma_0(g) \cap \Gamma_0(gh)$. Then $(N^- U A) n_\infty^{gh}(\omega) = (N^- U A) n_\infty^g(\omega) h$. Consequently, $h\, n_\infty^{gh}(\omega)^{-1} = n_\infty^g(\omega)^{-1} r$ for some $r \in N^- U A$.*

Proof. Let $h = kan$ be the Iwasawa decomposition $G = KAN^+$. By assumption, $\phi_\cdot(\omega) g \in \Lambda$ and $\phi_\cdot(\omega) gh \in \Lambda$. By Lemma 8.2, $h \in n_\infty^g(\omega)^{-1} N^- U A N^+$. This implies

$$k \in n_\infty^g(\omega)^{-1} N^- U A N^+;$$

hence, $\phi_\cdot(\omega) gk \in \Lambda$. This implies that $(N^- U A) n_\infty^{gk}(\omega) = (N^- U A) \eta_\infty^{gk}(\omega) = (N^- U A) \eta_\infty^g(\omega) k = (N^- U A) n_\infty^g(\omega) k$. Since $\phi_t gh = \phi_t gkan = k_t^{gk} a_t^{gk} n_t^{gk} an = k_t^{gk}(a_t^{gk} a)(a^{-1} n_t^{gk} an)$, we see that $n_t^{gh} = a^{-1} n_t^{gk} an$. Therefore,

$$(N^- U A) n_\infty^{gh}(\omega) = (N^- U A) a^{-1} n_\infty^{gk}(\omega) an = (N^- U A) n_\infty^g(\omega) kan$$
$$= (N^- U A) n_\infty^g(\omega) h. \qquad \square$$

As in Section 7.4, let $\{X_1, \ldots, X_d\}$ be an orthonormal basis of $\mathfrak{g}$ such that each X_i is contained in $\mathfrak{g}_\alpha$ for some root α or $\alpha = 0$, which will be denoted by α_i.

For $(g, \omega) \in \Gamma_0$, let $h_{ij}^g(t) = h_{ij}^g(t, \omega)$ be defined by

$$\mathrm{Ad}\big(n_t^g \big(n_\infty^g\big)^{-1}\big) X_j = \sum_{j=1}^{d} h_{ij}^g(t) X_i. \tag{8.16}$$

Then $h_{ii}^g(t) = 1$, and $h_{ij}^g(t) = 0$ if $i \neq j$ and $\alpha_i(H^+) \leq \alpha_j(H^+)$.

The next result follows directly from Theorem 7.6 by noting that the expression in (8.11) is unchanged when ϕ_t is replaced by $\phi_t g$.

Proposition 8.3. *Under the hypothesis (H), if $(g, \omega) \in \Gamma_0$, then, for any $\varepsilon > 0$,*

$$|h_{ij}^g(t, \omega)| \leq \exp\{-t[\alpha_i(H^+) - \alpha_j(H^+) - \varepsilon]\} \text{ for sufficiently large } t > 0. \tag{8.17}$$

8.3. Lyapunov Exponents and Stable Manifolds

From now on, Q will denote a closed subgroup of G of a lower dimension containing AN^+ and $M = G/Q$ unless explicitly stated otherwise. Since $G = KAN^+$, we may also regard M as K/L for $L = K \cap Q$; hence, M is compact and can be regarded as a homogeneous space of either G or K.

Let $\langle \cdot, \cdot \rangle$ be the inner product on $\mathfrak{g}$ induced by the Killing form, given by (5.3), and let $\| \cdot \|$ be the associated norm. Recall that $\langle \cdot, \cdot \rangle$ is $\mathrm{Ad}(K)$-invariant. Therefore, its restriction on $\mathfrak{k}$ induces a K-invariant Riemannian metric on $M = K/L$ (see the discussion in Section 2.3). Let $\| \cdot \|_x$ and $\mathrm{dist}(x, y)$ denote, respectively, the Riemannian norm on the tangent space at $x \in M$ and the Riemannian distance between two points x and y in M.

Let $\mathfrak{q}$ be the Lie algebra of Q. Then $\mathfrak{a} \oplus \mathfrak{n}^+ \subset \mathfrak{q}$. By the simultaneous diagonalization of the commutative family $\{\mathrm{ad}(H); H \in \mathfrak{a}\}$ of the linear operators restricted on $\mathfrak{q}$, which are symmetric with respect to $\langle \cdot, \cdot \rangle$, we obtain the direct sum decomposition

$$\mathfrak{q} = \left[\sum_{\alpha > 0} (\mathfrak{g}_{-\alpha} \cap \mathfrak{q}) \right] \oplus (\mathfrak{u} \cap \mathfrak{q}) \oplus \mathfrak{a} \oplus \mathfrak{n}^+, \qquad (8.18)$$

which is orthogonal under $\langle \cdot, \cdot \rangle$. The following proposition provides the corresponding decomposition at the group level. Note that, by the Bruhat decomposition (5.15), any $g \in G$ may be written as $g = nvp$ with $n \in N^-$, $v \in U'$, and $p \in AN^+$.

Proposition 8.4. *Let $g = nvp$ with $n \in N^-$, $v \in U'$, and $p \in AN^+$. Then $g \in Q$ if and only if $n \in Q$ and $v \in Q$. Consequently, by Theorem 5.4,*

$$Q = \bigcup_{v \in U' \cap Q} (N^- \cap Q) v (U \cap Q) AN^+, \qquad (8.19)$$

where any two sets in the union are either identical or disjoint, and $(N^- \cap Q)(U \cap Q)AN^+$ is a diffeomorphic product. Moreover, $N^- \cap Q = \exp(\mathfrak{n}^- \cap \mathfrak{q})$.

Proof. Since $AN^+ \subset Q$, if $nvp \in Q$, then $nv \in Q$. Let a_j be a positive contracting sequence in A. Then $a_j^{-1} v = v b_j$ for some $b_j \in A$ and $Q \ni a_j nv b_j = a_j na_j^{-1} v \to v$. This proves that $v \in Q$; hence, $n \in Q$. To prove $N^- \cap Q = \exp(\mathfrak{n}^- \cap \mathfrak{q})$, by Corollary 5.1, it suffices to show that $N^- \cap Q$ is connected. Any element of $N^- \cap Q$ may be written as e^Y for some $Y \in \mathfrak{n}^-$. Let $a \colon \mathbb{R}_+ \to A$ be a continuous and positive contracting function. Then $N^- \cap Q \ni a(t)e^Y a(t)^{-1} \to e$. This shows that e^Y is connected to e. $\square$

Let $\pi: G \to G/Q$ be the natural projection and let $o = \pi(e)$ be the point eQ in G/Q. Define

$$\Gamma = \{(x, \omega) \in M \times \Omega; \ \exists g \in G \text{ such that } x = go \text{ and } (g, \omega) \in \Gamma_0\}. \quad (8.20)$$

Let Φ_t be the skew-product flow associated to the stochastic flow ϕ_t on $M = G/Q$, given by $\Phi_t(x, \omega) = (\phi_t(\omega)x, \theta_t\omega)$. For any $x \in M$ and $\omega \in \Omega$, let $\Gamma(x)$ and $\Gamma(\omega)$ be respectively the x-section and the ω-section of Γ. Recall that Ω_0 is the projection of Γ_0 to Ω and that $P(\Omega_0) = 1$. It is easy to see that Ω_0 is also the projection of Γ to Ω. The following proposition follows immediately from Proposition 8.2.

Proposition 8.5. Γ *is* $\{\Phi_t\}$-*invariant and, for* $\omega \in \Omega_0$,

$$\phi_t(\omega)\Gamma(\omega) = \Gamma(\theta_t\omega). \quad (8.21)$$

Moreover, under the hypothesis (H), $P[\Gamma(x)] = 1$ *for any* $x \in M$, *and, for* $(g, \omega) \in \Gamma_0$,

$$\Gamma(\omega) = g\, n_\infty^g(\omega)^{-1}\pi(N^- U A N^+) = g\, n_\infty^g(\omega)^{-1}N^- U o. \quad (8.22)$$

By Propositon 5.19 and (8.22), for $\omega \in \Omega_0$, $\Gamma(\omega)$ is an open subset of $M = G/Q$ with a lower dimension complement. Note that this expression for $\Gamma(\omega)$ should be independent of the choice of $g \in \Gamma_0(\omega)$.

For $g, h \in G$ and $X \in \mathfrak{g}$, we may write gXh for $DL_g \circ DR_h(X)$. Note that any tangent vector in $T_{go}(G/Q)$ can be expressed as $D\pi(gZ)$ for some $Z \in \mathfrak{g}$. Since $h\pi(g) = \pi(hg)$ for $g, h \in G$, $Dh \circ D\pi(gZ) = D\pi(hgZ)$ for $Z \in \mathfrak{g}$.

Theorem 8.1. *Assume the hypothesis (H) and regard* ϕ_t *as a stochastic flow on* $M = G/Q$. *Let* α *be a negative root or zero. For any* $(x, \omega) \in \Gamma$, *choose* $(g, \omega) \in \Gamma_0$ *with* $x = go$. *Then*

$$\forall Y \in \text{Ad}\big(n_\infty^g(\omega)^{-1}\big)[\mathfrak{g}_\alpha - (\mathfrak{g}_\alpha \cap \mathfrak{q})],$$

$$\lim_{t \to \infty} \frac{1}{t} \log \|D\phi_t(\omega)D\pi(gY)\|_{\phi_t(\omega)x} = \alpha(H^+),$$

where $[\mathfrak{g}_\alpha - (\mathfrak{g}_\alpha \cap \mathfrak{q})]$ *is the set difference. Consequently, the Lyapunov exponents of the stochastic flow* ϕ_t *on* $M = G/Q$ *are given by* $\alpha(H^+)$, *where* α *ranges over all negative roots and zero such that* $\mathfrak{g}_\alpha$ *is not entirely contained in* $\mathfrak{q}$. *Therefore, all the exponents are nonpositive, and they are all negative if and only if* $\mathfrak{u} \subset \mathfrak{q}$.

Proof. Let $\{X_1, \ldots, X_d\}$ be the basis of $\mathfrak{g}$ chosen as before. We may assume that each $\mathfrak{g}_\alpha \cap \mathfrak{q}$ is spanned by some X_js; hence, so is a subspace of $\mathfrak{g}_\alpha$ complementary to $\mathfrak{g}_\alpha \cap \mathfrak{q}$. For simplicity, we will omit writing ω in this proof. Then, for $X_j \in [\mathfrak{g}_\alpha - (\mathfrak{g}_\alpha \cap \mathfrak{q})]$,

$$D\phi_t \, D\pi \left[g\mathrm{Ad}((n_\infty^g)^{-1}) X_j \right]$$
$$= D\pi \left[\phi_t g \mathrm{Ad}((n_\infty^g)^{-1}) X_j \right] = D\pi \left[k_t^g a_t^g n_t^g \mathrm{Ad}((n_\infty^g)^{-1}) X_j \right]$$
$$= D\pi \left[k_t^g \mathrm{Ad}(a_t^g) \mathrm{Ad}(n_t^g (n_\infty^g)^{-1}) X_j \right] = Dk_t^g \circ D\pi \left[\sum_i h_{ij}^g(t) \mathrm{Ad}(e^{\log a_t^g}) X_i \right]$$
$$= Dk_t^g \circ D\pi \left[\sum_i h_{ij}^g(t) e^{\mathrm{ad}\left(\log a_t^g\right)} X_i \right] = Dk_t^g \, D\pi \left[\sum_i h_{ij}^g(t) e^{\alpha_i \left(\log a_t^g\right)} X_i \right],$$

where the $h_{ij}^g(t)$ are given by (8.16). Since $(1/t) \log a_t^g \to H^+$, the term with $i = j$ in this expression has the exponential rate $\alpha_j(H^+)$ as $t \to \infty$. By (8.17), the exponential rates of the other terms are not greater. This proves the theorem by the orthogonality of the X_is. $\square$

Let $\lambda_1 > \lambda_2 > \cdots > \lambda_l$ be the distinct Lyapunov exponents of ϕ_t on $M = G/Q$ and let

$$\mathfrak{n}_i = \sum_{\alpha(H^+) \leq \lambda_i} \mathfrak{g}_\alpha, \tag{8.23}$$

where α ranges over all the roots and zero satisfying $\alpha(H^+) \leq \lambda_i$. Since $\lambda_i \leq 0$, the summation in (8.23) includes only $\mathfrak{g}_\alpha$ for $\alpha \leq 0$. Using $[\mathfrak{g}_\alpha, \mathfrak{g}_\beta] \subset \mathfrak{g}_{\alpha+\beta}$, it is easy to prove the following proposition.

Proposition 8.6.

(a) $\mathfrak{n}_i$ *is a Lie subalgebra of* $\mathfrak{g}$.
(b) *If* $\lambda_i < 0$, *then* $\mathfrak{n}_i$ *is an ideal of* $\mathfrak{n}^-$.
(c) *If* $\lambda_1 = 0$, *then* $\mathfrak{n}_1 = \mathfrak{n}^- \oplus \mathfrak{g}_0$.

Note that, if $Y \in \mathrm{Ad}((n_\infty^g(\omega)^{-1})(\mathfrak{g}_\alpha \cap \mathfrak{q})$, then $D\pi(gY) = g D\pi(Y) = 0$.

By Theorem 8.1, for $(x, \omega) \in \Gamma$ and $1 \leq i \leq l$, the subspace $V_i(x, \omega)$ of $T_x M$ associated to the exponent λ_i, defined by (8.6), is equal to

$$V_i(x, \omega) = D\pi \left[g\mathrm{Ad}(n_\infty^g(\omega)^{-1}) \mathfrak{n}_i \right]. \tag{8.24}$$

The multiplicity of λ_i is given by

$$d_i = \sum_{\alpha(H^+) = \lambda_i} \dim(\mathfrak{g}_\alpha) - \sum_{\alpha(H^+) = \lambda_i} \dim(\mathfrak{g}_\alpha \cap \mathfrak{q}). \tag{8.25}$$

Let N_i be the connected Lie subgroup of G with Lie algebra $\mathfrak{n}_i$. Because the exponential map is surjective on a connected nilpotent Lie group, $N_i = \exp(\mathfrak{n}_i)$ if $\lambda_i < 0$. In this case, N_i is a closed normal subgroup of N^-. If $\lambda_1 = 0$, then $N_1 = N^- U_0 A$, where U_0 is the identity component of U, is a closed Lie subgroup of G.

Lemma 8.4. *For* $r \in N^- U A$ *and* $1 \le i \le l$, $r\mathfrak{n}_i = \mathfrak{n}_i r$ *and* $r N_i = N_i r$.

Proof. It suffices to prove the first equality. Since $\mathfrak{n}_i$ is a direct sum of some $\mathfrak{g}_\alpha$, where α is a negative root or zero, and $\mathfrak{g}_\alpha = a\mathfrak{g}_\alpha a^{-1} = u\mathfrak{g}_\alpha u^{-1}$ for $a \in A$ and $u \in U$, we need only to prove the claim for $r = n \in N^-$. We have $n = e^Y$ for some $Y \in \mathfrak{n}^-$. Note that $e^Y \mathfrak{n}_i (e^Y)^{-1} = e^{\text{ad}(Y)} \mathfrak{n}_i$. The space $\mathfrak{n}_i$ is either an ideal of $\mathfrak{n}^-$ or equal to $\mathfrak{n}^- \oplus \mathfrak{g}_0$ with $i = 1$. In either case, $e^{\text{ad}(Y)} \mathfrak{n}_i \subset \mathfrak{n}_i$. $\qquad\square$

Let X^* be the vector field on M induced by $X \in \mathfrak{g}$ defined by (2.3). Since $X^*(x)$ is the tangent vector to the curve $t \mapsto e^{tX} x$ at $t = 0$ for $x \in M$, for the sake of simplicity, we will write Xx for $X^*(x) \in T_x M$. Moreover, for $g \in G$, regarded as a map on M, and $v \in T_x M$, we will write gv for $Dg(v) \in T_{gx} M$. Then $D\pi(gXh) = gXho$ for $X \in \mathfrak{g}$ and $g, h \in G$. Since $N^+ \subset Q$, and $D\pi[g\text{Ad}(q)Y] = D\pi(gqY) = gqXo$ for $g \in G$, $q \in Q$, and $Y \in \mathfrak{g}$, we see that $\text{Ad}(n_\infty^g(\omega)^{-1})$ in (8.24) may be replaced by $n_\infty^g(\omega)^{-1}$. It follows that, for $(x, \omega) \in \Gamma$,

$$V_i(x, \omega) = D\pi \left[g\, n_\infty^g(\omega)^{-1} \mathfrak{n}_i \right] = g\, n_\infty^g(\omega)^{-1} \mathfrak{n}_i o, \qquad (8.26)$$

where $g \in G$ is chosen to satisfy $x = go$ and $(g, \omega) \in \Gamma_0$.

As the subspace of $T_x M$ associated to the Lyapunov exponent λ_i, $V_i(x, \omega)$ given by (8.26) should be independent of the choice of g. This can also be verified directly as follows: A different choice of g in (8.26) can be expressed as $g' = gq$ for some $q \in Q$ such that $(gq, \omega) \in \Gamma_0$. By Lemma 8.3, $q(n_\infty^{gq})^{-1} = (n_\infty^g)^{-1} r$ for some $r \in N^- U A$. From this, it follows that $r \in Q$ and, by Lemma 8.4,

$$gq\left(n_\infty^{gq}\right)^{-1} \mathfrak{n}_i o = g\left(n_\infty^g\right)^{-1} r\mathfrak{n}_i o = g\left(n_\infty^g\right)^{-1} \mathfrak{n}_i r o = g\left(n_\infty^g\right)^{-1} \mathfrak{n}_i o.$$

This shows that the expression in (8.26) is independent of the choice of g.

By Proposition 8.5, the set Γ is invariant under the skew-product flow Φ_t. We now show that the subspace $V(x, \omega)$ as a function of $(x, \omega) \in \Gamma$ is invariant under the stochastic flow ϕ_t in the sense that, $\forall t \in \mathbb{R}_+$,

$$\phi_t(\omega)V_i(x, \omega) = V_i \circ \Phi_t(x, \omega) = V_i(\phi_t(\omega)x, \theta_t\omega). \qquad (8.27)$$

To show this, use the cocycle property (8.1) to see that $n_{t+s}^g(\omega) = n_s^h(\theta_t\omega)$, where $h = \phi_t(\omega)g$. It follows that

$$
\begin{aligned}
\phi_t(\omega)V_i(x,\omega) &= \phi_t(\omega)g\, n_\infty^g(\omega)^{-1}\mathfrak{n}_i o = h\, n_\infty^h(\theta_t\omega)^{-1}\mathfrak{n}_i o \\
&= V_i(ho, \theta_t\omega) = V_i(\phi_t(\omega)x, \theta_t\omega).
\end{aligned}
$$

Theorem 8.2. *Assume the hypothesis (H). Let $(x,\omega) \in \Gamma$ and let λ_i be a negative Lyapunov exponent of the stochastic flow ϕ_t on $M = G/Q$.*

(a) *Choose $g \in G$ such that $x = go$ and $(g,\omega) \in \Gamma_0$. Then*

$$
M_i(x,\omega) = \pi\left[g\, n_\infty^g(\omega)^{-1}N_i\right] = g\, n_\infty^g(\omega)^{-1}N_i o \qquad (8.28)
$$

is the maximal stable manifold of λ_i at (x,ω).

(b) *If $y \in M_i(x,\omega)$, then $(y,\omega) \in \Gamma$ and $M_i(y,\omega) = M_i(x,\omega)$.*

(c) *$M_i(x,\omega)$ is invariant under the stochastic flow ϕ_t in the sense that, for any $t \in \mathbb{R}_+$, $\phi_t(\omega)M_i(x,\omega) = M_i(\phi_t(\omega)x, \theta_t\omega)$.*

Proof. Note that $T_x M_i(x,\omega) = V_i(x,\omega)$ is a direct consequence of (8.26) and (8.28). To prove $M_i(x,\omega)$ is a stable manifold of λ_i at (x,ω), one needs to prove that, for any λ satisfying $\lambda_i < \lambda < 0$ and any compact subset C of $M_i(x,\omega)$, there exists $c > 0$ such that (8.8) holds for any $y \in C$. We may assume $C = g(n_\infty^g)^{-1}e^D o$ and $y = g(n_\infty^g)^{-1}e^Y o$, where D is a compact subset of $\mathfrak{n}_i$ and $Y \in D$. Let $\{X_1, \ldots, X_d\}$ be the basis of $\mathfrak{g}$ as given before. Assume the roots α_i including zero are ordered in (7.29). There is an integer σ between 1 and d such that $X_\sigma, X_2, \ldots, X_d$ form a basis of $\mathfrak{n}_i$. Then $Y = \sum_{j=\sigma}^d c_j X_j$, where the coefficients c_j remain bounded when Y is contained in D. Let $h_{jm}^g(t)$ be defined by (8.16). Then, omitting ω, we have $\phi_t x = \phi_t go = k_t^g o$ and

$$
\begin{aligned}
\phi_t y &= \phi_t g\left(n_\infty^g\right)^{-1}e^Y o = k_t^g a_t^g n_t^g \left(n_\infty^g\right)^{-1}e^Y o \\
&= k_t^g \exp\left[\sum_{j=\sigma}^d \mathrm{Ad}(a_t^g)\mathrm{Ad}(n_t^g\left(n_\infty^g\right)^{-1})c_j X_j\right] o \\
&= k_t^g \exp\left[\sum_{j=\sigma}^d \sum_{m=1}^d c_j h_{mj}^g(t)e^{\alpha_m\left(\log a_t^g\right)} X_m\right] o.
\end{aligned}
$$

Because $(1/t)\log a_t^g \to H^+$ and $\alpha_j(H^+) \leq \lambda_i < \lambda$ for $\sigma \leq j \leq d$, by (8.17), the expression inside $\exp[\cdots]$ is bounded by $ce^{\lambda t}$ for some constant $c > 0$. This proves (8.8) because, if z is a fixed point in M, then for any Z contained in a compact subset of $\mathfrak{g}$, $\mathrm{dist}(z, e^Z z) \leq c_1\|Z\|$ for some constant $c_1 > 0$.

If n_i is replaced by $\mathfrak{n}^-$ in the preceding computation, then it follows that $a_t^g n_t^g (n_\infty^g)^{-1} no \to o$ as $t \to \infty$ uniformly for n contained in a compact subset of N^-.

To prove that $M_i(x, \cdot)$ is the maximal stable manifold of λ_i at (x, ω), we need to show that, if $z \in \Gamma(\omega)$ and $z \notin M_i(x, \cdot)$, then $\liminf_{t \to \infty}(1/t)$ $\log \mathrm{dist}(\phi_t x, \phi_t z) > \lambda_i$. It suffices to consider the following two cases:

> (i) $z = g(n_\infty^g)^{-1} nuo$, where $n \in N^-$ and $u \in (U - Q)$, and
> (ii) $z = g(n_\infty^g)^{-1} no$, where $n \in [N^- - (N_i \cup Q)]$.

In case (i),

$$a_t^g n_t^g (n_\infty^g)^{-1} nuo = \exp\left[\mathrm{Ad}(a_t^g n_t^g (n_\infty^g)^{-1}) \log n\right] a_t^g n_t^g (n_\infty^g)^{-1} uo$$
$$= \exp\left[\mathrm{Ad}(a_t^g)\mathrm{Ad}(n_t^g (n_\infty^g)^{-1}) \log n\right] uo.$$

By (8.17), $\mathrm{Ad}(a_t^g)\mathrm{Ad}(n_t^g (n_\infty^g)^{-1}) \log n \to 0$. It follows that $a_t^g n_t^g (n_\infty^g)^{-1} nuo \to uo$, $\mathrm{dist}(\phi_t x, \phi_t z) \to \mathrm{dist}(k_t^g o, k_t^g uo) = \mathrm{dist}(o, uo) > 0$, and

$$\liminf_{t \to \infty} \frac{1}{t} \log \mathrm{dist}(\phi_t x, \phi_t z) = 0 > \lambda_i.$$

In case (ii), $n = \exp(\sum_{j=\tau}^d c_j X_j)$, where τ is a positive integer $< \sigma$ such that $X_\tau, X_{\tau+1}, \ldots, X_d$ form a basis of $\mathfrak{n}^-$ and $c_p \neq 0$ for some $p < \sigma$. We may assume that a subset of $\{X_1, \ldots, X_d\}$ form a basis of $\mathfrak{q}$ and $X_p \notin \mathfrak{q}$. In fact, we may further assume that p is the smallest integer in $\{\tau, \tau + 1, \ldots, \sigma - 1\}$ such that $c_p \neq 0$ and $X_p \notin \mathfrak{q}$. Then

$$\phi_t z = k_t^g \exp\left[\sum_{j=\tau}^d c_j \mathrm{Ad}(a_t^g)\mathrm{Ad}(n_t^g (n_\infty^g)^{-1}) X_j\right] o = k_t^g \exp\left[\sum_{m=1}^d C_m(t) X_m\right] o,$$

where $C_m(t) = \sum_{j=\tau}^d c_j h_{mj}^g(t) e^{\alpha_m (\log a_t^g)}$. By (8.17) and the convergence of $(1/t) \log a_t^g$ to H^+, we can show that, for any m,

$$\limsup_{t \to \infty} \frac{1}{t} \log |C_m(t)| \leq \alpha_\tau(H^+) < 0.$$

Therefore, $C_m(t) \to 0$ as $t \to \infty$. However, because $n_t^g (n_\infty^g)^{-1} \in \mathfrak{n}^+ \subset \mathfrak{q}$, if $X_m \notin \mathfrak{q}$ and if $X_j \in \mathfrak{q}$, then $h_{mj}^g(t) = 0$. It follows that, if $X_m \notin \mathfrak{q}$, then

$$\limsup_{t \to \infty} \frac{1}{t} \log |C_m(t)| \leq \alpha_p(H^+).$$

Since $h_{jj}^g(t) = 1$, and $h_{mj}^g(t) = 0$ if $j \neq m$ and $\alpha_m(H^+) \leq \alpha_j(H^+)$, we have

$$C_p(t) = c_p e^{\alpha_p\left(\log a_t^g\right)} + \sum_{p < j \leq d, \, \alpha_j(H^+) < \alpha_p(H^+)} c_j h_{pj}^g(t) e^{\alpha_p\left(\log a_t^g\right)}.$$

It follows that $\lim_{t \to \infty} (1/t) \log |C_p(t)| = \alpha_p(H^+) > \lambda_i$.

Let E be the subset of $\{1, 2, \ldots, d\}$ such that $\{X_j; \, j \in E\}$ is a basis of q and let $F = \{1, 2, \ldots, d\} - E$. If $\xi = (\xi_1, \ldots, \xi_d) \in \mathbb{R}^d$ is contained in a small neighborhood V of 0 in $\mathbb{R}^d$, then

$$\exp\left(\sum_{j=1}^{d} \xi_j X_j\right) = \exp\left(\sum_{j \in F} f_j(\xi) X_j\right) \exp\left(\sum_{j \in E} f_j(\xi) X_j\right),$$

where $f_j(\xi)$ are analytic functions defined for $\xi \in V$ and $f_j(\xi) = \xi_j + O(|\xi|^2)$ for $1 \leq j \leq d$. Since $f_j(\xi) = 0$ for $j \in F$ if $\xi_F = 0$, where $\xi_F = \{\xi_j; \, j \in F\} \in \mathbb{R}^F$, it follows that $f_j(\xi) = \xi_j + O(|\xi_F| \cdot |\xi|)$ for $j \in F$. We have

$$\phi_t z = k_t^g \exp\left[\sum_{j \in F} f_j(C_1(t), \ldots, C_d(t)) X_j\right] o$$

and

$$\lim_{t \to \infty} \frac{1}{t} \log |f_p(C_1(t), \ldots, C_d(t))| = \lim_{t \to \infty} \frac{1}{t} \log |C_p(t)| = \alpha_p(H^+) > \lambda_i.$$

This implies that

$$\liminf_{t \to \infty} (1/t) \log \, \text{dist}(\phi_t x, \phi_t z) > \lambda_i.$$

Part (a) is proved.

To prove (b), note that, if $y \in M_i(x, \omega)$, then $y = ghQ$, where $h = (n_\infty^g)^{-1} n$ for some $n \in N_i$. By Lemma 8.2, $\phi_\cdot(\omega) gh = [\phi_\cdot(\omega) g] h \in \Lambda$. This implies that $(gh, \omega) \in \Gamma_0$ and $(y, \omega) \in \Gamma$. By Lemma 8.3, $gh(n_\infty^{gh})^{-1} = g(n_\infty^g)^{-1} r$ for some $r \in N^- U A$. It follows that $ghQ = gh(n_\infty^{gh})^{-1} Q = g(n_\infty^g)^{-1} rQ$, $(n_\infty^g)^{-1} rQ = hQ = (n_\infty^g)^{-1} nQ$, $rQ = nQ$, and

$$M_i(y, \cdot) = gh\left(n_\infty^{gh}\right)^{-1} N_i o = g\left(n_\infty^g\right)^{-1} r N_i o = g\left(n_\infty^g\right)^{-1} N_i r o$$
$$= g\left(n_\infty^g\right)^{-1} N_i n o = g\left(n_\infty^g\right)^{-1} N_i o = M_i(x, \cdot).$$

Part (c) is proved in the same way as (8.27). $\qquad\square$

Let $1 \leq i \leq l$ with $\lambda_i < 0$. Then $N_i \subset N^-$. For $(g, \omega) \in \Gamma_0$ and $\bar{r} = rN_i \in (N^-U)/N_i$, let

$$M_i(\bar{r}, g, \omega) = g \, n_\infty^g(\omega)^{-1}\pi(rN_i) = g \, n_\infty^g(\omega)^{-1}rN_i o. \tag{8.29}$$

Then by (8.22),

$$\bigcup_{\bar{r}} M_i(\bar{r}, g, \omega) = \Gamma(\omega), \tag{8.30}$$

where $\bar{r}$ ranges over $(N^-U)/N_i$.

Proposition 8.7. *Let* $(g, \omega) \in \Gamma_0$. *If* $\bar{r} = rN_i \in (N^-U)/N_i$, *then* $M_i(\bar{r}, g, \omega) = M_i(x, \omega)$ *for* $x = g \, n_\infty^g(\omega)^{-1}ro$. *Consequently, the family of the stable manifolds* $\{M_i(x, \omega); x \in \Gamma(\omega)\}$ *coincides with* $\{M_i(\bar{r}, g, \omega); \bar{r} \in (N^-U)/N_i\}$.

Proof. Let $h = n_\infty^g(\omega)^{-1}r$. By Lemma 8.2, $(gh, \omega) \in \Gamma_0$, and by Lemma 8.3, $h \, n_\infty^{gh}(\omega)^{-1} = n_\infty^g(\omega)^{-1}r'$ for some $r' \in N^-UA$. Then $n_\infty^g(\omega)^{-1}rQ = hQ = h \, n_\infty^{gh}(\omega)^{-1}Q = n_\infty^g(\omega)^{-1}r'Q$, so $rQ = r'Q$ and $M_i(x, \omega) = gh \, n_\infty^{gh}(\omega)^{-1} N_i o = g \, n_\infty^g(\omega)^{-1}r'N_i o = g \, n_\infty^g(\omega)^{-1}N_i r'o = g \, n_\infty^g(\omega)^{-1}N_i ro = g \, n_\infty^g(\omega)^{-1} rN_i o = M_i(\bar{r}, g, \omega)$. $\square$

Let M be an arbitrary manifold. A family of submanifolds $\{M_\sigma; \sigma \in \Sigma\}$ of M of dimension k is said to be a foliation of M, or M is said to be foliated by $\{M_\sigma; \sigma \in \Sigma\}$, if any $x \in M$ has a coordinate neighborhood V with coordinates $x_1, \ldots, x_d$ such that each subset of V determined by

$$x_{k+1} = c_1, \quad x_{k+2} = c_2, \ldots, \quad x_d = c_{d-k}$$

is equal to $M_\sigma \cap V$ for some $\sigma \in \Sigma$, where $c_1, c_2, \ldots, c_{d-k}$ are arbitrary constants. The submanifolds M_σ form a disjoint union of M and are called the leaves of the foliation.

Lemma 8.5. *Let* H_1 *and* H_2 *be two closed Lie subgroups of a Lie group* H *with* $H_1 \subset H_2$ *and let* p *be the point* eH_1 *in* H/H_1. *Then* $\{hH_2p; h \in H\}$ *is a foliation of* H/H_1.

Proof. The set H_2p may be identified with H_2/H_1 and is a submanifold of H/H_1 when equipped with the manifold structure of the homogeneous space. We first show that H_2p is a topological submanifold of H. Since H_2 is a closed Lie subgroup of H, it is a topological Lie subgroup of H. Any open subset of H_2 has the form $O \cap H_2$ for some open subset O of H. It follows that any open subset of $H_2 p$ has the form $(O \cap H_2) p$. It is clear that $(O \cap H_2) p \subset (Op) \cap (H_2 p)$. If $x \in (Op) \cap (H_2 p)$, then $x = hp = h'p$ for $h \in O$ and $h' \in H_2$. Since $H_1 \subset H_2$, $h \in h'H_1 \subset H_2$. This implies that $x \in (O \cap H_2) p$ and hence $(O \cap H_2) p = (Op) \cap (H_2 p)$. Since any open subset of H/H_1 has the form Op, this proves the claim that $H_2 p$ is a topological submanifold of H/H_1.

To prove the lemma, it suffices to show that there is a coordinate neighborhood V of p in H/H_1 that has the property required in the definition of foliation because then hV is such a neighborhood of hp for any $h \in H$. Let $Y_1, \ldots, Y_d$ be a basis of the Lie algebra of H such that $X_1, \ldots, X_b$ form a basis of the Lie algebra of H_2 and $X_{k+1}, \ldots, X_b$ form a basis of the Lie algebra H_1 for some integers $k < b < d$. Let $n = k + d - b$. Consider the map $\psi \colon \mathbb{R}^n \to H/H_1$ given by $x = (x_1, \ldots, x_n) \mapsto h_x h_x^2 p$, where $h_x = \exp(\sum_{i=1}^{n-k} x_{k+i} Y_{b+i})$ and $h_x^2 = \exp(\sum_{i=1}^{k} x_i Y_i)$. The map ψ is regular at $x = 0$ with $\psi(0) = p$. There is a neighborhood W of 0 in $\mathbb{R}^n$ such that $\psi \colon W \to \psi(W)$ is a diffeomorphism. Note that, because $H_1 \subset H_2$, for any $h, h' \in H$, $hH_2 p$ and $h'H_2 p$ are either identical or disjoint. Since $H_2 p$ is a topological submanifold of H/H_1, by choosing W small enough we may assume that, if $x \in W$ with $h_x \neq e$, then $h_x H_2 p \neq H_2 p$. There is a neighborhood W_1 of 0 in $\mathbb{R}^n$ such that $W_1 \subset W$ and, for $x, y \in W_1$, $h_y^{-1} h_x h_x^2 p = h_z h_z^2 p$ for some $z \in W$. Let $V = \psi(W_1)$. To finish the proof, it is enough to show that, if $x, y \in W_1$ with $h_x \neq h_y$, then $h_x H_2 p \neq h_y H_2 p$. Otherwise, $h_x h_x^2 p \in h_y H_2 p$ and $h_z h_z^2 p = h_y^{-1} h_x h_x^2 p \in H_2 p$. It follows that $h_z = e$ and $h_x h_x^2 p = h_y h_z^2 p$. This is impossible because $\psi \colon W_1 \to V$ is a diffeomorphism. $\square$

As a direct consequence of Lemma 8.5, if H' is a closed Lie subgroup of a Lie group H, then the family of the left cosets, $\{hH'; h \in H\}$, is a foliation of H. The same is true for the family of the right cosets.

Proposition 8.8. *For $1 \leq i \leq l$ with $\lambda_i < 0$, $\{rN_i o; r \in N^- U\}$ is a foliation of $\pi(N^- U) = N^- U o$. Moreover, if $i < l$ and $r \in N^- U$, then $rN_i o$ is foliated by $\{sN_{i+1} o; s \in rN_i\}$.*

Proof. Note that N_i is a closed normal subgroup of N^-. It follows that $N_i(N^- \cap Q)$ is a Lie subgroup of N^-. It is connected because both factors N_i and $N^- \cap Q$ are connected (see Proposition 8.4 for the connectedness of $N^- \cap Q$). By Corollary 5.1, $N_i(N^- \cap Q)$ is a closed subgroup of N^-. By Lemma 8.4 and the compactness of U, it is easy to show that $N_i(N^- \cap Q)(U \cap Q)$ is a closed subgroup of $N^- U$. By Proposition 8.4,

$(N^- U) \cap Q = (N^- \cap Q)(U \cap Q)$. Now Lemma 8.5 may be applied with $H = N^- U$, $H_1 = (N^- U) \cap Q$, and $H_2 = N_i H_1$ to conclude that $\{r N_i o;\ r \in N^- U\}$ is a foliation of $\pi(N^- U) = N^- U o$. To prove the second statement, note that N_i contains N_{i+1} as a closed normal subgroup and Lemma 8.5 may be applied with $H = N_i$, $H_1 = N_i \cap Q$, and $H_2 = N_{i+1} H_1$ to show that $N_i o$ is foliated by $\{s N_{i+1} o;\ s \in N_i\}$. $\qquad\square$

As a direct consequence of Proposition 8.8, we see that, under the hypothesis (H), if $(g, \omega) \in \Gamma_0$ and $\lambda_i < 0$, then $\{M_i(\bar{r}, g, \omega);\ \bar{r} \in (N^- U)/N_i\}$ is a foliation of the open set $\Gamma(\omega)$ given by (8.22). Moreover, if $i < l$, then each $M_i(\bar{r}, g, \omega)$ is foliated by $M_{i+1}(\bar{s}, g, \omega)$ with $\bar{s} = s N_{i+1} \in r N_i / N_{i+1}$. By Proposition 8.7, we obtain the following result:

Corollary 8.1. *Assume the hypothesis (H). Let $\omega \in \Omega_0$ and $\lambda_i < 0$. Then $\{M_i(x, \omega);\ x \in \Gamma(\omega)\}$, the family of stable manifolds of the exponent λ_i, is a foliation of $\Gamma(\omega)$. Moreover, if $i < l$, then each stable manifold $M_i(x, \omega)$ is foliated by $\{M_{i+1}(y, \omega);\ y \in M_i(x, \omega)\}$, the family of the stable manifolds of the exponent λ_{i+1} contained in $M_i(x, \omega)$.*

Applying the Iwasawa decomposition $G = K A N^-$ to $g(n_\infty^g)^{-1}$ in (8.22) and (8.29), we see that, for $\omega \in \Omega_0$

$$\Gamma(\omega) = \eta(\omega) N^- U o \qquad (8.31)$$

and, for $x \in \Gamma(\omega)$ and $\lambda_i < 0$,

$$M_i(x, \omega) = \eta(\omega) r N_i o \text{ for some } r \in N^- U, \qquad (8.32)$$

where η is a K-valued random variable. Recall that we have chosen a K-invariant Riemannian metric on $M = K/L$. This means that $\Gamma(\omega)$ and the family of stable manifolds $\{M_i(x, \omega);\ x \in \Gamma(\omega)\}$ are the images of $N^- U o$ and $\{r N_i o;\ r \in N^- U\}$, respectively, under the same isometric transformation: $M \ni x \mapsto \eta(\omega)x \in M$. Therefore, up to a random isometric transformation, we may regard $N^- U o$ as $\Gamma(\omega)$ and $r N_i o,\ r \in N^- U$, as the stable manifolds of λ_i, the former being an open dense subset of M foliated by the latter.

8.4. Stationary Points and the Clustering Behavior

As in the last section, ϕ_t is a right Lévy process in G with $\phi_0 = e$ considered as a stochastic flow on $M = G/Q$, where Q is a closed subgroup of G containing $A N^+$. Assume the hypothesis (H) holds. Let $(H^+)^*$ be the vector field on $M = G/Q$ induced by the rate vector H^+. Then $e^{t H^+}$ is the flow of the vector field $(H^+)^*$. Recall that M is equipped with a K-invariant Riemannian metric.

By the Cartan decomposition $\phi_t = \xi_t a_t^+ \eta_t$, for large time t, the stochastic flow ϕ_t on M may be regarded as approximately composed of the following three transformations: a fixed random isometric transformation η_∞, followed by the flow of the vector field $(H^+)^*$, and then followed by a "moving" random isometric transformation ξ_t. Because only the second transformation causes geometric distortion on M, the asymptotic behavior of the stochastic flow ϕ_t is largely determined by the flow of the single vector field $(H^+)^*$.

In general, a point $x \in M$ is called a stationary point of a vector field X on M if $X(x) = 0$. This is equivalent to saying that x is a fixed point of the flow ψ_t of X. A stationary point x will be said to attract a subset W of M if $\forall y \in W$, $\psi_t(y) \to x$ as $t \to \infty$. It will be said to repel W if there is a neighborhood V of x such that $\forall y \in W - \{x\}$, $\psi_t(y) \notin V$ when $t > 0$ is sufficiently large. A subset W of M is called invariant under the flow ψ_t if $\forall y \in W$ and $t > 0$, $\psi_t(y) \in W$. A stationary point x of X will be called attracting if there is an open neighborhood V of x that is a disjoint union of positive dimensional submanifolds V_α such that each V_α is invariant under ψ_t and contains exactly one stationary point that attracts V_α. It will be called repelling if there is an open neighborhood V of x such that the set of nonstationary points in V is an open dense subset of V that is repelled by x. It will be called a saddle point if there are an open neighborhood V of x and a submanifold W containing x such that $0 < \dim(W) < \dim(M)$, and x is attracting for the vector field X restricted to W and repels $V - W$. In this case, the saddle point x will be said to attract inside W. These definitions may not be standard and do not cover all possible types of stationary points, but they are sufficient for our purpose here. If a stationary point is not isolated and is one of the three types defined here, then all the nearby stationary points must be of the same type.

Recall that o is the point eQ in G/Q and $\mathfrak{q}$ is the Lie algebra of Q.

Proposition 8.9. *(a) The stationary points of $(H^+)^*$ on $M = G/Q$ are precisely $x = vo$ for $v \in U'$.*

(b) For $v \in U'$, vo is attracting if and only if $vo = uo$ for some $u \in U$.

(c) For $v \in U'$, vo is repelling if and only if $\mathrm{Ad}(v^{-1})\mathfrak{n}^- \subset \mathfrak{q}$.

(d) If $x = vo$ for $v \in U'$ is neither attracting nor repelling, then it is a saddle point that attracts inside the complement of $N^- U o$ in M.

Proof. Any point $x \in G/Q$ is contained in $N^- vo$ for some $v \in U'$. Because U' normalizes A and, for $Y \in \mathfrak{n}^-$, $\mathrm{Ad}(\exp(tH^+))Y \to 0$ as $t \to \infty$, it follows that, for $n \in N^-$, $N^- vo \ni \exp(tH^+)nvo \to vo$ as $t \to \infty$. From this it is easy to see that x is a stationary point of $(H^+)^*$ if and only if $x = vo$ for some $v \in U'$. Since $N^- uo$, $u \in U$, form a foliation of the dense open subset

N^-Uo of G/Q, it follows that uo is an attracting stationary point of $(H^+)^*$ for $u \in U$, and the stationary points not contained in Uo cannot be attracting. Now let $v \in U'$ with $vo \notin Uo$. Then N^-vo is a submanifold of G/Q of a lower dimension and is attracted by vo. It follows that vo is either a repelling or saddle point. It is repelling if N^-vo contains only the point vo, that is, $v \exp[\text{Ad}(v^{-1})\mathfrak{n}^-]o = vo$ or $\text{Ad}(v^{-1})\mathfrak{n}^- \subset \mathfrak{q}$. Otherwise, the complement of N^-Uo in M satisfies the condition stated for W in the definition of a saddle point. $\qquad\qquad\square$

A probability measure ρ on $M = G/Q$ will be called irreducible if it does not charge the complement of gN^-Uo in M for any $g \in G$. This generalizes the notion of irreducible measures on $G/(MAN^+)$ given in Section 6.4.

The space $M = G/Q$ may be identified with K/L with $L = Q \cap K$. Let $\pi_{K/L}\colon K \to K/L$ be the natural projection. Denote by ν_U the normalized Haar measure on U and let $\nu = \pi_{K/L}\, \nu_U$, where the measure ν_U on U is regarded as a measure on K supported by U. Then ν is a probability measure on K/L supported by $Uo = \pi_{K/L}(U)$ that may be identified with $U/(U \cap L)$. Moreover, it is U-invariant in the sense that $\forall u \in U$, $u\nu = \nu$. In fact, ν is the unique U-invariant probability measure on $\pi_{K/L}(U)$ (see [27, chapter I, theorem 1.9] and the remark after its proof). A probability measure ρ on K will be called right U-invariant if $R_u \rho = \rho$ for any $u \in U$, where $R_u\colon K \to K$ denotes the right translation $k \mapsto ku$ on K.

Proposition 8.10. *Assume the hypothesis (H). Let ρ be an irreducible probability measure on $M = G/Q$. Then, for any neighborhood W of Uo, $\phi_t \rho(\xi_t W) \to 1$ almost surely as $t \to \infty$. Moreover, if $\rho = \pi_{K/L}\, \rho'$, where ρ' is an irreducible and right U-invariant probability measure on K, then almost surely $\phi_t \rho - \xi_t \nu \to 0$ weakly as $t \to \infty$.*

Proof. As a neighborhood of Uo, W must contain $e^V Uo$, where V is a neighborhood of 0 in $\mathfrak{n}^-$. Since $(a_t^+)^{-1} e^V a_t^+ = \exp[\text{Ad}((a_t^+)^{-1})V] \uparrow N^-$ as $t \uparrow \infty$ and $aUo = Uo$ for $a \in A$,

$$\phi_t \rho(\xi_t W) \geq \rho(\phi_t^{-1} \xi_t e^V Uo) = \rho[\eta_t^{-1}(a_t^+)^{-1} e^V Uo] \to \rho(\eta_\infty^{-1} N^- Uo).$$

The last expression is equal to 1 by the irreducibility of ρ. This proves the first claim. To prove the second claim, fix an ω for which the conclusions of Theorem 6.5 hold. For simplicity, we will omit writing this ω in the rest of the proof. Recall that the natural action of G on $G/(AN^+)$ induces the $*$-action of G on $K = G/(AN^+)$. (See the discussion at the end of Section 5.3.) Let $J\colon \eta_\infty^{-1} N^- * U \to U$ be the map defined by $J\colon x = \eta_\infty^{-1} n * u \mapsto u$ for $n \in N^-$

and $u \in U$. Then

$$a_t^+ \eta_t x = a_t^+ (\eta_t \eta_\infty^{-1}) n * u = a_t^+ n_t * u_t = [(a_t^+) n_t (a_t^+)^{-1}] * u_t \to u = J(x)$$

as $t \to \infty$, where $n_t \in N^-$ and $u_t \in U$ are determined by $\eta_t \eta_\infty^{-1} n * u = n_t * u_t$ and thus satisfy $n_t \to n$ and $u_t \to u$. It follows that $a_t^+ \eta_t \rho' \to J\rho'$ weakly. Since $J(x)u = J(xu)$ for $x \in K$ and $u \in U$, it is easy to show from the right U-invariance of ρ' that $J\rho'$ is a right invariant probability measure on U; hence, $J\rho' = \nu_U$. This implies that

$$\phi_t \rho' - \xi_t \nu_U = \xi_t (a_t^+ \eta_t \rho' - \nu_U) \to 0$$

and $\phi_t \rho - \xi_t \nu = \phi_t \pi_{K/L} \rho' - \xi_t \pi_{K/L} \nu_U = \pi_{K/L}(\phi_t \rho' - \xi_t \nu_U) \to 0$ weakly. $\qquad \square$

This proposition shows a clustering property of the stochastic flow ϕ_t: It transforms an irreducible probability measure ρ on M into a random measure $\phi_t \rho$ that is more and more concentrated on the "moving" random set $\xi_t U o$, an isometric image of the set of attracting stationary points of $(H^+)^*$, as $t \to \infty$. Moreover, under an additional assumption, this random measure is approximately equal to $\xi_t \nu$, an isometric image of the unique U-invariant probability measure ν on $\pi_{K/L}(U)$, for large $t > 0$.

Recall that G_μ is the closed group generated by the Lévy process ϕ_t. It is assumed to be totally right irreducible in the hypothesis (H). However, this hypothesis is not assumed in the following proposition.

Proposition 8.11. *Assume G_μ is totally left irreducible. Then a stationary measure ρ of ϕ_t on $M = G/Q = K/L$ is irreducible.*

Proof. Let μ be the probability measure on G defined by (6.22). As a stationary measure, ρ is μ-invariant. We now show that there is a μ-invariant probability measure $\tilde\rho$ on $K = G/(AN^+)$ with $\pi_{K/L} \tilde\rho = \rho$. Let $M \ni x \mapsto k_x \in K$ be a measurable map with $\pi_{K/L}(k_x) = x$, let ν be an arbitrary probability measure on $L = Q \cap K$, and define a probability measure ρ' on K by $\rho'(f) = \int f(k_x \nu)\nu(d\nu)\rho(dx)$ for $f \in \mathcal{B}(K)_+$. Then $\pi_{K/L} \rho' = \rho$. It is easy to show that any weak limiting point $\tilde\rho$ of the sequence of the probability measures $(1/n)(\rho' + \sum_{i=1}^{n-1} \mu^{*i} * \rho')$ is μ-invariant and satisfies $\pi_{K/L}\tilde\rho = \rho$, where μ^{*i} denotes the i-fold convolution of μ. Note that $\pi_{K/U} \tilde\rho$ is a μ-invariant probability measure on K/U and, by Theorem 6.3, is irreducible. From this it follows that $\tilde\rho$ is irreducible on K and hence $\rho = \pi_{K/L} \tilde\rho$ is irreducible on K/L. $\qquad \square$

Proposition 8.12. *Assume the hypothesis (H) and the total left irreducibility of G_μ. Then there is a unique stationary measure ρ_K of ϕ_t on K that is right U-invariant.*

Proof. It is easy to see that, if ρ' is a stationary measure of ϕ_t on K, then $\rho_K = \int_U R_u \rho' v_U(du)$ is a right U-invariant stationary measure on K. To show that such a ρ_K is unique, suppose ρ is another right U-invariant stationary measure on K. By Theorem 6.4, there is a unique stationary measure on K/U. Since both $\pi_{K/U} \rho$ and $\pi_{K/U} \rho_K$ are stationary measures on K/U, they must be identical. For any $f \in \mathcal{B}(K)_+$, let $f_U(kU) = \int v_U(du) f(ku)$. Then $f_U \in \mathcal{B}(K/U)_+$, and, by the right U-invariance, $\rho(f) = \pi_{K/U} \rho(f_U) = \pi_{K/U} \rho_K(f_U) = \rho_K(f)$. This proves $\rho = \rho_K$. $\qquad\square$

Let $\rho = \pi_{K/L} \rho_K$. Then ρ is a stationary measure of ϕ_t on K/L. As a consequence of Propositions 8.10–8.12, we obtain the following result:

Corollary 8.2. *Under the hypothesis (H) and the total left irreducibility of G_μ, if ϕ_t has a unique stationary measure ρ on $M = G/Q$, then, almost surely, $\phi_t \rho - \xi_t \nu \to 0$ weakly as $t \to \infty$, where $\nu = \pi_{K/L} \nu_U$ and ν_U is the normalized Haar measure on U.*

8.5. *SL(d, ℝ)*-flows

In the remaining sections, we consider some examples of stochastic flows that are induced by right Lévy processes in two special matrix groups, namely, the special linear group $SL(d, \mathbb{R})$ and the connected Lorentz group $SO(1, d)_+$ for $d \geq 3$. The hypothesis (H) will be assumed to hold in all these examples. Recall that a right Lévy process in a Lie group G regarded as a stochastic flow on a manifold M is called a G-flow on M.

In Section 5.4, these two matrix groups have been discussed in some detail. We will use the notation introduced there; however, remember that m, M, and M' are now denoted by u, U, and U' respectively.

Let ϕ_t be a right Lévy process in $G = SL(d, \mathbb{R})$ with $\phi_0 = e = I_d$ satisfying the hypothesis (H). Its rate vector is given by

$$H^+ = \text{diag}(b_1, b_2, \ldots, b_d) \text{ with } b_1 > b_2 > \cdot > b_d \text{ and } \sum_{i=1}^d b_i = 0. \tag{8.33}$$

Note that, by Theorem 7.5 and (6.9), if ϕ_t is continuous and left K-invariant, then $b_1 - b_2 = b_2 - b_3 = \cdots = b_{d-1} - b_d$. If Q is a closed subgroup of G containing AN^+ with Lie subalgebra q, then $\mathfrak{g}_0 = \mathfrak{a} \subset \mathfrak{q}$. By Theorem 8.1, when ϕ_t is regarded as a stochastic flow on $M = G/Q$, its Lyapunov exponents are negative and are given by $-(b_i - b_j)$ for $i < j$ such that $E_{ji} \notin \mathfrak{q}$, where E_{ij} is the matrix that has 1 at place (i, j) and 0 elsewhere.

Although the connected general linear group $\bar{G} = GL(d, \mathbb{R})_+$ is not semisimple, it is isomorphic to the product group $\mathbb{R}_+ \times G$, where $G = SL(d, \mathbb{R})$, via the map $(\mathbb{R}_+ \times G) \ni (a, g) \mapsto ag \in \bar{G}$, and $M = G/Q$ may be identified with $\bar{G}/\bar{Q}$, where $\bar{Q} = \mathbb{R}_+ \times Q$. We will write $S(g)$ for the G-component of $g \in \bar{G}$ under the product structure $\bar{G} = \mathbb{R}_+ \times G$. It is easy to show that, if $\bar{g}_t$ is a right Lévy process in $\bar{G}$, then $\phi_t = S(\bar{g}_t)$ is a right Lévy process in G. For example, if $\bar{g}_t$ is a right invariant diffusion process in $\bar{G}$ satisfying the stochastic differential equation

$$d\bar{g}_t = \sum_{i=1}^{m} \bar{Y}_i \bar{g}_t \circ dW_t^i + \bar{Y}_0 \bar{g}_t dt \tag{8.34}$$

on $GL(d, \mathbb{R})_+$, where $\bar{Y}_0, \bar{Y}_1, \ldots, \bar{Y}_m$ are linear combinations of E_{jk}, that is, $\bar{Y}_i = \sum_{j,k=1}^{d} c_{ijk} E_{jk}$ for some constants c_{ijk}, and $W_t = \{W_t^1, \ldots, W_t^m\}$ is an m-dimensional standard Brownian motion, then $\phi_t = S(\bar{g}_t)$ is a right invariant diffusion process in G satisfying the stochastic differential equation

$$d\phi_t = \sum_{i=1}^{m} Y_i \phi_t \circ dW_t^i + Y_0 \phi_t dt, \tag{8.35}$$

where $Y_i = \bar{Y}_i - c_i I_d \in \mathfrak{g} = \mathfrak{sl}(d, \mathbb{R})$ for the unique constant c_i. The hypothesis (H) holds for the right Lévy process ϕ_t if $\bar{Y}_0, \bar{Y}_1, \ldots, \bar{Y}_m$ generate $\bar{\mathfrak{g}} = \mathfrak{gl}(d, \mathbb{R})$ and if a symmetric matrix with distinct eigenvalues may be found in the Lie algebra generated by $\bar{Y}_1, \ldots, \bar{Y}_m$.

Example 1. ($SL(d, \mathbb{R})$-flow on $SO(d)$). Consider ϕ_t as a stochastic flow on $M = K = SO(d)$, which may be identified with G/Q with $Q = AN^+$. Then, for any $i > j$, $E_{ij} \notin \mathfrak{q}$. The number l of the distinct Lyapunov exponents depends on the set $\{(b_i - b_j); i < j\}$, where the constants b_i are given in (8.33). If $b_1 - b_2 = b_2 - b_3 = \cdots = b_{d-1} - b_d = c$, then l reaches its smallest possible value $d - 1$ and the distinct exponents are $-c, -2c, \ldots, -(d - 1)c$ with multiplicities $(d - 1), (d - 2), \ldots, 1$, respectively. In the other extreme case, if all the numbers $(b_i - b_j)$ for $i < j$ are distinct, then l reaches its largest possible value $d(d - 1)/2$ and the distinct exponents are given by $-(b_i - b_j)$ for $i < j$, which are all simple.

By Proposition 5.20,

$$N^- * e = \{k \in SO(d); \ \det(k[i]) > 0 \text{ for } 1 \leq i \leq d\}.$$

Recall that, for any square matrix g, $g[i]$ denotes the submatrix formed by the first i rows and i columns of g. Since $N_1 = N^-$, by (8.32), the stable manifolds of the top exponent can be indexed by $u \in U$ and are given by

$$\eta(\omega)(uN^- * e) = \eta(\omega)(N^- * u) = \eta(\omega)(N^- * e)u \quad \text{for } u \in U \quad (8.36)$$

for P-almost all ω, where $\eta(\omega)$ is an $SO(d)$-valued random variable. Recall that U is the group of diagonal matrices with diagonal elements equal to ± 1 and with an even number of -1s; hence, it has 2^{d-1} elements.

By the expression of U' given by (5.19), any $v \in U'$ is a permutation matrix $v = \{\pm \delta_{i\,\sigma(j)}\}$ for some permutation σ on $\{1, 2, \ldots, d\}$, where the distribution of $\pm$ signs must satisfy $\det(v) = 1$. It follows that U' has $2^{d-1}d!$ elements. By Proposition 8.9, U' can be identified with the set of the stationary points of the vector field $(H^+)^*$ on $SO(d)$, and U with the set of the attracting stationary points. To determine whether a point $v \in U' - U$ is repelling, again by Proposition 8.9, we calculate $\text{Ad}(v^{-1})Y = v^{-1}Yv$, for $Y = \{Y_{ij}\} \in \mathfrak{n}^-$, to see whether it is contained in $\mathfrak{q}$, that is, whether it is an upper triangular matrix. Since $\text{Ad}(v^{-1})Y = \{Y_{\sigma(i)\sigma(j)}\}$, we see that for this to be an upper triangular matrix for any $Y \in \mathfrak{n}^-$, the permutation σ must be the one that sends 1 to d, 2 to $(d-1)$, $\ldots$, d to 1; that is, $v = \{\pm \delta_{i\,(d-j+1)}\}$. It follows that there are exactly 2^{d-1} repelling stationary points.

Example 2. ($SL(3, \mathbb{R})$-flow on $SO(3)$). We now take a closer look at the case of $d = 3$ in Example 1, that is, an $SL(3, \mathbb{R})$-flow on $SO(3)$. This stochastic flow has two or three distinct Lyapunov exponents. We will describe geometrically the stable manifolds for the top exponent λ_1 and the smallest exponent, which is either λ_2 or λ_3, and locate the stationary points of the vector field $(H^+)^*$.

Let e_1, e_2, e_3 be the standard basis of $\mathbb{R}^3$ and let x_1, x_2, x_3 be the coordinates associated to this basis. For $v \in \mathbb{R}^3$, let $R(v)$ denote the rotation on $\mathbb{R}^3$ by the angle $|v|$ around the axis that contains v in the positive direction. Then $R(0)$ is the identity map on $\mathbb{R}^3$, $R(v) = R(-v)$ if $|v| = \pi$, and

$$SO(3) = \{R(u); \ |u| \le \pi\}. \quad (8.37)$$

Therefore, $SO(3)$ may be identified with the ball $B_\pi(0)$ of radius π in $\mathbb{R}^3$ centered at the origin if any two antipodal points on its boundary are identified. For $k \in SO(3)$, using $k' = k^{-1}$ and the cofactors of the matrix k, it is easy to show that $\det(k[2]) = k_{33}$. Therefore,

$$N^- * e = \{k \in SO(3); \ k_{11} > 0 \text{ and } k_{33} > 0\}. \quad (8.38)$$

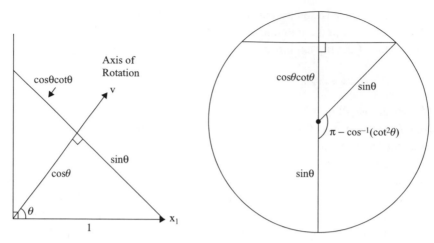

Figure 1. Determination of angle θ when $R(v)_{11} = 0$.

Let $u_i = R(\pi e_i)$ for $i = 1, 2, 3$ and let $u_0 = e = R(0)$. Then $U = \{u_0, u_1, u_2, u_3\}$. The sets $N^- * u_1 = (N^- * e)u_1, N^- * u_2 = (N^- * e)u_2$, and $N^- * u_3 = (N^- * e)u_3$ can be expressed by (8.38) when the two inequalities $k_{11} > 0$ and $k_{33} > 0$ are replaced by $k_{11} > 0$ and $k_{33} < 0, k_{11} < 0$ and $k_{33} < 0$, and $k_{11} < 0$ and $k_{33} > 0$, respectively.

By (8.36), the stable manifolds of the top Lyapunov exponent λ_1 are open subsets of $SO(3)$ that are, up to a left translation by some $\eta(\omega) \in SO(3)$, equal to $N^- * u_i$ for $i = 0, 1, 2, 3$.

For $v \in \mathbb{R}^3$ with $|v| \leq \pi$, let θ be the angle between e_1 and v. Note that $R(v)_{11}$ is the orthogonal projection of $R(v)e_1$ to e_1. Clearly, if $\theta < \pi/4$, then $R(v)_{11} > 0$. If $\pi/4 \leq \theta \leq \pi/2$, then, from Figure 1, it is easy to see that $R(v)_{11} > 0$ if and only if $|v| < \pi - \cos^{-1}(\cot^2 \theta)$. The inequality $R(v)_{11} > 0$ determines a region inside $B_\pi(0)$ enclosed by a surface that looks like a hyperboloid of one-sheet along the x_1-axis. Similarly, the inequality $R(v)_{33} > 0$ determines such a region along the x_3-axis. The set $N^- * e$ is the intersection of these two regions. Let A, B, C, and D be the regions inside the ball $B_\pi(0)$ determined by the sets $N^- * e, N^- * u_1, N^- * u_2$, and $N^- * u_3$, respectively. Their cross sections with the (x_1, x_3)-, (x_2, x_3)-, and (x_1, x_2)-planes are depicted in Figure 2. The stable manifolds of the top Lyapunov exponent λ_1 are obtained by the left translations of these four sets by a random $\eta(\omega) \in SO(3)$.

By the discussion at the end of Example 1, the vector field $(H^+)^*$ has $2^2 \cdot 3! = 24$ stationary points. There are four attracting stationary points, u_i

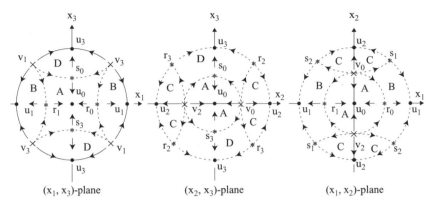

Figure 2. $SL(3, \mathbb{R})$-flow on $SO(3)$ represented by $B_\pi(0)$. Stationary points of $(H^+)^*$:
•, attracting; x, repelling; *, saddle.

for $i = 0, 1, 2, 3$, which are shown in Figure 2 as solid dots. There are four repelling stationary points,

$$v_0 = R\left(\frac{\pi}{2}e_2\right), \quad v_1 = R\left(\pi\frac{e_1 - e_3}{\sqrt{2}}\right), \quad v_2 = R\left(-\frac{\pi}{2}e_2\right), \quad \text{and}$$

$$v_3 = R\left(\pi\frac{e_1 + e_3}{\sqrt{2}}\right),$$

which are shown as crosses. The other stationary points are saddle points. Only eight of them,

$$r_0 = R\left(\frac{\pi}{2}e_1\right), \quad r_1 = R\left(-\frac{\pi}{2}e_1\right), \quad r_2 = R\left(\pi\frac{e_2 + e_3}{\sqrt{2}}\right),$$

$$r_3 = R\left(\pi\frac{e_2 - e_3}{\sqrt{2}}\right),$$

$$s_0 = R\left(\frac{\pi}{2}e_3\right), \quad s_1 = R\left(\pi\frac{e_1 + e_2}{\sqrt{2}}\right), \quad s_2 = R\left(\pi\frac{e_1 - e_2}{\sqrt{2}}\right), \quad \text{and}$$

$$s_3 = R\left(-\frac{\pi}{2}e_3\right),$$

are shown as asterisks. The remaining saddle points,

$$p_0 = R\left(\frac{2\pi}{3}\left(\frac{e_1 + e_2 + e_3}{\sqrt{3}}\right)\right), \quad p_1 = R\left(\frac{2\pi}{3}\left(\frac{-e_1 - e_2 + e_3}{\sqrt{3}}\right)\right),$$

$$p_2 = R\left(\frac{2\pi}{3}\left(\frac{e_1 - e_2 - e_3}{\sqrt{3}}\right)\right),$$

$$p_3 = R\left(\frac{2\pi}{3}\left(\frac{-e_1 + e_2 - e_3}{\sqrt{3}}\right)\right), \quad q_0 = R\left(\frac{2\pi}{3}\left(\frac{-e_1 - e_2 - e_3}{\sqrt{3}}\right)\right),$$

$$q_1 = R\left(\frac{2\pi}{3}\left(\frac{e_1 - e_2 + e_3}{\sqrt{3}}\right)\right),$$

$$q_2 = R\left(\frac{2\pi}{3}\left(\frac{e_1 + e_2 - e_3}{\sqrt{3}}\right)\right), \quad \text{and} \quad q_3 = R\left(\frac{2\pi}{3}\left(\frac{-e_1 + e_2 + e_3}{\sqrt{3}}\right)\right),$$

are not shown because they are not contained in any coordinate plane. Note that

$$v_i = v_0 u_i, \quad r_i = r_0 u_i, \quad s_i = s_0 u_i, \quad p_i = p_0 u_i, \quad \text{and} \quad q_i = q_0 u_i \quad (8.39)$$

for $i = 1, 2, 3$. These relationships can be verified easily by either writing down the corresponding matrices or examining the images of e_i under the linear maps on the both sides of the equations. The directions of the vector field $(H^+)^*$ are shown as arrows in Figure 2.

The stable manifolds of the smallest Lyapunov exponent are given by $\eta(\omega)(nN_s * u_i)$ for $n \in N^-$ and $i = 0, 1, 2, 3$, where $\eta(\omega) \in SO(3)$ and N_s is the one-dimensional Lie subgroup of N^- whose Lie algebra is the linear span of E_{31}. We will describe the one-dimensional sets $nN_s * u_i = (nN_s * e)u_i$ geometrically. For real a, b, c, let

$$n(a, b, c) = \begin{bmatrix} 1 & 0 & 0 \\ a & 1 & 0 \\ b & c & 1 \end{bmatrix} \quad \text{and} \quad h(a, b, c) = \begin{bmatrix} 1 & -a - bc & ac - b \\ a & 1 + b^2 - abc & -c \\ b & c + a^2 c - ab & 1 \end{bmatrix}.$$

Then $N^- = \{n(a, b, c); a, b, c \in \mathbb{R}\}$ and $N_s = \{n(0, t, 0); t \in \mathbb{R}\}$. Let v_1, v_2, v_3 and w_1, w_2, w_3 be the column vectors of the matrices $n(a, b, c)$ and $h(a, b, c)$, respectively. Then $w_1 = v_1$, $w_2 = -(a + bc)v_1 + (1 + a^2 + b^2)v_2$, and w_1, w_2, w_3 are mutually orthogonal. Moreover, $\det[h(0, 0, 0)] = 1$, so by orthogonality and continuity, $\det[h(a, b, c)] > 0$.

Let $k(a, b, c)$ be the matrix formed by the three column vectors $w_1/|w_1|$, $w_2/|w_2|$, and $w_3/|w_3|$, which are obtained from the column vectors of $n(a, b, c)$ by a Gram–Schmidt orthogonalization procedure. It follows that $k(a, b, c)$ is the K-component of $n(a, b, c)$ in the Iwasawa decomposition $G = KAN^+$, where $K = SO(3)$, and

$$N^- * e = \{k(a, b, c); \quad a, b, c \in \mathbb{R}\}. \quad (8.40)$$

By the expression for $h(a, b, c)$, it is easy to show that

$$k(a, b, c) \rightarrow \begin{bmatrix} 0 & 0 & \mp 1 \\ 0 & 1 & 0 \\ \pm 1 & 0 & 0 \end{bmatrix} \quad \text{as } b \rightarrow \pm \infty.$$

Since $n(a, b, c)n(0, t, 0) = n(0, t, 0)n(a, b, c) = n(a, b + t, c)$, we see that the set $nN_s * e$ for fixed $n \in N^-$ is a curve inside region A connecting the two repelling stationary points $v_0 = R(\frac{\pi}{2}e_2)$ and $v_2 = R(-\frac{\pi}{2}e_2)$ in Figure 2. More precisely, it is a smooth curve $\gamma : \mathbb{R} \rightarrow A$ such that $\gamma(t) \rightarrow v_0$ as $t \rightarrow -\infty$ and $\gamma(t) \rightarrow v_2$ as $t \rightarrow +\infty$. Note that

$$k(0, b, 0) = \begin{bmatrix} (1 + b^2)^{-1/2} & 0 & -b(1 + b^2)^{-1/2} \\ 0 & 1 & 0 \\ b(1 + b^2)^{-1/2} & 0 & (1 + b^2)^{-1/2} \end{bmatrix} = R(\alpha e_2),$$

where $\alpha = -\sin^{-1}[b(1 + b^2)^{-1/2}]$ with $|\alpha| < \pi/2$ for $b \in \mathbb{R}$. Therefore, the curve $N_s * e$ is the solid line segment connecting v_0 and v_2 inside region A depicted in Figure 2. The other curves, $nN_s * e$ for $n \neq e$, are not depicted because they are twisted and are not contained in any plane. All these curves connect the points v_0 and v_2, and together they form a foliation of region A. Similarly, regions B, C, and D are foliated by the families of curves

$$\{(nN_s * e)u_1; \ n \in N^-\}, \quad \{(nN_s * e)u_2; \ n \in N^-\}, \quad \text{and}$$
$$\{(nN_s * e)u_3; \ n \in N^-\}$$

respectively. Using (8.39), it is easy to see that the curves in the first and third families connect the two other repelling stationary points v_1 and v_3, and those in the second family connect the points v_0 and v_2. The curves $(N_s * e)u_1$, $(N_s * e)u_2$, and $(N_s * e)u_3$, shown in Figure 2, are respectively the solid arc connecting v_1 and v_3 inside region B, the solid line segment connecting v_0 and v_2 inside region C (identifying the antipodal points on the boundary), and the solid arc connecting v_1 and v_3 inside region D.

Example 3. $(SL(d)$-flow on $S^{d-1})$. Let Q be the closed subgroup of $G = SL(d, \mathbb{R})$ consisting of the matrices $g \in G$ such that $g_{11} > 0$ and $g_{i1} = 0$ for $2 \leq i \leq d$, and let $L = K \cap Q$. Then $L = \text{diag}\{1, SO(d - 1)\}$, which may be naturally identified with $SO(d - 1)$. Let $e_1, \ldots, e_d$ and $x_1, \ldots, x_d$ be respectively the standard basis and the standard coordinates on $\mathbb{R}^d$.

The group G acts on S^{d-1} as follows: Regard S^{d-1} as the unit sphere embedded in $\mathbb{R}^d$ and let gx be the usual matrix product for $g \in G$ and $x \in \mathbb{R}^d$, where x is regarded as a column vector. Define $g * x = gx/|gx|$ for $x \in S^{d-1}$.

It is easy to show that the map $S^{d-1} \ni x \mapsto g * x \in S^{d-1}$ for $g \in G$ defines an action of G on S^{d-1}. Let $o = e_1$ be regarded as a point in S^{d-1}. Then Q is the subgroup of G that fixes the point o and S^{d-1} may be identified with $G/Q = K/L$. Moreover, the $*$-action of G on S^{d-1} defined here coincides with the $*$-action of G on K/L defined earlier. We may regard o as the point representing the coset Q in G/Q or L in K/L.

Because all $\mathfrak{g}_\alpha \in \mathfrak{q}$ except when $\alpha = \alpha_{i1}$ for $i = 2, 3, \ldots, d$, the stochastic flow ϕ_t has $(d - 1)$ simple negative exponents:

$$\lambda_1 = -(b_1 - b_2) > \lambda_2 = -(b_1 - b_3) > \cdots > \lambda_{d-1} = -(b_1 - b_d).$$

In this example, Uo contains only two points: o and $-o$.

Let N_i be the connected subgroup of N^- whose Lie algebra is $\mathfrak{n}_i$ defined by (8.23) for $1 \leq i \leq l = d - 1$. Then any $g \in N_i$ is a lower triangular matrix whose first column is equal to $(1, 0, \ldots, 0, y_{i+1}, \ldots, y_d)'$, where there are $(i - 1)$ zeros and $(d - i)$ arbitrary real numbers $y_{i+1}, \ldots, y_d$. Let σ_i be the subspace $\{x = (x_1, \ldots, x_d) \in \mathbb{R}^d; \ x_i = 0\}$ of $\mathbb{R}^d$ and let Σ be the open hemisphere $\{x = (x_1, \ldots, x_d) \in S^{d-1}; \ x_1 > 0\}$. Then $(N^- U A N^+) * o = \Sigma \cup (-\Sigma)$ and

$$N_1 * o = \sum, \quad N_2 * o = \sum \cap \sigma_2, \ldots, \quad N_{d-1} * o = \sum \cap \sigma_2 \cap \cdots \cap \sigma_{d-1}.$$

For $1 \leq j \leq (d - 1)$, the intersection of S^{d-1} with a $(j + 1)$-dimensional subspace of $\mathbb{R}^d$ and the intersection of an open hemisphere of S^{d-1} with such a subspace are connected submanifolds of S^{d-1}, called respectively a j-dimensional great circle and j-dimensional great half circle on S^{d-1}. Note that a $(d - 1)$-dimensional great circle is S^{d-1} itself and a $(d - 1)$-dimensional great half circle is an open hemisphere on S^{d-1}. For $1 \leq i \leq (d - 1)$, $N_i * o$ is a $(d - i)$-dimensional great half circle. By (8.32), the stable manifolds of $\lambda_i = -(b_1 - b_{i+1})$ are given by $\eta(\omega) r N_i * o$ for $r \in N^- U$. Because the action of G on $\mathbb{R}^d$ maps the subspaces into subspaces, the $*$-action of G on S^{d-1} maps great circles into great circles and great half circles into great half circles. To summarize, we obtain the following result:

Proposition 8.13. *Let ϕ_t be a right Lévy process in $G = SL(d, \mathbb{R})$ with $\phi_0 = I_d$ satisfying the hypothesis (H). Regard ϕ_t as a stochastic flow on S^{d-1} as defined previously. Then it has $(d - 1)$ negative simple Lyapunov exponents $\lambda_1 > \lambda_2 > \cdots > \lambda_{d-1}$, whose stable manifolds can be described as follows: For P-almost all ω, there are two stable manifolds of λ_1, $\Sigma(\omega)$ and $-\Sigma(\omega)$, which are two open hemispheres separated by a $(d - 2)$-dimensional great circle. For each $i \in \{1, 2, \ldots, d - 1\}$, the stable manifolds of λ_i are $(d - i)$-dimensional great half circles that form a foliation of $\Sigma(\omega) \cup [-\Sigma(\omega)]$. Each*

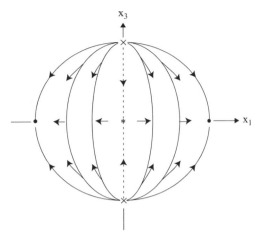

Figure 3. $SL(3, \mathbb{R})$-flow on S^2.

such great half circle, when $i < (d - 1)$, is itself foliated by a family of $(d - i - 1)$-dimensional great half circles that are stable manifolds of λ_{i+1}.

Because any $v \in U'$ is a permutation matrix, $v * o = \pm e_i$ for some $i = 1, 2, \ldots, d$. Therefore, the stationary points of the vector field $(H^+)^*$ on S^{d-1} are given by $\pm e_i$ for $1 \leq i \leq d$. By Proposition 8.9 and the discussion at the end of Example 1, it is easy to show that the attracting and repelling stationary points are respectively $\pm e_1$ and $\pm e_d$.

When $d = 3$, these stationary points are shown on one side of S^2 in Figure 3 (looking along the x_2-axis in $\mathbb{R}^3$) with solid dots, crosses, and asterisks representing attracting, repelling, and saddle points respectively. The arrows show the directions of the vector field $(H^+)^*$. The dashed line shown in Figure 3 is the great circle through the two repelling stationary points $(0, 0, \pm 1)$, the north and south poles, which separates the sphere S^2 into two hemispheres, and the solid lines connecting the two poles are great half circles. Up to a random isometric transformation $S^2 \ni x \mapsto \eta(\omega)x \in S^2$ for some $\eta(\omega) \in SO(3)$, the two hemispheres and the family of the great half circles are respectively the stable manifolds of λ_1 and λ_2.

Now let $\bar{g}_t$ be a right invariant diffusion process in $GL(d, \mathbb{R})_+$ satisfying stochastic differential equation (8.34) and $g_0 = I_d$. Then $\phi_t = S(\bar{g}_t)$ is a right invariant diffusion process in $G = SL(d, \mathbb{R})$. Both $\bar{g}_t$ and ϕ_t may be regarded as the stochastic flow on S^{d-1} generated by the stochastic differential equation

$$dx_t = \sum_{i=1}^m Y_i^*(x_t) \circ dW_t^i + Y_0^*(x_t)dt, \qquad (8.41)$$

to which Proposition 8.13 applies.

In particular, if $\bar{g}_t$ satisfies the stochastic differential equation

$$d\bar{g}_t = \sum_{i,j=1}^{d} E_{ij} \bar{g}_t \circ dW_t^{ij}, \tag{8.42}$$

where $W_t = \{W_t^{ij}\}$ is a d^2-dimensional standard Brownian motion, then both $\bar{g}_t$ and ϕ_t may be regarded as the stochastic flow generated by the stochastic differential equation

$$dx_t = \sum_{i,j=1}^{d} E_{ij}^{*}(x_t) \circ dW_t^{ij} \tag{8.43}$$

on S^{d-1}. It can be shown by a direct computation that the vector field E_{ij}^{*} on S^{d-1} induced by E_{ij} is equal to the orthogonal projection to S^{d-1} of the vector field $x_i(\partial/\partial x_j)$ on $\mathbb{R}^d$ given by

$$E_{ij}^{*}(x) = x_i \frac{\partial}{\partial x_j} + x_i x_j \sum_{k=1}^{d} x_k \frac{\partial}{\partial x_k} \quad \text{for } x \in S^{d-1}. \tag{8.44}$$

By the version of Theorem 6.2 for right Lévy processes, the rate vector H^+ is given by (6.13). From this it follows that the Lyapunov exponents of ϕ_t are given by $\lambda_1 = -1, \lambda_2 = -2, \ldots, \lambda_{d-1} = -(d-1)$, which are all simple.

Also by the right process version of Theorem 6.2, $\bar{g}_t$ is left $SO(d)$-invariant; hence, its one-point motion on $S^{d-1} = SO(d)/SO(d-1)$, or equivalently the solution x_t of the stochastic differential equation (8.43), is invariant under the action of $SO(d)$ on S^{d-1}. It is well known that such a diffusion process, up to a time scaling, is a Riemannian Brownian motion in S^{d-1} with respect to the usual metric, that is, the metric on S^{d-1} induced by the standard Euclidean metric on $\mathbb{R}^d$. To show that x_t is in fact a Riemannian Brownian motion in S^{d-1} with respect to this metric, using the $SO(d)$-invariance, it suffices to show that its generator L agrees with $(1/2)\Delta$ at the point $o = e_1$, where Δ is the Laplace–Beltrami operator on S^{d-1}. This can be verified by a direct computation using $L = (1/2)\sum_{i,j=1}^{d} E_{ij}^{*} E_{ij}^{*}$ and (8.44).

We have proved the following result:

Proposition 8.14. *Let ϕ_t be the stochastic flow generated by the stochastic differential equation (8.43) on S^{d-1}. Then it has $d-1$ simple Lyapunov exponents $\lambda_1 = -1, \lambda_2 - 2, \ldots, \lambda_{d-1} = -(d-1)$, and the associated stable manifolds are as described in Proposition 8.13. Moreover, its one-point motion is the Riemannian Brownian motion in S^{d-1} with respect to the usual metric.*

8.6. $SO(1, d)_+$-flow

Let $G = SO(1, d)_+$ be the identity component of the Lorentz group $SO(1, d)$ on $\mathbb{R}^{d+1}$ with standard coordinates $x_0, x_1, \ldots, x_d$ and assume $d \geq 3$. Recall that $K = \text{diag}\{1, SO(d)\}$, which can be identified with the special orthogonal group $SO(d)$ acting on $\mathbb{R}^d$ with coordinates $x_1, \ldots, x_d$, and $U = \text{diag}\{1, 1, SO(d - 1)\}$, which can be identified with $SO(d - 1)$ acting on $\mathbb{R}^{d-1}$ with coordinates $x_2, \ldots, x_d$. We may take $\mathfrak{a}$ to be the one-dimensional subspace spanned by ξ_y, where ξ_y is defined by (5.23) with $y = (1, 0, \ldots, 0)' \in \mathbb{R}^d$, and the Weyl chamber $\mathfrak{a}_+$ to be the set $\{c\xi_y; c > 0\}$. There are only two roots $\pm\alpha$ given by $\alpha(c\xi_y) = c$. Let ϕ_t be a right Lévy process in $G = SO(1, d)_+$ with $\phi_0 = I_{d+1}$ satisfying the hypothesis (H). Its rate vector H^+ is equal to $c_0\xi_y$ for some constant $c_0 > 0$.

Example 4. ($SO(1, d)_+$-flow on $SO(d)$). Let ϕ_t be regarded as a stochastic flow on $K = G/(AN^+)$. With the natural identification of $K = \text{diag}\{1, SO(d)\}$ with $SO(d)$, it may be regarded as a stochastic flow on $SO(d)$. It has two Lyapunov exponents: 0 and $-c_0$.

By Proposition 5.21, $N^- * e$ consists of all the diagonal matrices $\text{diag}(1, R(\sigma, \alpha))$, where $0 \leq \alpha < \pi$, σ is an oriented two-dimensional subspace of $\mathbb{R}^d$ containing the x_1-axis and $R(\sigma, \alpha)$ is the element of $SO(d)$ that rotates σ by an angle α while fixing its orthogonal complement in $\mathbb{R}^d$. By (8.32), the stable manifolds of the negative exponent $-c_0$, up to a left translation by some $\eta(\omega) \in SO(d)$, are given by

$$uN^- * e = N^- * u = (N^- * e)u \quad \text{for } u \in U.$$

By (5.29),

$$\mathfrak{n}^\pm = \left\{ \begin{bmatrix} 0 & 0 & z' \\ 0 & 0 & \pm z' \\ z & \mp z & 0 \end{bmatrix} ; \; z \in \mathbb{R}^{d-1} \right\}.$$

It is easy to show that, if $v = \text{diag}(1, -1, -1, I_{d-2})$, then $v^{-1}\mathfrak{n}^- v = \mathfrak{n}^+$. By (5.28) and Proposition 8.9, the attracting stationary points of the vector field $(H^+)^*$ on $SO(d)$ form the set $U = \text{diag}\{1, 1, SO(d - 1)\}$ and the repelling stationary points form the set $U' - U = \text{diag}\{1, -1, h SO(d - 1)\}$, where $h = \text{diag}(-1, I_{d-2})$. There are no saddle points.

Example 5. ($SO(1, 3)_+$-flow on $SO(3)$). Now assume $d = 3$ in Example 4. Recall that, for $v \in \mathbb{R}^3$, $R(v) \in SO(3)$ denotes the rotation in $\mathbb{R}^3$ around the axis v by the angle $|v|$. Using the ball $B_\pi(0)$ to represent $SO(3)$ as in Example 2

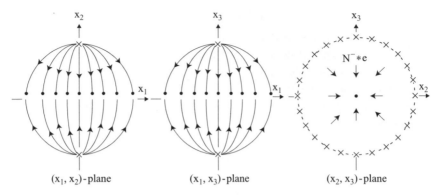

Figure 4. $SO(1, 3)_+$-flow on $SO(3)$ represented by $B_\pi(0)$.

(with antipodal points on its boundary identified), we see that the attracting and repelling stationary points of $(H^+)^*$ are precisely the sets

$$\{R(ae_1); |a| \le \pi\} \quad \text{and} \quad \{R(\pi(ae_2 + be_3)); (a, b) \in \mathbb{R}^2 \text{ with } a^2 + b^2 = 1\}$$

respectively, which are shown in Figure 4 as solid dots and crosses. As in Figures 2 and 3, the arrows show the directions of the vector field $(H^+)^*$.

With the identification of $K = \text{diag}\{1, SO(3)\}$ and $SO(3)$, $N^- * e$ can be identified with the set of the rotations $R(v)$, where v is contained in span(e_2, e_3) with $|v| < \pi$. We see that $N^- * e$ is the intersection of the interior of $B_\pi(0)$ with the subspace $x_1 = 0$.

Up to a left translation by some $\eta(\omega) \in SO(3)$, the stable manifolds of the negative exponent $-c_0$ are given by $N^- * u$ for $u \in U$. Any $u \in U$ is identified with $R(se_1)$ with $|s| \le \pi$. The set $N^- * u = (N^- * e)u$ consists of the rotations $R(v)R(se_1)$, where $v \in \text{Span}(e_2, e_3)$ with $|v| < \pi$. As a surface in $B_\pi(0)$, it is symmetric about the x_1-axis. To determine its shape, by symmetry, we may assume $v = te_2$ with $t \in [0, \pi)$. We have

$$R(te_2)R(se_1) = \begin{bmatrix} \cos t & 0 & \sin t \\ 0 & 1 & 0 \\ -\sin t & 0 & \cos t \end{bmatrix} \begin{bmatrix} 1 & 0 & 0 \\ 0 & \cos s & -\sin s \\ 0 & \sin s & \cos s \end{bmatrix}$$

$$= \begin{bmatrix} \cos t & \sin t \sin s & \sin t \cos s \\ 0 & \cos s & -\sin s \\ -\sin t & \cos t \sin s & \cos t \cos s \end{bmatrix}.$$

Let $R(w) = R(te_2)R(se_1)$. Then the rotation angle $r = |w|$ is given by

$$r = \cos^{-1}\left[\frac{1}{2}(\text{Trace}(w) - 1)\right] = \cos^{-1}\left[\frac{1}{2}(\cos t + \cos s + \cos t \cos s - 1)\right].$$

$$(8.45)$$

To show this, note that $w = hR(re_1)h^{-1}$ for some $h \in SO(3)$ and $\text{Trace}(w) = \text{Trace}[R(re_1)] = 1 + 2\cos r$.

To determine the direction of the rotation axis $w = (x_1, x_2, x_3)$, we can solve the equations

$$\begin{cases} x_1 \cos t & +x_2 \sin t \sin s & +x_3 \sin t \cos s & = x_1, \\ & x_2 \cos s & -x_3 \sin s & = x_2, \\ -x_1 \sin t & +x_2 \cos t \sin s & +x_3 \cos t \cos s & = x_3 \end{cases}$$

to obtain $x_1 = -x_3 \sin t/(1 - \cos t)$ and $x_2 = -x_3 \sin s/(1 - \cos s)$. Therefore, the angle θ between e_1 and w is given by

$$\theta = \tan^{-1} \frac{\sqrt{x_2^2 + x_3^2}}{|x_1|} = \tan^{-1}\left[\frac{\sqrt{2}(1 - \cos t)}{\sin t \sqrt{1 - \cos s}}\right] \quad \text{with } \theta = 0 \text{ at } t = 0.$$

(8.46)

At $t = 0$, $r = s$ and $\theta = 0$, and as $t \to \pi$, $r \to \pi$ and $\theta \to \pi/2$. Let C be the great circle on the boundary of $B_\pi(0)$ through the points $(0, \pi, 0)$ and $(0, 0, \pi)$. This is the set of repelling stationary points of $(H^+)^*$. The preceding computation shows that the set $N^- * u$ as a surface in $B_\pi(0)$ is symmetric about the x_1-axis and intersects the x_1-axis at $u = R(se_1)$. Moreover, the family of the surfaces $N^- * u$, $u \in U$, have the great circle C as the common boundary and form a foliation of the complement of C in $B_\pi(0)$. The left and middle portions of Figure 4 show the cross sections of some of these surfaces in the (x_1, x_2)- and (x_1, x_3)-planes as solid lines connecting two antipodal repelling stationary points on the boundary of $B_\pi(0)$, which are in fact the same point in $SO(3)$. The right portion shows the cross section of $B_\pi(0)$ in the (x_2, x_3)-plane. Its interior is $N^- * e$ and its boundary is C.

Example 6. ($SO(1, d)_+$-flow on S^{d-1}). For $G = SO(1, d)_+$ and $Q = UAN^+$, $G/Q = K/U = SO(d)/SO(d - 1)$ can be identified with the unit sphere S^{d-1} in $\mathbb{R}^d$. The natural action of G on G/Q, or the $*$-action of G on K/U, induces an action of G on S^{d-1}, under which the point $o = (1, 0, \ldots, 0)' \in S^{d-1}$ is fixed by UAN^+. This action can also be defined as follows: For $x \in S^{d-1}$, considered as a column vector, and $g \in G = SO(d, 1)_+$,

$$g(1, x')' = g\begin{pmatrix} 1 \\ x \end{pmatrix} = \begin{pmatrix} r \\ y \end{pmatrix} = (r, y')'$$

for some $y \in \mathbb{R}^d$, regarded as a column vector, and $r \in \mathbb{R}$. Because g preserves the quadratic form $x_0^2 - \sum_{i=1}^d x_i^2$ on $\mathbb{R}^{d+1}$, $r^2 - |y|^2 = 1 - |x|^2 = 0$; hence,

$y/r \in S^{d-1}$. It can be shown that

$$G \times S^{d-1} \ni (g, x) \mapsto g * x = \frac{y}{r} \in S^{d-1}$$

defines an action of G on S^{d-1}. We now show that this action is the same as the ∗-action of G on K/U. Note that any $x \in S^{d-1}$ may be written as $x = Bo$ for some $B \in SO(d)$ and $g(1, x')' = gk(1, o')'$ with $k = \text{diag}(1, B) \in K$. Under the Iwasawa decomposition $G = KAN^+$, $gk = \text{diag}(1, C)\exp(t\xi_o)n^+(z)$ for some $C \in SO(d)$, $t \in \mathbb{R}$, and $z \in \mathbb{R}^{d-1}$, where $\exp(t\xi_o)$ and $n^+(z)$ are given by (5.27) and (5.29). It follows that

$$g\begin{pmatrix} 1 \\ x \end{pmatrix} = \text{diag}(1, C)\exp(t\xi_o)n^+(z)\begin{pmatrix} 1 \\ o \end{pmatrix} = \text{diag}(1, C)\exp(t\xi_o)\begin{pmatrix} 1 \\ o \end{pmatrix}$$

$$= \begin{pmatrix} e^t \\ e^t Co \end{pmatrix}.$$

This proves $g * x = Co$ and shows that the ∗-action of G on S^{d-1} defined here is the same as the ∗-action of G on K/U defined earlier. We may write gx for $g * x$ for $g \in G$ and $x \in S^{d-1}$.

Via this action, a right Lévy process ϕ_t in $G = SO(1, d)_+$ with $g_0 = e$ can be regarded as a stochastic flow on S^{d-1}. By the Bruhat decomposition, $G = SO(1, d)_+$ is the disjoint union of N^-UAN^+ and N^-vUAN^+, where $v = \text{diag}(1, -1, -1, I_{d-2})$. Since $v^{-1}n^-v = n^+$, we see that the complement of $N^-o = N^-Uo$ in S^{d-1} is $\{-o\}$. Moreover, the vector field $(H^+)^*$ has only two stationary points, o and $-o$. The former is attracting and the latter repelling. There is only one Lyapunov exponent $-c_0$ of multiplicity $d - 1$. The associated stable manifold is the complement of a random point in S^{d-1}.

For $1 \leq i \leq d$, let $y = (0, \ldots, 0, 1, 0, \ldots, 0)' \in \mathbb{R}^d$ be the column vector that has 1 at the ith place and 0 elsewhere. A direct computation shows that the vector field $X_i = (\xi_y)^*$ on S^{d-1} induced by ξ_y is the orthogonal projection to S^{d-1} of the coordinate vector field $\partial/\partial x_i$ on $\mathbb{R}^d$ given by

$$X_i(x) = \frac{\partial}{\partial x_i} + x_i \sum_{k=1}^{d} x_k \frac{\partial}{\partial x_k} \quad \text{for } x \in S^{d-1}.$$

Since the rate vector H^+ of ϕ_t is equal to $c_0\xi_o$ for some constant $c_0 > 0$, the induced vector field $(H^+)^*$ on S^{d-1} is the orthogonal projection of the vector field $c_0\partial/\partial x_1$ on $\mathbb{R}^d$ to S^{d-1}. This shows again that $(H^+)^*$ has an attracting stationary point o and a repelling stationary point $-o$. The flow of $(H^+)^*$ sweeps all the other points on S^{d-1} toward o.

Let ϕ_t be the stochastic flow generated by the stochastic differential equation

$$dx_t = \sum_{i=1}^{d} c_i X_i(x_t) \circ dW_t^i \tag{8.47}$$

on S^{d-1}, where c_i are nonzero constants and $W_t = (W_t^1, \ldots, W_t^d)$ is a d-dimensional standard Brownian motion. Then ϕ_t is a right Lévy process in $G = SO(1, d)_+$ and satisfies the hypothesis (H). Therefore, all the preceding discussion holds.

When all $c_i = 1$ in (8.47), by the version of Theorem 6.1 for right Lévy processes and (6.10), it is easy to show $c_0 = (d - 1)/2$. Therefore, the Lyapunov exponent is $-(d - 1)/2$. This value and the limiting behavior of the stochastic flow on S^{d-1} described in this case were obtained in Baxendale [8] by a direct computational method. Also, by the right process version of Theorem 6.1, the process ϕ_t is left $SO(d)$-invariant; hence, its one-point motion x_t in S^{d-1} is $SO(d)$-invariant. This implies that, up to a time scaling, x_t is a Riemannian Brownian motion in S^{d-1} with respect to the usual metric. By a direct computation, one can show that the generator of x_t agrees with $(1/2)\Delta$ at the point $o = (1, 0, \ldots,)'$, where Δ is the Laplace–Beltrami operator on S^{d-1}. It follows that the one-point motion of ϕ_t is a Riemannian Brownian motion in S^{d-1} with respect to the usual metric on S^{d-1}.

8.7. Stochastic Flows on S^3

We now consider some examples of stochastic flows on S^3 obtained by lifting from $SO(3)$. For this purpose, we first review briefly the covering manifolds. Let M be a connected compact manifold and let $\tilde{M}$ be a compact covering manifold of M with covering projection $p: \tilde{M} \to M$ of order k. Then p is a local diffeomorphism in the sense that M is covered by a family of open sets U such that $p^{-1}(U)$ is a disjoint union of k open subsets $\tilde{U}_1, \ldots, \tilde{U}_k$ of $\tilde{M}$ and the restriction $p: \tilde{U}_i \to U$ is a diffeomorphism for each i. A diffeomorphism $h: \tilde{M} \to \tilde{M}$ is called a covering transformation or a deck transformation if $p \circ h = p$. All the deck transformations form a finite group D_p, called the deck group of the covering map p. We will assume the covering is regular; that is, D_p is transitive on $p^{-1}(x)$ for any $x \in M$. Then M may be naturally identified with the orbit space $\tilde{M}/D_p$ (see Boothby [11, chapter III, theorem 9.3]).

Because p is a local diffeomorphism, any vector field X on M induces a unique vector field $\tilde{X}$ on $\tilde{M}$ such that $Dp(\tilde{X}(\tilde{x})) = X(p(\tilde{x}))$ for $\tilde{x} \in \tilde{M}$. Moreover, $Dh(\tilde{X}(\tilde{x})) = \tilde{X}(h(\tilde{x}))$ for $\tilde{x} \in \tilde{M}$ and $h \in D_p$.

In general, a diffeomphism $f: M \to M$ may not be lifted to become a diffeomorphism $\tilde{f}: \tilde{M} \to \tilde{M}$ so that $p \circ \tilde{f} = f \circ p$. However, because a vector field X on M can be lifted to $\tilde{X}$ on $\tilde{M}$, a diffeomorphism f obtained from the flow of a vector field on M can be lifted to a diffeomorphism $\tilde{f}$ on $\tilde{M}$. This is also true for stochastic flows generated by stochastic differential equations. Let $X_0, X_1, \ldots, X_m$ be smooth vector fields on M. Consider the stochastic differential equation

$$dx_t = \sum_{j=1}^{m} X_j(x_t) \circ dW_t^j + X_0(x_t)dt \qquad (8.48)$$

on M, where $W_t = (W_t^1, \ldots, W_t^m)$ is an m-dimensional standard Brownian motion, and the corresponding stochastic differential equation

$$d\tilde{x}_t = \sum_{j=1}^{m} \tilde{X}_j(\tilde{x}_t) \circ dW_t^j + \tilde{X}_0(\tilde{x}_t)dt \qquad (8.49)$$

on $\tilde{M}$.

Let ϕ_t and $\tilde{\phi}_t$ be respectively stochastic flows generated by the stochastic differential equations (8.48) and (8.49). It is easy to show by the standard stochastic calculus that, for any $\tilde{x} \in \tilde{M}$, $x_t = p\tilde{\phi}_t(\tilde{x})$ is a solution of (8.48) with $x_0 = p\tilde{x}$, and, for $h \in D_p$, $\tilde{x}_t = h\tilde{\phi}_t(\tilde{x})$ is a solution of (8.49) with $\tilde{x}_0 = h\tilde{x}$. It follows from the uniqueness of solutions of stochastic differential equations that

$$\forall \tilde{x} \in \tilde{M}, \quad p[\tilde{\phi}_t(\tilde{x})] = \phi_t(p\tilde{x}) \quad \text{and} \quad \forall h \in D_p, \quad h\tilde{\phi}_t(\tilde{x}) = \tilde{\phi}_t(h\tilde{x}). \quad (8.50)$$

The stochastic flow $\tilde{\phi}_t$ on $\tilde{M}$ is said to cover the stochastic flow ϕ_t on M via the covering map p.

Now let $M = G/Q$ be as before and assume that $X_i = Y_i^*$ for Y_0, $Y_1, \ldots, Y_m \in \mathfrak{g}$. Then ϕ_t may be regarded as a right invariant diffusion process in G satisfying the stochastic differential equation

$$d\phi_t = \sum_{j=1}^{r} Y_j\phi_t \circ dW_t^j + Y_0\phi_t dt \qquad (8.51)$$

on G with $\phi_0 = e$. We will assume that the hypothesis (H) is satisfied. It is easy to see that ϕ_t and $\tilde{\phi}_t$ have the same Lyapunov spectrum and, if M' is a stable manifold of some negative Lyapunov exponent λ for ϕ_t, then each connected component of $p^{-1}(M')$ is a stable manifold of λ for $\tilde{\phi}_t$.

Let $\tilde{H}$ be the vector field on $\tilde{M}$ obtained by lifting $(H^+)^*$ on $M = G/Q$. Then the attracting (resp. repelling, saddle) stationary points of $(H^+)^*$ are covered by the attracting (resp. repelling, saddle) stationary points of $\tilde{H}$ via the covering map p.

Example 7. We now continue the discussion in Example 2. A right invariant diffusion process ϕ_t in $G = SL(3, \mathbb{R})$ determined by a stochastic differential equation of the form (8.51) may be regarded as a stochastic flow on $SO(3)$. The four elements $u_0 = e, u_1 = R(\pi e_1), u_2 = R(\pi e_2)$, and $u_3 = R(\pi e_3)$ in U correspond to the identity map and $180°$ rotations around the three coordinate axes e_1, e_2, and e_3 in $\mathbb{R}^3$, respectively.

It is well known that the three-dimensional sphere S^3 is a regular covering space of $SO(3)$ of order two. By the preceding discussion, ϕ_t induces a stochastic flow $\tilde{\phi}_t$ on S^3 that covers ϕ_t. The stable manifolds of $\tilde{\phi}_t$ and the stationary points of the vector field $\tilde{H}$ on S^3 can be lifted from those of ϕ_t and of $(H^+)^*$ on $SO(3)$ via the regular covering map $p : S^3 \to SO(3)$ to be described as follows.

Regard S^3 as the unit sphere embedded in $\mathbb{R}^4$ and regard $\mathbb{R}^4$ as the quaternion algebra, which is a vector space with basis $\{1, I, J, K\}$ satisfying the product rule:

$$I^2 = J^2 = K^2 = -1, \quad IJ = -JI = K, \quad JK = -KJ = I, \quad \text{and}$$
$$KI = -IK = J.$$

The Euclidean norm is compatible with the quaternion product in the sense that $|xy| = |x| \, |y|$ for $x, y \in \mathbb{R}^4$. Any nonzero $x = (a + bI + cJ + dK)$ has an inverse $x^{-1} = (a - bI - cJ - dK)/|x|^2$. Regard $\mathbb{R}^3$ as the subspace of $\mathbb{R}^4$ spanned by $\{I, J, K\}$, the space of pure quaternions. Then, for any nonzero $x \in \mathbb{R}^4$, the map $\mathbb{R}^4 \ni y \mapsto xyx^{-1} \in \mathbb{R}^4$ leaves both $|\cdot|$ and $\mathbb{R}^3$ invariant; hence, it is an element of $SO(3)$. The covering map is defined by

$$p : S^3 \to SO(3) \text{ given by } p(x)y = xyx^{-1} \text{ for } y \in \mathbb{R}^3. \tag{8.52}$$

It is easy to see that the four points u_0, u_1, u_2, and u_3 in U are covered by ± 1, $\pm I, \pm J$, and $\pm K$, respectively, in quaternion notation. Written in the usual Cartesian coordinates, these eight points are $(\pm 1, 0, 0, 0), (0, \pm 1, 0, 0,), (0, 0, \pm 1, 0)$, and $(0, 0, 0, \pm 1)$.

A direct computation shows that

$$p[(\cos t) + (\sin t)I] = R(2tI).$$

More generally, for $H = x_1 I + x_2 J + x_3 K$ with $x_1^2 + x_2^2 + x_3^2 = 1$,

$$p[(\cos t) + (\sin t)H] = R(2tH). \tag{8.53}$$

To prove this, note that S^3 is a Lie group via the quaternion structure and the covering map $p: S^3 \to SO(3)$ is also a Lie group homomorphism. There is $y \in S^3$ such that $y I y^{-1} = H$. Then

$$
\begin{aligned}
p[(\cos t) + (\sin t)H] &= p\{y[(\cos t) + (\sin t)I]y^{-1}\} \\
&= p(y)p[(\cos t) + (\sin t)I]p(y)^{-1} \\
&= p(y)R(2tI)p(y)^{-1} = R(2tH).
\end{aligned}
$$

Recall that $SO(3)$ can be represented by the ball $B_\pi(0)$ in $\mathbb{R}^3$ when the antipodal points on the boundary of $B_\pi(0)$ are identified. The stable manifolds of the top Lyapunov exponent λ_1 of the stochastic flow ϕ_t on $SO(3)$ are given, up to a left translation by some $SO(3)$-valued random variable η, by $N^- * u_0$, $N^- * u_1$, $N^- * u_2$ and $N^- * u_3$, which are respectively the regions A, B, C and D depicted in Figure 2. The stable manifolds of the smallest exponent, which is either λ_2 or λ_3, are given, up to a left translation by η, by $nN_s * u_i$ for $n \in N^-$ and $i = 0, 1, 2, 3$, where N_s is a one-dimensional Lie subgroup of N^-.

We now present a graphic representation of S^3 to illustrate the stochastic flow $\tilde{\phi}_t$ on S^3. Consider S^3 as the unit sphere in $\mathbb{R}^4$, with the latter equipped with the standard coordinates x_0, x_1, x_2, x_3 and regarded as the quaternion algebra. The two closed hemispheres on S^3, determined by $x_0 \geq 0$ and $x_0 \leq 0$ respectively, can be represented by two copies of the unit ball $B_1(0)$ in $\mathbb{R}^3$ with coordinates x_1, x_2, x_3. Figure 5 depicts this representation. The upper half of the figure displays the cross-sections of the first copy of $B_1(0)$ with the three coordinate planes, whereas the lower half displays those of the second copy. The boundary points of the first copy of $B_1(0)$ are identified with the corresponding points of the second copy.

Using (8.53), it is easy to show that

$$
\forall \tilde{x} = (x_0 + x_1 I + x_2 J + x_3 K) \in S^3, \quad p(\tilde{x}) = R(2tv), \tag{8.54}
$$

where

$$
v = \operatorname{sgn}(x_0) \frac{x_1 I + x_2 J + x_3 K}{\sqrt{x_1^2 + x_2^2 + x_3^2}} \quad \text{and}
$$

$$
\sin t = \sqrt{x_1^2 + x_2^2 + x_3^2} \quad \text{for} \quad t \in [0, \pi/2].
$$

When $x_0 = 0$, the sign $\operatorname{sgn}(x_0)$ of x_0 can be taken to be either $+$ or $-$ because $R(\pi v) = R(-\pi v)$.

A direct computation shows that the I-component of $p(\tilde{x})I$ is $x_0^2 + x_1^2 - x_2^2 - x_3^2$ and the K-component of $p(\tilde{x})K$ is $x_0^2 + x_3^2 - x_1^2 - x_2^2$. By (8.38), we see that the set $p^{-1}(N^- * u_0) = p^{-1}(N^- * e)$ is determined by the

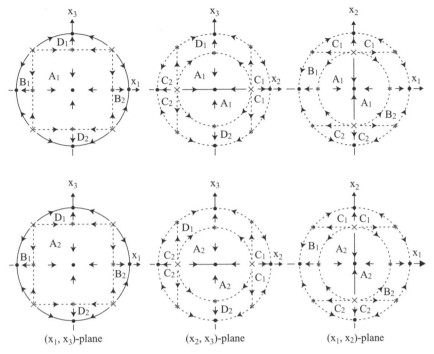

Figure 5. Induced flow on S^3 represented by two copies of $B_1(0)$.

inequalities $x_0^2 + x_1^2 > x_2^2 + x_3^2$ and $x_0^2 + x_3^2 > x_1^2 + x_2^2$. Using $\sum_{i=0}^3 x_i^2 = 1$, we see that they become $x_2^2 + x_3^2 < \frac{1}{2}$ and $x_1^2 + x_2^2 < \frac{1}{2}$. Therefore,

$$p^{-1}(N^- * e) = \left\{ x_0 + x_1 I + x_2 J + x_3 K \in S^3; \right.$$
$$\left. x_2^2 + x_3^2 < \frac{1}{2} \text{ and } x_1^2 + x_2^2 < \frac{1}{2} \right\}. \qquad (8.55)$$

When the two inequalities $<$ and $<$ in this expression are replaced by $<$ and $>$, $>$ and $>$, and $>$ and $<$, we obtain the corresponding expressions for the sets $p^{-1}(N^- * u_1)$, $p^{-1}(N^- * u_2)$, and $p^{-1}(N^- * u_3)$, respectively.

The inequality $x_2^2 + x_3^2 < \frac{1}{2}$ determines a cylindrical region inside each copy of $B_1(0)$ around the x_1-axis and $x_1^2 + x_2^2 < \frac{1}{2}$ determines such a region around the x_3-axis. The set $p^{-1}(N^- * u_0)$ is the intersection of these two regions and has two connected components A_1 and A_2 as depicted in Figure 5. Similarly, the sets $p^{-1}(N^- * u_1)$, $p^{-1}(N^- * u_2)$ and $p^{-1}(N^- * u_3)$ are represented by the regions B_1 and B_2, C_1 and C_2, and D_1 and D_2 in Figure 5, respectively. These sets are, up to a left translation by some random $\tilde{\eta} \in S^3$ (which may be regarded as a random isometric transformation on S^3), the stable manifolds of the top Lyapunov exponent λ_1 of $\tilde{\phi}_t$. Each of these sets is foliated by one-dimensional stable manifolds of the smallest exponent, but only one of them is shown in Figure 5 as a solid line segment or a solid arc. The directions and stationary points of the vector field $\tilde{H}$ on S^3 are also shown. As before, the attracting, repelling and saddle points are shown respectively as solid dots, crosses and asterisks.

Example 8. Continue the discussion in Example 5. Let $d = 3$. The right invariant diffusion process ϕ_t in $SO(1, 3)_+$ determined by (8.51) is regarded as a stochastic flow on $SO(3)$. Via the regular covering map $p: S^3 \to SO(3)$ described in Example 7, it induces a stochastic flow $\tilde{\phi}_t$ on S^3 that covers ϕ_t.

Recall that the set of the attracting stationary points of the vector field $(H^+)^*$ on $SO(3)$ is $U = \{R(te_1); t \in [0, 2\pi)\}$, the set of the repelling stationary points of $(H^+)^*$ is the set $U' - U = \{R(be_2 + ce_3); \sqrt{b^2 + c^2} = \pi\}$, and there is no saddle point. Here, for simplicity, diag$\{1, SO(3)\}$ is identified with $SO(3)$. As in Example 7, regard S^3 as the unit sphere in $\mathbb{R}^4$ and regard $\mathbb{R}^4$ as the quaternion algebra with basis $\{1, I, J, K\}$. Note that $e_1 = I$, $e_2 = J$, and $e_3 = K$. Recall the definitions of great circles and great half circles given in Example 3 in Section 8.5. Using (8.54), it is easy to show that $p^{-1}(U)$, the set of attracting stationary points of $\tilde{H}$ on S^3, is the one-dimensional great circle on S^3 determined by the equations $x_2 = x_3 = 0$ and that $p^{-1}(U' - U)$, the set of repelling stationary points of $\tilde{H}$, is the one-dimensional great circle on S^3 determined by $x_0 = x_1 = 0$.

Recall that, up to a left translation by some $\eta(\omega) \in SO(d)$, the stable manifolds of the negative Lyapunov exponent $-c_0$ of ϕ_t on $SO(3)$ are the sets $N^- * u, u \in U$. Let

$$M_t = \{R(be_2 + ce_3)R(2te_1); \quad \sqrt{b^2 + c^2} < \pi\}$$

for $t \in [-\pi/2, \pi/2]$. Note that $M_t = (N^- * e)u = N^- * u$ for $u = R(2te_1)$. The inverse images of these sets under the covering map p are, up to a left translation by some $\tilde{\eta}(\omega) \in S^3$, the stable manifolds of the exponent $-c_0$ of $\tilde{\phi}_t$ on S^3. Because

$$(a + bJ + cK)(\cos t + \sin t I)$$
$$= a \cos t + a \sin t I + (b \cos t + c \sin t)J + (-b \sin t + c \cos t)K,$$

it is easy to show using (8.54) that, for $t \in (-\pi/2, \pi/2)$, the set M_t is covered by

$$C_t^+ = \{\tilde{x} \in S^3; \quad x_0 \sin t = x_1 \cos t \text{ and } x_0 > 0\}$$

and

$$C_t^- = \{\tilde{x} \in S^3;\ x_0 \sin t = x_1 \cos t \text{ and } x_0 < 0\},$$

where $\tilde{x} = x_0 + x_1 I + x_2 J + x_3 K$. Note that $M_{\pi/2} = M_{-\pi/2}$ is covered by

$$C_{\pi/2}^+ = \{\tilde{x} \in S^3;\ x_0 = 0 \text{ and } x_1 > 0\} \quad \text{and}$$
$$C_{\pi/2}^- = \{\tilde{x} \in S^3;\ x_0 = 0 \text{ and } x_1 < 0\}.$$

For each $t \in [-\pi/2,\ \pi/2]$, the sets C_t^+ and C_t^+ are two disjoint two-dimensional great half circles that are contained in the same two-dimensional great circle

$$C_t = \{\tilde{x} \in S^3;\ x_0 \sin t = x_1 \cos t\}.$$

The common boundary of C_t^+ and C_t^- is the one-dimensional great circle on S^3 determined by the equations $x_0 = x_1 = 0$, that is, the set of repelling stationary points of $\tilde{H}$. Note that each $C_t^\pm$ contains exactly one attracting stationary point $\pm(\cos t + \sin t\ I)$ of $\tilde{H}$.

Appendix A
Lie Groups

Basic definitions and facts about Lie groups are recalled in this appendix. The reader is referred to the standard textbooks such as Warner [60] or the first two chapters of Helgason [26] for more details.

A.1. Lie Groups

A Lie group G is a group and a manifold such that both the product map $G \times G \ni (g, h) \mapsto gh \in G$ and the inverse map $G \ni g \mapsto g^{-1} \in G$ are smooth. Let e be the identity element of G, the unique element of G satisfying $eg = ge = g$ for any $g \in G$. It is well known that a Lie group is in fact analytic in the sense that the underlying manifold structure together with the product and inverse maps are analytic.

The Lie algebra $\mathfrak{g}$ of a Lie group G is the tangent space $T_e G$ of G at the identity element e of G. For $g \in G$, let L_g and $R_g \colon G \to G$ be, respectively, the left and right translations on G. A vector field $\tilde{X}$ on G is called left invariant if it is L_g-related to itself, that is, if $DL_g(\tilde{X}) = \tilde{X}$, for any $g \in G$, where DL_g is the differential of L_g. Similarly, we define a right invariant vector field on G using R_g. Any left or right invariant vector field $\tilde{X}$ is determined by its value at e, $X \in \mathfrak{g}$, and may be written as $\tilde{X}(g) = DL_g(X)$ or $\tilde{X}(g) = DR_g(X)$ for $g \in G$. Such a left or right invariant vector field will be denoted by X^l or X^r, respectively. The Lie bracket of two vector fields Z_1 and Z_2 on a manifold is defined to be the vector field $[Z_1, Z_2] = Z_1 Z_2 - Z_2 Z_1$. For any two elements X and Y of $\mathfrak{g}$, $[X^l, Y^l]$ is left invariant. The Lie bracket $[X, Y]$ is defined to be the element of $\mathfrak{g}$ corresponding to this left invariant vector field; that is, $[X, Y]^l = [X^l, Y^l]$. It satisfies the following properties: For any $X, Y, Z \in \mathfrak{g}$ and $a, b \in \mathbb{R}$,

(a) $[aX + bY, Z] = a[X, Z] + b[Y, Z]$,
(b) $[X, Y] = -[Y, X]$, and
(c) $[X, [Y, Z]] + [Y, [Z, X]] + [Z, [X, Y]] = 0$.

The identity (c) is called the Jacobi identity.

In general, any linear space $\mathfrak{g}$ equipped with a Lie bracket satisfying (a), (b), and (c) is called a Lie algebra. For example, the space of all the vector fields on a manifold is a Lie algebra with Lie bracket of vector fields.

The exponential map exp: $\mathfrak{g} \to G$ is defined by $\exp(X) = \psi(1)$, where $\psi(t)$ is the solution of the ordinary differential equation $(d/dt)\psi(t) = X^l(\psi(t))$ satisfying the initial condition $\psi(0) = e$, and is a diffeomorphism of an open neighborhood of 0 in $\mathfrak{g}$ onto an open neighborhood of e in G. We may write e^X for $\exp(X)$.

A group homomorphism $F: G \to G'$ between two Lie groups G and G' is called a Lie group homomorphism (resp. isomorphism) if it is a smooth (resp. diffeomorphic) map. A linear map $f: \mathfrak{g} \to \mathfrak{g}'$ between two Lie algebras $\mathfrak{g}$ and $\mathfrak{g}'$ is called a Lie algebra homomorphism if $[f(X), f(Y)] = f([X, Y])$ for $X, Y \in \mathfrak{g}$. Such an f will be called a Lie algebra isomorphism if it is a bijection. Let $\mathfrak{g}$ and $\mathfrak{g}'$ be, respectively, the Lie algebras of G and G'. If $F: G \to G'$ is a Lie group homomorphim (resp. isomorphism), then $DF: \mathfrak{g} \to \mathfrak{g}'$ is a Lie algebra homomorphism (resp. isomorphism) and $F(e^X) = e^{DF(X)}$ for any $X \in \mathfrak{g}$. Two Lie groups (resp. Lie algebras) are called isomorphic if there is a Lie group (resp. Lie algebra) isomorphism between them. A Lie group (resp. Lie algebra) isomorphism from G (resp. $\mathfrak{g}$) onto itself will be called a Lie group (resp. Lie algebra) automorphism on G (resp. $\mathfrak{g}$).

For $g \in G$, the map $c_g: G \ni h \mapsto ghg^{-1} \in G$ is a Lie group automorphism on G, called a conjugation map on G. Its differential map $\mathrm{Ad}(g): \mathfrak{g} \to \mathfrak{g}$, given by $\mathrm{Ad}(g)X = DL_g \circ DR_{g^{-1}}(X)$, is a Lie algebra automorphism on $\mathfrak{g}$ and satisfies $ge^X g^{-1} = e^{\mathrm{Ad}(g)X}$ for any $X \in \mathfrak{g}$. It can be shown that $[X, Y] = (d/dt)\mathrm{Ad}(e^{tX})Y \mid_{t=0}$ for $Y \in \mathfrak{g}$.

A Lie group G is called abelian if any two elements of G commute, that is, if $gh = hg$ for $g, h \in G$. A Lie algebra $\mathfrak{g}$ is called abelian if $[X, Y] = 0$ for $X, Y \in \mathfrak{g}$. It is easy to show that the Lie algebra of an abelian Lie group is abelian and that a connected Lie group with an abelian Lie algebra is abelian.

A subset H of G is called a Lie subgroup of G if it is a subgroup and a submanifold of G such that both the product and inverse maps are smooth under the submanifold structure. It is well known that if H is a closed subgroup of G, then there is a unique manifold structure on H under which H is a Lie subgroup and a topological subspace of G. It is easy to show that the identity component of G, that is, the connected component of G containing e, is an open and closed Lie subgroup of G.

For any two subsets $\mathfrak{h}$ and $\mathfrak{k}$ of a Lie algebra $\mathfrak{g}$, let $[\mathfrak{h}, \mathfrak{k}]$ denote the subspace of $\mathfrak{g}$ spanned by $[X, Y]$ for $X \in \mathfrak{h}$ and $Y \in \mathfrak{k}$. A linear subspace $\mathfrak{h}$ of a Lie algebra $\mathfrak{g}$ is called a Lie subalgebra, or simply a subalgebra, of $\mathfrak{g}$, if $[\mathfrak{h}, \mathfrak{h}] \subset \mathfrak{h}$. If G is a Lie group with Lie algebra $\mathfrak{g}$, then the Lie algebra $\mathfrak{h}$ of a Lie subgroup

H of G is a subalgebra of $\mathfrak{g}$. Conversely, given any subalgebra $\mathfrak{h}$ of $\mathfrak{g}$, there is a unique connected Lie subgroup H whose Lie algebra is $\mathfrak{h}$, called the Lie subgroup generated by $\mathfrak{h}$. A subalgebra $\mathfrak{h}$ of $\mathfrak{g}$ is called an ideal if $[\mathfrak{g}, \mathfrak{h}] \subset \mathfrak{h}$. In this case, the Lie subgroup H generated by $\mathfrak{h}$ is a normal subgroup of G; that is, $gHg^{-1} = H$ for $g \in G$. For $\Gamma \subset \mathfrak{g}$, the Lie algebra generated by Γ is defined to be the smallest Lie subalgebra of $\mathfrak{g}$ containing Γ and is denoted by $\mathrm{Lie}(\Gamma)$.

The center of G is $Z = \{h \in G; hg = gh$ for any $g \in G\}$. This is a closed normal subgroup of G with Lie algebra $\mathfrak{z} = \{X \in \mathfrak{g}; [X, Y] = 0$ for any $Y \in \mathfrak{g}\}$, which is called the center of $\mathfrak{g}$.

Let G_1 and G_2 be two Lie groups. Then $G = G_1 \times G_2$ is a Lie group with the product group and product manifold structures, called the product Lie group of G_1 and G_2. Its Lie algebra is isomorphic to $\mathfrak{g} = \mathfrak{g}_1 \oplus \mathfrak{g}_2$, where $\mathfrak{g}_i$ is the Lie algebra of G_i for $i = 1, 2$ with $[\mathfrak{g}_1 \oplus \{0\}, \{0\} \oplus \mathfrak{g}_2] = \{0\}$. Therefore, each $\mathfrak{g}_i$ can be regarded as an ideal of $\mathfrak{g}$ and each G_i as a normal subgroup of G. Similarly, we define the product of more than two Lie groups.

A Riemannian metric $\{\langle \cdot, \cdot \rangle_g; g \in G\}$ on G is called left invariant if $\langle DL_g(X), DL_g(Y) \rangle_{gh} = \langle X, Y \rangle_h$ for any $g, h \in G$ and $X, Y \in T_h G$. A right invariant metric is defined by replacing DL_g by DR_g. If $\langle \cdot, \cdot \rangle$ is an inner product on $\mathfrak{g}$, then it determines a unique left (resp. right) invariant Riemannian metric $\langle \cdot, \cdot \rangle_g^l$ (resp. $\langle \cdot, \cdot \rangle_g^r$), defined by

$$\langle DL_g(X), DL_g(Y) \rangle_g^l = \langle X, Y \rangle \quad (\text{resp. } \langle DR_g(X), DR_g(Y) \rangle_g^r = \langle X, Y \rangle)$$

for $X, Y \in \mathfrak{g}$ and $g \in G$. Under this metric, L_g (resp. R_g) is an isometry on G for any $g \in G$.

As an example of Lie group, let $GL(d, \mathbb{R})$ be the group of all the $d \times d$ real invertible matrices. Since such a matrix represents a linear automorphism on $\mathbb{R}^d$, $GL(d, \mathbb{R})$ is called the general linear group on $\mathbb{R}^d$. With the matrix multiplication and inverse, and with the natural identification of $GL(d, \mathbb{R})$ with an open subset of the d^2-dimensional Euclidean space $\mathbb{R}^{d^2}$, $GL(d, \mathbb{R})$ is a Lie group of dimension d^2. The identity element e is the $d \times d$ identity matrix I_d. The Lie algebra of $GL(d, \mathbb{R})$ is the space $\mathfrak{gl}(d, \mathbb{R})$ of all the $d \times d$ real matrices. The exponential map is given by the usual matrix exponentiation $e^X = I + \sum_{k=1}^{\infty} X^k / k!$ and the Lie bracket is given by $[X, Y] = XY - YX$ for $X, Y \in \mathfrak{gl}(d, \mathbb{R})$. Note that $GL(d, \mathbb{R})$ has two connected components and its identity component $GL(d, \mathbb{R})_+$ is the group of all $g \in GL(d, \mathbb{R})$ with positive determinant $\det(g)$. We have

$$\forall X \in \mathfrak{gl}(d, \mathbb{R}), \quad \det(e^X) = \exp[\mathrm{Trace}(X)],$$

where $\mathrm{Trace}(X) = \sum_i X_{ii}$ is the trace of X.

Similarly, the group $GL(d, \mathbb{C})$ of $d \times d$ complex invertible matrices is a Lie group, called the complex general linear group on $\mathbb{C}^d$. Its Lie algebra $\mathfrak{gl}(d, \mathbb{C})$ is the space of all the $d \times d$ complex matrices with Lie bracket $[X, Y] = XY - YX$.

Let V be a d-dimensional linear space and let $GL(V)$ denote the set of the linear bijections: $V \to V$. By choosing a basis of V, $GL(V)$ may be identified with the general linear group $GL(d, \mathbb{R})$. Therefore, $GL(V)$ is a Lie group and its Lie algebra is the space $\mathfrak{gl}(V)$ of all the linear endomorphsims on V with Lie bracket $[X, Y] = XY - YX$.

A.2. Action of Lie Groups and Homogeneous Spaces

A left action of a group G on a set S is a map $F: G \to S$ satisfying $F(gh, x) = F(g, F(h, x))$ and $F(e, x) = x$ for $g, h \in G$ and $x \in S$. When G is a Lie group, M is a manifold, and F is smooth, we say G acts smoothly on M on the left by F and we often omit the word "smoothly." A right action of G on M is defined similarly with $F(gh, x) = F(h, F(g, x))$. For simplicity, we may write gx for the left action $F(g, x)$ and xg for the right action. Because left actions are encountered more often, therefore, an action in this book will mean a left action unless explicitly stated otherwise.

The subset $Gx = \{gx; g \in G\}$ of M is called an orbit of G on M. It is clear that, if $y \in Gx$, then $Gx = Gy$. The action of G on M will be called effective if $F(g, \cdot) = \mathrm{id}_M \implies g = e$, and it will be called transitive if any orbit of G is equal to M. If G acts effectively on M, it will be called a Lie transformation group on M. By Helgason [26, chapter II, proposition 4.3], if G is transitive on a connected manifold M, then its identity component is also transitive on M.

Any $X \in \mathfrak{g}$ induces a vector field X^* on M given by $X^* f(x) = (d/dt) f(e^{tX} x) \mid_{t=0}$ for any $f \in C^1(M)$ and $x \in M$. We have (see in [26, chapter II, theorem 3.4]) $[X, Y]^* = -[X^*, Y^*]$ for $X, Y \in \mathfrak{g}$. The Lie algebra $\mathfrak{g}^* = \{X^*; X \in \mathfrak{g}\}$ is finite dimensional. A vector field Y on M is called complete if any solution of the equation $(d/dt)x(t) = Y(x(t))$ extends to all time t. It is clear that X^* is complete. In general, if Γ is a collection of complete vector fields on M such that $\mathrm{Lie}(\Gamma)$ is finite dimensional, then there is a Lie group G acting effectively on M with Lie algebra $\mathfrak{g}$ such that $\mathrm{Lie}(\Gamma) = \mathfrak{g}^*$ (see [Palais 48, chapter IV, theorem III]). Note that any vector field on a compact manifold is complete.

Let H be a closed subgroup of G. The set of left cosets gH for $g \in G$ is denoted by G/H and is called a homogeneous space of G. It is equipped with the quotient topology, the smallest topology under which the natural projection $G \to G/H$ given by $g \mapsto gH$ is continuous. By [26, chapter II, theorem 4.2], there is a unique manifold structure on G/H under which the natural action of G on G/H, defined by $g'H \mapsto gg'H$, is smooth. Similarly, we may consider the right coset space $H \backslash G = \{Hg; g \in G\}$ on which G acts naturally via the right action $Hg' \mapsto Hg'g$.

Suppose a Lie group G acts on a manifold M. Fix $p \in M$. Let $H = \{g \in G; gp = p\}$. Then H is a closed subgroup of G, called the isotropy subgroup of G at p. By theorem 3.2 and proposition 4.3 in [26, chapter II] if the action of G on M is transitive, then the map $gH \mapsto gp$ is a diffeomorphism from G/H onto M; therefore, M may be identified with G/H.

Let G be a Lie group and H be a closed subgroup. If N is a Lie subgroup of G, then for any $x = gH \in G/H$, the orbit Nx may be naturally identified with $N/(N \cap H)$ via the map $nx \mapsto n(N \cap H)$ and, hence, it is equipped with the manifold structure of a homogeneous space. By [26, chapter II, proposition 4.4], Nx is in fact a submanifold of G/H. Moreover, if H is compact and N is closed, then Nx is a closed topological submanifold of G/H.

Let G be a Lie group with Lie algebra $\mathfrak{g}$ and let H be a closed subgroup with Lie algebra $\mathfrak{h}$. Via the natural action of G on G/H, any $g \in G$ is a diffeomorphism: $G/H \to G/H$. Let $\pi: G \to G/H$ be the natural projection. Since $h \in H$ fixes the point $\pi(e) = eH$ in G/H, Dh is a linear bijection $T_{eH}(G/H) \to T_{eH}(G/H)$ and

$$Dh \circ D\pi = D\pi \circ \mathrm{Ad}(h) : \mathfrak{g} \to T_{eH}(G/H).$$

The homogeneous space G/H is called reductive if there is a subspace $\mathfrak{p}$ of $\mathfrak{g}$ such that $\mathfrak{g} = \mathfrak{h} \oplus \mathfrak{p}$ (direct sum) and $\mathfrak{p}$ is $\mathrm{Ad}(H)$-invariant in the sense that $\mathrm{Ad}(h)\mathfrak{p} \subset \mathfrak{p}$ for any $h \in H$. Let $\langle \cdot, \cdot \rangle$ be an inner product on $\mathfrak{p}$ that is $\mathrm{Ad}(H)$-invariant in the sense that $\langle \mathrm{Ad}(h)X, \mathrm{Ad}(h)Y \rangle = \langle X, Y \rangle$ for $h \in H$ and $X, Y \in \mathfrak{p}$. Then it induces a Riemannian metric $\{\langle \cdot, \cdot \rangle_x; x \in G/H\}$ on G/H, given by

$$\forall g \in G \text{ and } X, Y \in \mathfrak{p}, \quad \langle Dg \circ D\pi(X), Dg \circ D\pi(Y) \rangle_{gH} = \langle X, Y \rangle.$$

By the $\mathrm{Ad}(H)$-invariance of $\langle \cdot, \cdot \rangle$, this metric is well defined and is G-invariant in the sense that

$$\forall g, u \in G \quad \text{and} \quad X, Y \in T_{uH}(G/H), \quad \langle Dg(X), Dg(Y) \rangle_{guH} = \langle X, Y \rangle_{uH}.$$

Appendix B

Stochastic Analysis

This appendix contains a brief summary of stochastic processes and stochastic analysis, including Markov processes, Feller processes, Brownian motion, martingales, Itô and Stratonovich stochastic integrals, stochastic differential equations, and Poisson random measures. The reader is referred to standard textbooks such as those by Kallenberg [34] or Revuz and Yor [52] for more details.

B.1. Stochastic Processes

Let $(\Omega, \mathcal{F}, P)$ be the probability space on which all the stochastic processes will be defined. A random variable z taking values in a measurable space $(S, \mathcal{S})$ by definition is a measurable map $z \colon \Omega \to S$. Its distribution is the probability measure $zP = P \circ z^{-1}$ on $(S, \mathcal{S})$. A stochastic process or simply a process x_t, $t \in \mathbb{R}_+ = [0, \infty)$, taking values in S, is a family of such random variables and may be regarded as a measurable map $x \colon \Omega \to S^{\mathbb{R}_+}$, where $S^{\mathbb{R}_+}$ is equipped with the product σ-algebra $\mathcal{S}^{\mathbb{R}_+}$. The probability measure xP on $(S^{\mathbb{R}_+}, \mathcal{S}^{\mathbb{R}_+})$ is called the distribution of the process. The process x_t may also be written as $x_t(\omega)$ or $x(t, \omega)$ to indicate its dependence on $\omega \in \Omega$. We normally identify two random variables that are equal P-almost surely and two processes that are equal as functions of time t almost surely. A version of a process x_t is another process y_t such that $x_t = y_t$ almost surely for each $t \in \mathbb{R}_+$.

The expectation $E(X)$ of a real-valued random variable X is just the integral $P(X) = \int X dP$ whenever it is defined. We may write $E[X; B] = \int_B X dP$ for $B \in \mathcal{F}$. When $E(|X|) < \infty$ or when $X \geq 0$ P-almost surely, the conditional expectation of X given a σ-algebra $\mathcal{G} \subset \mathcal{F}$, denoted by $E[X \mid \mathcal{G}]$, is defined to be the P-almost surely unique $\mathcal{G}$-measurable random variable Y such that $\int_A Y dP = \int_A X dP$ for any $A \in \mathcal{G}$.

A filtration on $(\Omega, \mathcal{F}, P)$ is a family of σ-algebras $\mathcal{F}_t$ on Ω, $t \in \mathbb{R}_+$, such that $\mathcal{F}_s \subset \mathcal{F}_t \subset \mathcal{F}$ for $0 \leq s \leq t < \infty$. A probability space equipped with a filtration is called a filtered probability space and may be denoted by $(\Omega, \mathcal{F}, \{\mathcal{F}_t\}, P)$. The filtration is called right continuous if $\mathcal{F}_{t+} = \bigcap_{u>t} \mathcal{F}_u$ is equal to $\mathcal{F}_t$. It is called complete if $(\Omega, \mathcal{F}_\infty, P)$ is a complete probability space and if $\mathcal{F}_t$ contains all the P-null sets in $\mathcal{F}_\infty = \sigma(\bigcup_{t \geq 0} \mathcal{F}_t)$, where $\sigma(\mathcal{C})$ for a collection $\mathcal{C}$ of sets or functions denotes the smallest σ-algebra under which all the elements in $\mathcal{C}$ are measurable, called the σ-algebra generated by $\mathcal{C}$. The completion of a filtration $\{\mathcal{F}_t\}$ is the complete filtration $\{\bar{\mathcal{F}}_t\}$, where $\bar{\mathcal{F}}_t = \sigma\{\mathcal{F}_t, \mathcal{N}\}$ and $\mathcal{N}$ is the collection of all the P-null sets in $\mathcal{F}_\infty$.

A filtration $\{\mathcal{F}_t\}$ is said to be generated by a collection of processes x_t^λ, $\lambda \in \Lambda$, and random variables z^α, $\alpha \in A$, if $\mathcal{F}_t = \sigma\{z^\alpha \text{ and } x_u^\lambda; \alpha \in A, \lambda \in \Lambda, \text{ and } 0 \leq u \leq t\}$. The filtration generated by a single process is called the natural filtration of that process.

A $\{\mathcal{F}_t\}$-stopping time τ is a random variable taking values in $\overline{\mathbb{R}_+} = [0, \infty]$ such that the set $[\tau \leq t]$ belongs to $\mathcal{F}_t$ for any $t \in \mathbb{R}_+$. For a stopping time τ,

$$\mathcal{F}_\tau = \{B \in \mathcal{F}_\infty; \qquad B \cap [\tau \leq u] \in \mathcal{F}_u \text{ for any } u \in \mathbb{R}_+\}$$

is a σ-algebra, which is equal to $\mathcal{F}_t$ if $\tau = t$. If $\tau \leq \sigma$ are two stopping times, then $\mathcal{F}_\tau \subset \mathcal{F}_\sigma$.

A process x_t is said to be adapted to a filtration $\{\mathcal{F}_t\}$ if x_t is $\mathcal{F}_t$-measurable for each $t \in \mathbb{R}_+$. It is always adapted to its natural filtration $\{\mathcal{F}_t^0\}$ given by $\mathcal{F}_t^0 = \sigma\{x_s; 0 \leq s \leq t\}$.

A process x_t will be called continuous, right continuous, etc. if for P-almost all ω, the path $t \mapsto x_t(\omega)$ is such a function. A process is called càdlàg (continu à droite, limites à gauche) if almost all its paths are right continuous with left limits. It is called càglàd if almost all its paths are left continuous with right limits.

A probability kernel on $(S, \mathcal{S})$ is a map $K: S \times \mathcal{S} \to \mathbb{R}_+$ such that $K(x, \cdot)$ is a probability measure for each fixed $x \in S$ and $x \mapsto K(x, B)$ is $\mathcal{S}$-measurable for each fixed $B \in \mathcal{S}$. For $f \in \mathcal{S}_+$ (the space of nonnegative $\mathcal{S}$-measurable functions), we may write $Kf(x) = \int f(y)K(x, dy)$ and thus may regard K as an operator on a function space. Let $\{P_t; t \in \mathbb{R}_+\}$ be a semigroup of probability kernels on $(S, \mathcal{S})$; that is, $P_{s+t} = P_t P_s$ and $P_0(x, \cdot) = \delta_x$ (unit mass at point x). A process x_t taking values in S will be called a Markov process with transition semigroup P_t if the following Markov property holds:

$$\forall t > s \text{ and } \forall f \in \mathcal{S}_+, \qquad E[f(x_t) \mid \mathcal{F}_s^0] = P_{t-s}f(x_s),$$

P-almost surely. The distribution of a Markov process is determined completely by the transition semigroup P_t and the initial distribution ν, the distribution of x_0. In fact,

$$P[f_1(x_{t_1})f_2(x_{t_2})\cdots f_k(x_{t_k})] = \int v(dx)\int P_{t_1}(x,dz_1)f_1(z_1)$$

$$\int P_{t_2-t_1}(z_1,dz_2)f_2(z_2)\cdots \int P_{t_k-t_{k-1}}(z_{k-1},dz_k)f_k(z_k)$$

for $t_1 < t_2 < \cdots < t_k$ and $f_1, f_2, \ldots, f_k \in \mathcal{S}_+$. The Markov process will often be assumed to be càdlàg when S is a topological space equipped with the Borel σ-algebra $\mathcal{S}$. In this case, we may take Ω to be the canonical sample space, that is, the space of all the càdlàg maps $\omega: \mathbb{R}_+ \to S$ with $\mathcal{F}$ being the σ-algebra generated by all the maps $\omega \mapsto \omega(t)$ for $t \in \mathbb{R}_+$ and $x_t(\omega) = \omega(t)$. Then P will be be denoted by $P_{(v)}$ for the given initial distribution v of the process x_t. For $x \in S$ and $v = \delta_x$, $P_{(v)}$ may be written as $P_{(x)}$.

A probability measure v on S will be called a stationary measure of the semigroup P_t or the Markov process x_t if $vP_t = v$ for $t \in \mathbb{R}_+$, where vP_t is the measure $\int v(dx)P_t(x,\cdot)$. A process x_t will be called stationary if its distribution is invariant under the time shift, that is, if the processes x_t and x_{t+h} have the same distribution for any fixed $h > 0$. A Markov process is stationary if and only if its initial distribution is a stationary measure.

Let x_t be a càdlàg Markov process and let Ω be the canonical sample space. Define $r_t: \Omega \to \Omega$ by $(r_t\omega)(s) = \omega(t+s)$. Then $x_{t+s} = x_s \circ r_t$ and $r_t^{-1}\mathcal{F}_{u,v}^0 = \mathcal{F}_{u+t,v+t}^0$ for any $s, t, u, v \in \mathbb{R}_+$ with $u < v$, where $\mathcal{F}_{u,v}^0 = \sigma\{x_w; u \le w \le v\}$. The Markov property may now be written as $E[f(x_u \circ r_t) \mid \mathcal{F}_t^0] = P_u f(x_t)$ for $f \in \mathcal{S}_+$ and $u, t \in \mathbb{R}_+$. The process x_t is said to have the strong Markov property if the constant time t here can be replaced by an $\{\mathcal{F}_t^0\}$-stopping time τ, that is, $E[f(x_u \circ r_\tau)1_{[\tau<\infty]} \mid \mathcal{F}_\tau^0] = P_u f(x_\tau)$ on $[\tau < \infty]$, where $\mathcal{F}_t^0$ is the natural filtration of x_t, and for any set A, 1_A is its indicator function. The strong Markov property can be stated in the following apparently stronger but in fact equivalent form: For any $\xi \in (\mathcal{F}_\infty)_+$, $u \in \mathbb{R}_+$, and $\{\mathcal{F}_t\}$-stopping time τ,

$$E[\xi \circ r_\tau 1_{[\tau<\infty]} \mid \mathcal{F}_\tau] = P_{(x_\tau)}(\xi) \quad \text{on } [\tau < \infty],$$

where the filtration $\mathcal{F}_t$ is the completion of the natural filtration $\mathcal{F}_t^0$.

An important class of Markov processes comprises the Feller processes. Let S be a locally compact Hausdorff space with a countable base of open sets. The space $C_0(S)$ of the continuous functions on S vanishing at ∞ is a Banach space under the sup norm $\|f\| = \sup_{x \in S}|f(x)|$. A semigroup of probability kernels P_t on S is called a Feller semigroup if $C_0(S)$ is invariant under P_t (that is, $P_t f \in C_0(S)$ for $f \in C_0(S)$ and $t \in \mathbb{R}_+$) and $P_t f \to f$ in $C_0(S)$ as $t \to 0$ (which may be relaxed to the apparently weaker but in this case equivalent pointwise convergence). Given such a semigroup and a probability measure v

on S, there is a càdlàg Markov process x_t with P_t as transition semigroup and ν as the initial distribution. Such a process is called a Feller process. A Feller process has the strong Markov property and its completed natural filtration is right continuous.

The generator of the Feller semigroup P_t or the Feller process x_t on $C_0(S)$ is a linear operator on $C_0(S)$ defined by

$$\forall f \in D(L), \qquad Lf = \lim_{t \to 0} \frac{P_t f - f}{t},$$

where the domain $D(L)$ is the space of $f \in C_0(S)$ for which this limit exists in $C_0(S)$. The generator L determines the semigroup P_t completely and is a closed operator with $D(L)$ dense in $C_0(S)$. The reader is referred to a standard text such as that by Kallenberg [34, chapter 17] for more details. Note that, with a slight modification, in this discussion we may replace $C_0(S)$ by $C_u(S)$, the space of functions on S that are uniformly continuous under the one-point compactification topology. Then $D(L)$ will contain all the constant functions on S.

B.2. Stochastic Integrals

A d-dimensional $\{\mathcal{F}_t\}$-Brownian motion $B_t = (B_t^1, \ldots, B_t^d)$ is a continuous process in $\mathbb{R}^d$, adapted to $\{\mathcal{F}_t\}$, such that for any $t > s$, $B_t - B_s$ is independent of $\mathcal{F}_s$ and has the normal distribution with mean zero and covariance matrix $a_{jk}(t - s)$. The nonnegative definite symmetric $d \times d$ matrix $\{a_{jk}\}$ is called the covariance matrix of B_t. When the filtration $\mathcal{F}_t$ is the natural filtration, B_t will be simply called a Brownian motion. If $a_{jk} = \delta_{jk}$, it will be called a standard Brownian motion.

Given a filtration $\{\mathcal{F}_t\}$, a real-valued process M_t is called a martingale if it is adapted, $E[|M_t|] < \infty$ for any $t \in \mathbb{R}_+$, and $E[M_t \mid \mathcal{F}_s] = M_s$ for $s < t$. If $\{\mathcal{F}_t\}$ is right continuous, then any martingale has a càdlàg version. From now on, a martingale will be always assumed to be càdlàg. A martingale M_t is called a L^2-martingale if $E[M_t^2] < \infty$ for $t \in \mathbb{R}_+$. As a special case of Doob's norm inequality (see, for example, [34, proposition 6.16]), if M_t is an L^2-martingale, then

$$E[(M_t^*)^2] \leq 4E(M_t^2), \tag{B.1}$$

where $M_t^* = \sup_{0 \leq s \leq t} |M_s|$.

A process H_t stopped by a stopping time τ is the process $H_t^\tau = H_{t \wedge \tau}$, where $t \wedge \tau = \min(t, \tau)$. It is said to have a certain property locally or to be a local process of a certain type if $\exists$ stopping times $\tau_n \uparrow \infty$ P-almost surely such that the stopped processes $H_t^{\tau_n}$ possess the property or are the processes of that type. If $\mathcal{C}$ is a class of processes defined by a certain property, then $\mathcal{C}_{\text{loc}}$ will denote the class of processes that possess the property locally.

Regard a process as a function on $\mathbb{R}_+ \times \Omega$. Let $\mathcal{P}$ be the smallest σ-algebra on $\mathbb{R}_+ \times \Omega$ with respect to which all real-valued, adapted, and left continuous processes are measurable. A process is called predictable if it is measurable with respect to $\mathcal{P}$. Let M_t and N_t be two (local) L^2-martingales. By the Doob–Meyer decomposition, there is a unique predictable càdlàg process $\langle M, N \rangle_t$ of finite variation such that $\langle M, N \rangle_0 = 0$ and $M_t N_t - \langle M, N \rangle_t$ is a (local) martingale. Note that $\langle M \rangle_t = \langle M, M \rangle_t$ is increasing, and if $\langle M \rangle_t = 0$ for all $t > 0$, then $M_t = M_0$ almost surely for all $t > 0$.

A d-dimensional Brownian motion with covariance matrix a_{ij} is a d-tuple of continuous L^2-martingales $B_t = (B_t^1, \dots, B_t^d)$ with $B_0 = 0$ and $\langle B^i, B^j \rangle_t = a_{ij} t$.

Let M_t be a local L^2-martingale and let H_t be an adapted càglàd (not càdlàg) process. The Itô stochastic integral $(H \cdot M)_t = \int_0^t H_s dM_s$ is defined to be the limit in probability P of the sum

$$\sum_{i=1}^{k} H(t_{i-1})[M(t_i) - M(t_{i-1})] \tag{B.2}$$

over the partition $\Delta: 0 = t_0 < t_1 < \cdots < t_k = t$ as the mesh of partition, given by $|\Delta| = \max_{1 \le i \le k} |t_i - t_{i-1}|$, tends to zero. The process $(H \cdot M)_t$ is also a local L^2-martingale. It is continuous if M_t is continuous. It is an L^2-martingale with $E[(H \cdot M)_t^2] = E[\int_0^t H_s^2 d\langle M \rangle_s]$ if $E[\int_0^t H_s^2 d\langle M \rangle_s] < \infty$. If (K_t, N_t) is another pair of an adapted càglàd process and a local L^2-martingale, then $\langle H \cdot M, K \cdot N \rangle_t = \int_0^t H_s K_s d\langle M, N \rangle_s$, where the convention $\int_s^t = \int_{(s, t]}$ is used. We also have $\int_0^t K_s d(H \cdot M)_s = \int_0^t K_s H_s dM_s$, which may be written concisely as $K_t(H_t dM_t) = (K_t H_t) dM_t$.

A process X_t is called a semi-martingale if it can be written as $X_t = M_t + A_t$, where M_t is a local L^2-martingale and A_t is an adapted càdlàg process of finite variation. For an adapted càglàd process H_t, the Itô stochastic integral $(H \cdot X)_t = \int_0^t H_s dX_s$ is defined to be $\int_0^t H_s dM_s + \int_0^t H_s dA_s$, where the second integral is a pathwise Lebesgue–Stieljes integral. Note that $(H \cdot X)_t$ is the limit in probability P of the expression in (B.2) with M_t replaced by X_t as the mesh of the partition tends to zero. The process $(H \cdot X)_t$ is also a semi-martingale.

Let X_t and Y_t be two adapted càdlàg processes, and let

$$S_t^{\Delta} = \sum_{i=1}^{\infty} [X(t_i \wedge t) - X(t_{i-1} \wedge t)][Y(t_i \wedge t) - Y(t_{i-1} \wedge t)],$$

where $\Delta: 0 = t_0 < t_1 < t_2 < \cdots$ is a partition of $\mathbb{R}_+ = [0, \infty)$. The quadratic covariance process $[X, Y]_t$ is defined to be an adapted càdlàg process A_t of finite variation such that $\sup_{0 \le t \le T} |A_t - S_t^\Delta|$ converges to 0 in probability P for any $T > 0$ as $|\Delta| \to 0$, provided that such a process A_t exists. See Protter [50, section V5] for more details but note that, in [50], $[X, Y]$ is defined to be the present $[X, Y]$ plus $X_0 Y_0$. The continuous part $[X, Y]_t^c$ of $[X, Y]_t$ is defined by

$$[X, Y]_t = [X, Y]_t^c + \sum_{0 < s \le t} \Delta_s[X, Y],$$

where $\Delta_t Z = Z_t - Z_{t-}$ and $\Delta_0 Z = 0$ for any càdlàg process Z_t. Note that $\Delta_t[X, Y] = (\Delta_t X)(\Delta_t Y)$; hence, $[X, Y]_t = [X, Y]_t^c$ if X_t and Y_t do not jump at the same time. Note also that if either X_t or Y_t has finite variation, then $[X, Y]_t^c = 0$.

If X_t and Y_t are semi-martingales, then $[X, Y]_t$ is defined and we have the following integration-by-parts formula:

$$X_t Y_t = X_0 Y_0 + \int_0^t X_{s-} dY_s + \int_0^t Y_{s-} dX_s + [X, Y]_t.$$

In this case, if H and K are adapted càglàd processes, then $[H \cdot X, K \cdot Y] = (HK) \cdot [X, Y]$.

If M_t and N_t are (local) L^2-martingales, then $[M, N]_t$ is the unique adapted càdlàg process A_t of finite variation such that $A_0 = 0$, $M_t N_t - A_t$ is a (local) martingale, and $\Delta_t A = (\Delta_t M)(\Delta_t N)$ for $t > 0$. In this case, $[M, N]_t - \langle M, N \rangle_t$ is a (local) martingale of finite variation. If M_t and N_t do not jump at the same time, then $[M, N]_t = \langle M, N \rangle_t$.

The Stratonovich stochastic integral of an adapted càdlàg process H_t with respect to a semi-martingale X_t is defined as

$$\int_0^t H_{s-} \circ dX_s = \int_0^t H_{s-} dX_s + \frac{1}{2}[H, X]_t^c,$$

provided $[H, X]$ exists. If H_t and X_t do not jump at the same time, then $\int_0^t H_{s-} \circ dX_s$ is equal to the limit in probability P of the expression

$$\sum_{i=1}^k \frac{1}{2}[H(t_{i-1}) + H(t_i)][X(t_i) - X(t_{i-1})]$$

as the mesh of the partition tends to zero. The reader is referred to [50, section V5] for more details on Stratonovich integrals.

The Stratonovich stochastic integral $V_t = \int_0^t Y_{s-} \circ dZ_s$ is always defined for two semi-martingales Y_t and Z_t, and it is also a semi-martingale. If X_t is another semi-martingale, then $\int_0^t X_{s-} \circ dV_s = \int_0^t (X_{s-} Y_{s-}) \circ dZ_s$, which may be written concisely as $X_{t-} \circ (Y_{t-} \circ dZ_t) = (X_{t-} Y_{t-}) \circ dZ_t$.

Let $X_t = (X_t^1, \ldots, X_t^d)$ be a d-tuple of semi-martingales and $f, g \in C^1(\mathbb{R}^d)$. Then $[f(X), g(X)]_t$ is defined and is given by

$$[f(X), g(X)]_t = \sum_{i,j=1}^{d} \int_0^t f_i(X_s) g_j(X_s) d[X^i, X^j]_s^c + \sum_{0 < s \le t} \Delta_s f(X) \Delta_s g(X),$$
(B.3)

where $f_i = (\partial/\partial x_i) f$. In particular, the Stratonovich stochastic integral $\int_0^t f(X_{s-}) \circ dY_s$ is defined for any $f \in C^1(\mathbb{R}^d)$ and semi-martingale Y_t. If $f \in C^2(\mathbb{R}^d)$, then $f(X_t)$ is a semi-martingale and

$$f(X_t) = f(X_0) + \sum_{i=1}^{d} \int_0^t f_i(X_{s-}) \circ dX_s^i$$

$$+ \sum_{0 < s \le t} [\Delta_s f(X) - \sum_{i=1}^{d} f_i(X_{s-})(\Delta_s X^i)].$$
(B.4)

This is called the Itô formula. When X_t is continuous, the Itô formula looks like the usual rule of calculus: $f(X_t) = f(X_0) + \int_0^t \sum_{i=1}^{d} f_i(X_s) \circ dX_s^i$.

A càdlàg process x_t taking values in a d-dimensional manifold M is called a semi-martingale in M if $\forall f \in C_c^\infty(M)$, the space of smooth functions on M with compact supports, $f(x_t)$ is a semi-martingale. In this case, if $f \in C^1(M)$ and if Y_t is a semi-martingale, then the Stratonovich integral $\int_0^t f(x_{s-}) \circ dY_s$ is defined. Moreover, if $f \in C^2(M)$, then $f(x_t)$ is a semi-martingale. To show this, let $x^1, \ldots, x^d$ be local coordinates on an open subset U of M extended to be functions in $C_c^\infty(M)$. Then $X_t = (x^1(x_t), \ldots, x^d(x_t))$ is a d-tuple of semi-martingales. If $f \in C^1(M)$ (resp. $f \in C^2(M)$) is supported by U, then $f(x_t) = \tilde{f}(X_t)$ for some $\tilde{f} \in C^1(\mathbb{R}^d)$ (resp. $\tilde{f} \in C^2(\mathbb{R}^d)$). The claim follows from the discussion in the last paragraph. To prove this for a general $f \in C^1(M)$ (resp. $f \in C^2(M)$), choose countably many coordinate neighborhoods U_n of M that cover M. Let $\{\phi_n\}$ be a partition of unity subordinate to $\{U_n\}$. By this we mean that $\phi_n \in C_c^\infty(M)_+$, $\sum_n \phi_n = 1$, $\mathrm{supp}(\phi_n) \subset U_n$, and $\{\mathrm{supp}(\phi_n)\}$ is locally finite in the sense that any point of M has a neighborhood that intersects only finitely many $\mathrm{supp}(\phi_n)$. Then the claim holds for each $f_n = f \phi_n$. The local finiteness can be used to prove the claim for $f = \sum_n f_n$.

Let M be a d-dimensional manifold, let $Y_0, Y_1, \ldots, Y_m$ be smooth vector fields on M, and let B_t be an m-dimensional Brownian motion with covariance matrix $\{a_{ij}\}$. A solution of the Stratonovich stochastic differential equation

$$dx_t = \sum_{i=1}^{m} Y_i(x_t) \circ dB_t^i + Y_0(x_t) dt$$
(B.5)

is a semi-martingale x_t in M such that, $\forall f \in C_c^\infty(M)$,

$$f(x_t) = f(x_0) + \sum_{i=1}^{m} \int_0^t Y_i f(x_s) \circ dB_s^i + \int_0^t Y_0 f(x_s) ds.$$

If $M = \mathbb{R}^d$ and if the vector fields Y_i satisfy the uniform Lipschitz condition on $\mathbb{R}^d$, then given any x_0 independent of the Brownian motion B_t, there is a unique solution x_t. This can be proved rather easily by the usual successive approximation method (see Ikeda and Watanabe [33, IV.3]). The existence of the unique solution also holds on a compact manifold M because such a manifold can be embedded in a Euclidean space $\mathbb{R}^N$ and Y_i can be extended to be vector fields on $\mathbb{R}^N$ with compact supports. This is also true if M is a Lie group and the Y_is are left (resp. right) invariant vector fields on M (see Kunita [36, theorem 4.8.7]). In all these cases, x_t is a continuous Feller process, called a diffusion process, in M, and its generator L restricted to $C_c^\infty(M)$ is given by

$$\forall f \in C_c^\infty(M), \qquad Lf = \frac{1}{2} \sum_{i,j=1}^{m} a_{ij} Y_i Y_j f + Y_0 f. \tag{B.6}$$

Note that the restriction of L to $C_c^\infty(M)$ determines L completely (see Remark 1.2 in [33, Chapter V]). A d-dimensional Brownian motion B_t with covariance matrix a_{ij} is a diffusion process in $\mathbb{R}^d$ with generator restricted to $C_c^2(\mathbb{R}^d)$ given by $\sum_{i,j=1}^{d} a_{ij} \partial_i \partial_j$, where $\partial_i = \partial/\partial x_i$. In particular, a d-dimensional standard Brownian motion is a diffusion process in $\mathbb{R}^d$ with generator restricted to $C_c^2(\mathbb{R}^d)$ given by $(1/2)\Delta$, where $\Delta = \sum_{i=1}^{d} \partial_i \partial_i$ is the Laplace operator on $\mathbb{R}^d$.

In general, the solution x_t of the stochastic differential equation (B.5) exists uniquely for a given initial condition x_0, but it may have a finite lifetime. It may not be a Feller process, but it is still a strong Markov process and is called a diffusion process in M with generator L given by (B.6). The reader is referred to standard texts, such as those by Elworthy [17, 18] or Ikeda and Watanabe [33], for more details.

B.3. Poisson Random Measures

This section is devoted to a discussion of the stochastic integrals defined with respect to Poisson random measures following essentially the exposition in Ikeda and Watanabe [33, II.3]. A Poisson distribution ν of rate $c > 0$ is a probability measure on the set $Z_+ = \{0, 1, 2, \ldots\}$ defined by $\nu(\{k\}) = e^{-c} c^k / k!$ for $k \in Z_+$. If $c = 0$ or $c = \infty$, then it is defined to be δ_0 or δ_∞. A Z_+-valued càdlàg process N_t with $N_0 = 0$ is called a Poisson process of rate $c > 0$ if $\forall s < t$, $N_t - N_s$ is independent of $\sigma\{N_u; 0 \le u \le s\}$ and has Poisson distribution of rate $c(t - s)$.

A random measure ξ on a measurable space $(S, \mathcal{S})$ is a map $\xi: \Omega \times \mathcal{S} \to [0, \infty]$ such that $\forall \omega \in \Omega$, $\xi(\omega, \cdot)$ is a measure on $(S, \mathcal{S})$ and $\forall B \in \mathcal{S}$, $\xi(\cdot, B)$ is $\mathcal{F}$-measurable. Let μ be a σ-finite measure on $(S, \mathcal{S})$. A random measure ξ on $(S, \mathcal{S})$ is called a Poisson random measure with intensity measure μ if for disjoint $B_1, \ldots, B_k \in \mathcal{S}$, $\xi(\cdot, B_1), \ldots, \xi(\cdot, B_k)$ are independent random variables with Poisson distributions of rates $\mu(B_1), \ldots, \mu(B_k)$. Then $\mu = E\xi$; that is, $\mu(B) = E[\xi(\cdot, B)]$ for $B \in \mathcal{S}$.

A Poisson random measure N on $\mathbb{R}_+ \times S$, which is equipped with the product σ-algebra $\mathcal{B}(\mathbb{R}_+) \times \mathcal{S}$, where $B(\mathbb{R}_+)$ is the Borel σ-algebra on $\mathbb{R}_+$, is called homogeneous if its intensity measure is given by $E[N(dt\,ds)] = dt\,\nu(ds)$ for some σ-finite measure ν on $(S, \mathcal{S})$, called the characteristic measure. In this work, a Poisson random measure on $\mathbb{R}_+ \times S$ is always assumed to be homogeneous.

A positive random variable T is said to have an exponential distribution of rate $c > 0$ if $P(T > t) = e^{-ct}$ for any $t \in \mathbb{R}_+$. A nontrivial Poisson random measure N on $\mathbb{R}_+ \times S$ with a finite characteristic measure ν can be constructed from a sequence of independent exponential random variables τ_n of a common rate $\nu(S)$ and an independent sequence of independent S-valued random variables σ_n with a common distribution $\nu/\nu(S)$ by setting

$$N([0, t] \times B) = \#\{n > 0; \ T_n \le t \text{ and } \sigma_n \in B\}$$

for $t \in \mathbb{R}_+$ and $B \in \mathcal{S}$, where $T_n = \sum_{i=1}^{n} \tau_i$. Note that $T_n \uparrow \infty$ almost surely. In this case, the Poisson random measure N is said to be determined by the two sequences $\{\tau_n\}$ and $\{\sigma_n\}$, or by the two sequences $\{T_n\}$ and $\{\sigma_n\}$. We may also say that two sequences are determined by N.

Let F be the space of all the functions $g: \mathbb{R}_+ \times S \times \Omega \to [-\infty, \infty]$ that have the following two properties:

(i) $\forall t \in \mathbb{R}_+$, $(x, \omega) \mapsto g(t, x, \omega)$ is $\mathcal{S} \times \mathcal{F}_t$-measurable;
(ii) $\forall (x, \omega) \in S \times \Omega$, $t \mapsto g(t, x, \omega)$ is left continuous.

Note that for $f \in F$, $E[\int_0^t \int_S |f(u, x, \cdot)| N(du\,dx)] = E[\int_0^t \int_S |f(u, x, \cdot)| \, du\,\nu(dx)]$. For simplicity, $f(t, x, \omega)$ may be written as $f(t, x)$.

Let x_t be a process taking values in a linear space and let $T \in \mathbb{R}_+$ be fixed. The process $x_t^{(T)} = x_{T+t} - x_T$ is called the process x_t shifted by time T. For a Poisson random measure N, regarded as a measure-valued process $N_t = N([0, t] \times \cdot)$, the shifted process is $N_t^{(T)} = N([T, T+t] \times \cdot)$, which is a Poisson random measure identical in distribution to N.

A Poisson random measure N on $\mathbb{R}_+ \times S$ is said to be associated to the filtration $\{\mathcal{F}_t\}$, or is called an $\{\mathcal{F}_t\}$-Poisson random measure, if it is adapted to $\{\mathcal{F}_t\}$, when regarded as a process N_t as defined here, and for any $T \in \mathbb{R}_+$, the shifted process $N_t^{(T)}$ is independent of $\mathcal{F}_T$. Let N be an $\{\mathcal{F}_t\}$-Poisson random measure on $\mathbb{R}_+ \times S$ with characteristic measure ν. The compensated random measure $\tilde{N}$ of N is defined by $\tilde{N}(dtdx) = N(dtdx) - dt\nu(dx)$. Then, for any $B \in S$ with $\nu(B) < \infty$, $\tilde{N}_t(B) = \tilde{N}([0, t] \times B)$ is an L^2-martingale. Define

$$F^\alpha(N) = \left\{ g \in F; \quad E\left[\int_0^t \int_S |g(u, x)|^\alpha du\nu(dx)\right] < \infty \right\}$$

for $\alpha > 0$. Then for $g \in F^1(N)$, $\int_0^t \int_S g(u, x)N(du\,dx)$ is well defined and for $g \in F^1(N) \cap F^2(N)$,

$$\int_0^t \int_S g(u, x)\tilde{N}(du\,dx) = \int_0^t \int_S g(u, x)N(du\,dx) - \int_0^t \int_S g(u, x)du\nu(dx)$$

is an L^2-martingale satisfying

$$E\left\{ \left[\int_0^t \int_S g(u, x)\tilde{N}(du\,dx)\right]^2 \right\} = E\left[\int_0^t \int_S g(u, x)^2 du\nu(dx)\right]. \quad \text{(B.7)}$$

For $g \in F^2(N)$, choose $g_n \in F^1(N) \cap F^2(N)$ such that

$$E\left[\int_0^t \int_S |g_n(u, x) - g(u, x)|^2 du\nu(dx)\right] \to 0 \text{ as } n \to \infty$$

and define the stochastic integral $\int_0^t \int_S g(u, x)\tilde{N}(du\,dx)$ as the limit in $L^2(P)$ of $\int_0^t \int_S g_n(u, x)\tilde{N}(du\,dx)$. Then it is still an L^2-martingale satisfying (B.7). Note that in general $\int_0^t \int_S g(u, x)N(du\,dx)$ may not be defined for $g \in F^2(N)$. Note also that $\int_0^t \int_S g(u, x)\tilde{N}(ds\,dx)$ is equal to the limit in $L^2(P)$ of the expression

$$\sum_{i=1}^k \int_S g_n(t_{i-1}, x)\tilde{N}((t_{i-1}, t_i] \times dx) \quad \text{(B.8)}$$

as the mesh of the partition, $0 = t_0 < t_1 < \cdots t_k = t$, tends to zero and $n \to \infty$.

Let $F_{\text{loc}}^2(N)$ be the space of functions $f \in F$ such that there are stopping times $T_n \uparrow \infty$ satisfying $f_n \in F^2(N)$, where $f_n(t, x) = f(t \wedge T_n, x)$. Then the stochastic integral $\int_0^t \int_S f(u, x)\tilde{N}(du\,dx)$ may be defined by setting it equal to $\int_0^t \int_S f_n(u, x)\tilde{N}(du\,dx)$ for $t < T_n$. Then it is a local L^2-martingale.

We will denote $(g \cdot \tilde{N})_t = \int_0^t \int_S g(u, x)\tilde{N}(du\,dx)$. It can be shown that, for $g, h \in F_{\mathrm{loc}}^2(N)$,

$$\langle (g \cdot \tilde{N}), (h \cdot \tilde{N}) \rangle_t = \int_0^t \int_S g(u, x)h(u, x)du\,v(dx) \qquad \text{(B.9)}$$

and

$$[(g \cdot \tilde{N}), (h \cdot \tilde{N})]_t = \int_0^t \int_S g(u, x)h(u, x)N(du\,dx). \qquad \text{(B.10)}$$

Given a filtration $\{\mathcal{F}_t\}$, a family of processes $\{x_t^\lambda; \lambda \in \Lambda\}$, adapted to $\{\mathcal{F}_t\}$ and taking values possibly on different linear spaces, are called independent under $\{\mathcal{F}_t\}$ if the random variables x_0^λ are independent and, for all $T \in \mathbb{R}_+$, the shifted processes $x_t^{\lambda(T)}$ and $\mathcal{F}_T$ are independent.

Let B_t be a $\{\mathcal{F}_t\}$-Brownian motion and let N be a $\{\mathcal{F}_t\}$-Poisson random measure, independent under $\{\mathcal{F}_t\}$. In this case, if H_t is an adapted càglàd process satisfying $E(\int_0^t H_s^2 ds) < \infty$ and $g \in F^2(N)$, then

$$\langle (H \cdot B), (g \cdot \tilde{N}) \rangle_t = [(H \cdot B), (g \cdot \tilde{N})]_t = 0. \qquad \text{(B.11)}$$

To show this, note that since both $X_t = (H \cdot B)_t$ and $Y_t = (g \cdot \tilde{N})_t$ are L^2-martingales, and the former is continuous, it suffices to prove that $X_t Y_t$ is a martingale. This can be accomplished by a standard argument using the L^2-convergence of (B.2) and (B.8). We note that given a Brownian motion B_t and a Poisson random measure N on $\mathbb{R}_+ \times S$, if they are independent, then they are independent under the filtration $\{\mathcal{F}_t\}$ generated by $\{B_t\}$ and N (regarded as a measure-valued process N_t defined earlier) and by possibly other independent processes and random variables.

Bibliography

[1] Applebaum, D., "Lévy processes in stochastic differential geometry," in *Lévy Processes: Theory and Applications*, ed. by O. Barnsdorff-Nielsen, T. Mikosch, and S. Resnick, pp. 111–139, Birkhäuser (2000).

[2] Applebaum, D., "Compound Poisson processes and Lévy processes in groups and symmetric spaces," *J. Theor. Probab.* 13, pp. 383–425 (2000).

[3] Applebaum, D. and Kunita, H., "Lévy flows on manifolds and Lévy processes on Lie groups," *J. Math. Kyoto Univ.* 33, pp. 1105–1125 (1993).

[4] Arnold, L., *Random Dynamical Systems*, Springer-Verlag (1998).

[5] Azencott, R. et al., "Géodésiques et diffusions en temps petit," *Asterisque* 84–85 (1981).

[6] Babullot, M., "Comportement asymptotique due mouvement Brownien sur une variété homogène à courbure négative ou nulle," *Ann. Inst. H. Poincaré* (Prob. et Stat.) 27, pp. 61–90 (1991).

[7] Baxendale, P. H., "Brownian motions in the diffeomorphism group. I," *Compositio Math.* 53, pp. 19–50 (1984).

[8] Baxendale, P. H., "Asymptotic behaviour of stochastic flows of diffeomorphisms: Two case studies," *Probab. Theory Rel. Fields* 73, pp. 51–85 (1986).

[9] Bertoin, J., *Lévy Processes*, Cambrige Univ. Press (1996).

[10] Bourbaki, N., *Lie Groups and Lie Algebras*, Chap. 1–3, Springer-Verlag (1989).

[11] Boothby, W. M., *An introduction to Differentiable Manifolds and Riemannian Geometry*, 2nd ed., Academic Press (1986).

[12] Bröcker, T. and Dieck, T., *Representations of Compact Lie Groups*, Springer-Verlag (1985).

[13] Carverhill, A. P., "Flows of stochastic dynamical systems: Ergodic theory," *Stochastics* 14, pp. 273–317 (1985).

[14] Chatelin, F., *Eigenvalues of Matrices*, Wiley (1993).

[15] Diaconis, P., *Group Representations in Probability and Statistics*, IMS, Hayward, CA (1988).

[16] Dynkin, E. B., "Nonnegative eigenfunctions of the Laplace–Betrami operators and Brownian motion in certain symmetric spaces," *Dokl. Akad. Nauk SSSR* 141, pp. 1433–1426 (1961).

[17] Elworthy, K. D., *Stochastic Differential Equations on Manidfolds*, Cambridge Univ. Press (1982).

[18] Elworthy, K. D., "Geometric aspects of diffusions on manifolds," in *Ecole d'Eté de Probabilités de Saint-Flour XVII, July 1987,* ed. P.L. Hennequin, Lecture Notes Math. 1362, pp. 276–425 (1988).

[19] Furstenberg, H., "A Poisson formula for semi-simple Lie groups," *Ann. Math.* 77, pp. 335–386 (1962).

[20] Furstenberg, H., "Noncommuting random products," *Trans. Am. Math. Soc.* 108, pp. 377–428 (1963).

[21] Furstenberg, H. and Kesten, H., "Products of random matrices," *Ann. Math. Statist.* 31, pp. 457–469 (1960).

[22] Gangolli, R., "Isotropic infinitely divisible measures on symmetric spaces," *Acta Math.* 111, pp. 213–246 (1964).

[23] Gol'dsheid, I. Ya. and Margulis, G. A., "Lyapunov indices of a product of random matrices" (in Russian), *Uspekhi Mat. Nauk* 44:5, pp. 13–60 (1989). English translation in *Russian Math. Survey* 44:5, pp. 11–71 (1989).

[24] Guivarc'h, Y. and Raugi, A., "Frontière de Furstenberg, propriétés de contraction et convergence," *Z. Wahr. Gebiete* 68, pp. 187–242 (1985).

[25] Guivarc'h, Y. and Raugi, A., "Propriétés de contraction d'un semi-groupe de matrices inversibles. Coefficients de Liapunoff d'un produit de matrices aléatoires indépendantes," *Israel J. Math.* 65, pp. 165–196 (1989).

[26] Helgason, S., *Differential Geometry, Lie Groups, and Symmetric Spaces*, Academic Press (1978).

[27] Helgason, S., *Groups and Geometric Analysis*, Academic Press (1984).

[28] Heyer, H., *Probability Measures on Locally Compact Groups*, Springer-Verlag (1977).

[29] Högnäs, G. and Mukherjea, A., *Probability Measures on Semigroups*, Plenum Press (1995).

[30] Holevo, A. S., "An analog of the Ito decomposition for multiplicative processes with values in a Lie group," in *Quantum Probability and Applications V*, ed. by L. Accardi and W. von Waldenfels, Lecture Notes Math. 1442, pp. 211–215 (1990).

[31] Hsu, E., *Stochastic Analysis on Manifolds*, Am. Math. Soc. (2002).

[32] Hunt, G. A., "Semigroups of measures on Lie groups," *Trans. Am. Math. Soc.* 81, pp. 264–293 (1956).

[33] Ikeda, N. and Watanabe, S., *Stochastic Differential Equations and Diffusion Processes*, 2nd ed., North-Holland-Kodansha (1989).

[34] Kallenberg, O., *Foundations of Modern Probability*, Springer-Verlag (1997).

[35] Kobayashi, S. and Nomizu, K., *Foundations of Differential Geometry*, vols. I and II, Interscience Publishers (1963 and 1969).

[36] Kunita, H., *Stochastic Flows and Stochastic Differential Equations*, Cambrige Univ. Press (1990).

[37] Liao, M., "Brownian motion and the canonical stochastic flow on a symmetric space," *Trans. Am. Math. Soc.* 341, pp. 253–274 (1994).

[38] Liao, M., "Invariant diffusion processes in Lie groups and stochastic flows," *Proc. Symp. Pure Math.* vol. 57, Am. Math. Soc., pp. 575–591 (1995).

[39] Liao, M., "Lévy processes in semi-simple Lie groups and stability of stochastic flows," *Trans. Am. Math. Soc.* 350, pp. 501–522 (1998).

[40] Liao, M., "Stable manifolds of stochastic flows," *Proc. London Math. Soc.* 83, pp. 493–512 (2001).

[41] Liao, M., "Dynamical properties of Lévy processes in Lie groups," *Stochastics Dynam.* 2, pp. 1–23 (2002).

[42] Liao, M., "Lévy processes and Fourier analysis on compact Lie groups," to appear in *Ann. Probab.* (2004).

[43] Malliavin, M. P. and Malliavin, P., "Factorizations et lois limites de la diffusion horizontale audessus d'un espace Riemannien symmetrique," *Lecture Notes Math.* 404, pp. 164–271 (1974).

[44] Moore, C. C., "Compactification of symmetric spaces," *Am. J. Math.* 86, pp. 201–218 (1964).

[45] Norris, J. R., Rogers, L. C. G., and Williams, D., "Brownian motion of ellipsoids," *Trans. Am. Math. Soc.* 294, pp. 757–765 (1986).

[46] Orihara, A., "On random ellipsoids," *J. Fac. Sci. Univ. Tokyo Sect. IA Math.* 17, pp. 73–85 (1970).

[47] Palais, R. S., "A global formulation of the Lie theory of transformation groups," *Mem. Am. Math. Soc.* 22 (1957).

[48] Prat, J.-J., "Étude asympototique et convergence angulaire du mouvement brownien sure une variété à courbure négative," *C. R. Acad. Sci. Paris Sér. A* 280, pp. 1539–1542 (1975).

[49] Protter, P., *Stochastic Integration and Differential Equations: A New Approach*, Springer-Verlag (1990).

[50] Ramaswami, S., "Semigroups of measures on Lie groups," *J. Indiana Math. Soc.* 38, pp. 175–189 (1974).

[51] Raugi, A., "Fonctions harmoniques et théorèmes limites pour les marches gléatoires sur les groupes," *Bull. Soc. Math. France,* mémoire 54 (1997).

[52] Revuz, D. and Yor, M., *Continuous Martingales and Brownian Motion*, 2nd ed., Springer-Verlag (1994).

[53] Rosenthal, J. S., "Random rotations: Characters and random walks on SO(N)," *Ann. Probab.* 22, pp. 398–423 (1994).

[54] Sato, K., "Lévy processes and infinitely devisible distributions," translated from 1990 Japanese original and revised by the author, Cambridge Studies Adv. Math. 68, Cambridge Univ. Press (1999).

[55] Siebert, E., "Fourier analysis and limit theorems for convolution semigroups on a locally compact groups," *Adv. Math.* 39, pp. 111–154 (1981).

[56] Taylor, J. C., "The Iwasawa decomposition and the limiting behavior of Brownian motion on a symmetric space of non-compact type," in *Geometry of Random Motion*, ed. by R. Durrett and M. A. Pinsky, Contemp. Math. 73, Am. Math. Soc., pp. 303–332 (1988).

[57] Taylor, J. C., "Brownian motion on a symmetric space of non-compact type: Asymptotic behavior in polar coordinates," *Can. J. Math.* 43, pp. 1065–1085 (1991).

[58] Tutubalin, V. N., "On limit theorems for a product of random matrices," *Theory Probab. Appl.* 10, pp. 25–27 (1965).

[59] Virtser, A. D., "Central limit theorem for semi-simple Lie groups," *Theory Probab. Appl.* 15, pp. 667–687 (1970).

[60] Warner, F. W., *Foundations of Differential Manifolds and Lie Groups,* Springer-Verlag (1971).

[61] Warner, G., *Harmonic Analysis on Semi-simple Lie Groups I,* Springer-Verlag (1972).

[62] Yosida, K., *Functional Analysis,* 6th ed., Springer-Verlag (1980).

Index